AF531834

Encyclopaedia of

NUMBER THEORY

Encyclopaedia of
NUMBER THEORY

Volume 1

R.K. Pandey

ANMOL PUBLICATIONS PVT. LTD.
NEW DELHI - 110 002 (INDIA)

ANMOL PUBLICATIONS PVT. LTD.
H.O.: 4374/4B, Ansari Road, Daryaganj,
New Delhi-110 002 (India)
Ph.: 23278000, 23261597
B.O.: No. 1015, Ist Main Road, BSK IIIrd Stage
IIIrd Phase, IIIrd Block,
Bangalore - 560 085 (India)
Visit us at: www.anmolpublications.com

Encyclopaedia of Number Theory

First Edition, 2009

ISBN 978-81-261-3910-1 (Set)

PRINTED IN INDIA

Printed at Mehra Offset Press, Delhi.

Contents

	Preface	*vii*
	List of Symbols	*ix*
1.	Introduction to Integers	1
2.	Arithmetic of Integers	16
3.	Prime Numbers	35
4.	Modular Square Roots	59
5.	Arithmetic of Polynomials	79
6.	Complex Numbers	96
7.	Complex Topology	157
8.	Sequences Series	201
9.	Logarithmic Functions	223
10.	Operations of Vector	261

Preface

Natural Number is the original Mathematical concept and the most fundamental. Speculations about the nature and properties of whole numbers doubtless constitute the oldest form of mathematical thought. It is known that the sumerians and Babylonians as well as the ancient Egyptian had a fair knowledge of the properties of Natural Numbers. But first in connection with the Greeks is possible to speak of a proper theory of Numbers. Pythagoras and his pupils pursued extensive studies in the field of integers.

The Author has followed the practice of first presenting required background material, which is then used to develop the results. Relevant material is included in each chapter form other sources improves readability and makes the presentation "self-contained" to a large extent. In addition, at the end of each chapter a comprehensive Exercise is provided so that the interested reader can explore in more depth.

This volume presupposes familiarity only with the most elementary concepts of Arithmetic (Divisibility Properties, Greatest Common Divisor, etc.), is an expanded version of a series of lectures for graduate students on elementary number theory.

List of Symbols

$\pm$	$\not\subset$	$\subset$	$\subseteq$	$\in$	$\notin$	$\Re$	∂	$\neq$	$\approx$	$\angle$	∇
Σ	Π	Ω	ϕ	α	β	χ	ε	γ	η	λ	μ
ν	$\perp$	π	θ	ρ	σ	τ	$\Re$	ω	ξ	ψ	ζ
Υ		$\exists$	$\int$	$\iint$	$\iiint$	$\times$	$\sqrt{}$	$\because$	$\geq$	$\oint$	$\Rightarrow$

Chapter 1

Introduction to Integers

The integers are the set of numbers consisting of the natural numbers including 0 (0, 1, 2, 3,...) and their negatives (0, – 1, – 2, – 3,...). They are numbers that can be written without a fractional or decimal component, and fall within the set {... –2, – 1, 0, 1, 2,...}. For example, 65, 7, and – 756 are integers; 1.6 and 1½ are not integers. In other terms, integers are the numbers one can count with items such as apples or fingers, and their negatives, including 0.

More formally, the integers are the only integral domain whose positive elements are well-ordered, and in which order is preserved by addition. Like the natural numbers, the integers form a countably infinite set. The set of all integers is often denoted by a boldface Z (or blackboard bold ℤ, Unicode U+2124 ℤ), which stands for *Zahlen* (German for *numbers*).

Fig. Symbol Often Used to Denote the Set of Integers

In algebraic number theory, these commonly understood integers, embedded in the field of rational numbers, are referred to as rational integers to distinguish them from the more broadly defined algebraic integers.

ALGEBRAIC PROPERTIES

Like the natural numbers, Z is closed under the operations of addition and multiplication, that is, the sum and product of any two integers is an integer. However, with the inclusion of the negative natural numbers, and, importantly, zero, Z (unlike the natural numbers) is also closed under subtraction. Z is not closed under the operation of division, since the quotient of two integers (*e.g.*, 1 divided by 2), need not be an integer.

The following lists some of the basic properties of addition and multiplication for any integers *a*, *b* and *c*.

	Addition	*Multiplication*
Closure:	$a+b$ is an integer	$a \times b$ is an integer
Associativity:	$a+(b+c) = (a+b)+c$	$a \times (b \times c) = (a \times b) \times c$
Commutativity:	$a+b = b+a$	$a \times b = b \times a$
Existence of an identity element:	$a+0 = a$	$a \times 1 = a$
Existence of inverse elements:	$a+(-a) = 0$	
Distributivity:	$a \times (b+c) = (a \times b)+(a \times c)$	
No zero divisors:		if $ab = 0$, then either $a = 0$ or $b = 0$ (or both)

In the language of abstract algebra, the first five properties listed above for addition say that Z under addition is an abelian group. As a group under addition, Z is a cyclic group, since every nonzero integer can be written as a finite sum 1 + 1 +... 1 or (–1) + (–1) +... + (–1). In fact, Z under addition is the *only* infinite cyclic group, in the sense that any infinite cyclic group is isomorphic to Z.

The first four properties listed above for multiplication say that Z under multiplication is a commutative monoid. However, note that not every integer has a multiplicative inverse; e.g. there is no integer x such that $2x = 1$, because the left hand side is even, while the right hand side is odd. This means that Z under multiplication is not a group.

All the rules from the above property table, except for the last, taken together say that Z together with addition and multiplication is a commutative ring with unity. Adding the last property says that Z is an integral domain. In fact, Z provides the motivation for defining such a structure.

The lack of multiplicative inverses, which is equivalent to the fact that Z is not closed under division, means that Z is *not* a field. The smallest field containing the integers is the field of rational numbers. This process can be mimicked to form the field of fractions of any integral domain.

Although ordinary division is not defined on Z, it does possess an important property called the division algorithm: that is, given two integers a and b with $b \neq 0$, there exist unique integers q and r such that $a = q \times b + r$ and $0 \leq r < | b |$, where $| b |$ denotes the absolute value of b. The integer q is called the *quotient* and r is called the *remainder*, resulting from division of a by b. This is the basis for the Euclidean algorithm for computing greatest common divisors.

Again, in the language of abstract algebra, the above says that Z is a Euclidean domain. This implies that Z is a principal ideal domain and any positive integer can be written as the products of primes in an essentially unique way. This is the fundamental theorem of arithmetic.

ORDER-THEORETIC PROPERTIES

Z is a totally ordered set without upper or lower bound. The ordering of Z is given by

$$... < -2 < -1 < 0 < 1 < 2 < ...$$

An integer is *positive* if it is greater than zero and *negative* if it is less than zero. Zero is defined as neither negative nor positive.

The ordering of integers is compatible with the algebraic operations in the following way:

- If $a < b$ and $c < d$, then $a + c < b + d$
- If $a < b$ and $0 < c$, then $ac < bc$. (From this fact, one can show that if $c < 0$, then $ac > bc$.)

It follows that Z together with the above ordering is an ordered ring.

CONSTRUCTION

The integers can be constructed from the natural numbers by defining equivalence classes of pairs of natural numbers N × N under an equivalence relation, "~", where

$$(a, b) \sim (c, d)$$

precisely when

$$a + d = b + c.$$

Taking 0 to be a natural number, the natural numbers may be considered to be integers by the embedding that maps n to $[(n,0)]$, where $[(a,b)]$ denotes the equivalence class having (a,b) as a member.

Addition and multiplication of integers are defined as follows:

$$[(a, b)] + [(c, d)] = [(a + c, b + d)].$$

$$[(a, b)] \cdot [(c, d)] := [(ac + bd, ad + bc)].$$

It is easily verified that the result is independent of the choice of representatives of the equivalence classes.

Typically, $[(a,b)]$ is denoted by

$$\begin{cases} n, & \text{if } a \geq b \\ -n, & \text{if } a < b, \end{cases}$$

where

$$n = | a - b |.$$

If the natural numbers are identified with the corresponding integers (using the embedding mentioned above), this convention creates no ambiguity.

This notation recovers the familiar representation of the integers as {..., –3,–2,–1, 0, 1, 2, 3, ...}.

Some examples are:

$$0 = [(0, 0)] = [(1, 1)] = ... = [(k, k)]$$

$$1 = [(1, 0)] = [(2, 1)] = ... = [(k + 1, k)]$$

$$-1 = [(0, 1)] = [(1, 2)] = ... = [(k, k+1)]$$
$$2 = [(2, 0)] = [(3, 1)] = ... = [(k+2, k)]$$
$$-2 = [(0, 2)] = [(1, 3)] = ... = [(k, k+2)]$$

INTEGERS IN COMPUTING

An integer (sometimes known as an "int", from the name of a datatype in the C programming language) is often a primitive datatype in computer languages. However, integer datatypes can only represent a subset of all integers, since practical computers are of finite capacity. Also, in the common two's complement representation, the inherent definition of sign distinguishes between "negative" and "non-negative" rather than "negative, positive, and 0". (It is, however, certainly possible for a computer to determine whether an integer value is truly positive.)

Variable-length representations of integers, such as bignums, can store any integer that fits in the computer's memory. Other integer datatypes are implemented with a fixed size, usually a number of bits which is a power of 2 (4, 8, 16, *etc.*) or a memorable number of decimal digits (*e.g.*, 9 or 10).

In contrast, theoretical models of digital computers, such as Turing machines, typically do not have infinite (but only *unbounded finite*) capacity.

CARDINALITY

The cardinality of the set of integers is equal to $\aleph_0$. This is readily demonstrated by the construction of a bijection, that is, a function that is injective and surjective from $\mathbb{Z}$ to $\mathbb{N}$. Consider the function

$$\begin{cases} 2x+1, & \text{if } x \geq 0 \\ 2|x|, & \text{if } x < 0 \end{cases}$$

If the domain is restricted to $\mathbb{Z}$ then each and every member of $\mathbb{Z}$ has one and only one corresponding member of and by the definition of cardinal equality the two sets have equal cardinality.

DIVISIBILITY

Let us consider the integers $\mathbb{Z} = \{..., -2, -1, -1, 0, 1, 2, ...\}$. For $a, b \in \mathbb{Z}$, we say that a divides b if $az = b$ for some $z \in \mathbb{Z}$. If a divides b, we write $a \mid b$, and we may say that a is a divisor of b, or that b is a multiple of a, or that b is divisible by a. If a does not divide b, then we write $a \mid b$. A few facts about divisibility:

Theorem. For all $a, b, c \in \mathbb{Z}$, we have

(*i*) $a \mid a$, $1 \mid a$, and $a \mid 0$;

(*ii*) $0 \mid a$ if and only if $a = 0$;

(*iii*) $a \mid b$ if and only if $-a \mid b$ if and only if $a \mid -b$;

(*iv*) $a \mid b$ and $a \mid c$ implies $a \mid (b + c)$;

(v) $a \mid b$ and $b \mid c$ implies a j c.

Proof. All of these properties can be easily derived from the definition of divisibility, using elementary algebraic properties of the integers. For example, $a \mid a$ because we can write $a.1 = a$; $1 \mid a$ because we can write $1.a = a$; $a \mid 0$ because we can write $a.0 = 0$. We make a simple observation: if $a \mid b$ and $b \neq 0$, then $1 \leq |a| \leq |b|$. Indeed, if $az = b \neq 0$ for some integer z, then $a \neq 0$ and $z \neq 0$; it follows that $|a| \geq 1, |z| \geq 1$, and so $|a| \leq |a||z| = |b$

Theorem. For all $a, b \in \mathbb{Z}$, we have $a \mid b$ and $b \mid a$ if and only if $a = \pm b$. In particular, for every $a \in \mathbb{Z}$, we have $a \mid 1$ if and only if $a = \pm 1$.

Proof: If $a = \pm b$, then $a \mid b$ and $b \mid a$. So let us assume that $a \mid b$ and $b \mid a$, and prove that $a = \pm b$. If either of a or b are zero, then the other must be zero as well. So assume that neither is zero. By the above observation, $a \mid b$ implies $|a| \leq |b|$, and $b \mid a$ implies $|b| \leq |a|$; thus, $|a| = |b|$, and so $a = \pm b$. This proves the first statement.

The second statement follows from the first by setting $b := 1$, and noting that $1 \mid a$. The product of any two non-zero integers is again non-zero. This implies the usual cancellation law: if a, b, and c are integers such that $a \neq 0$ and $ab = ac$, then we must have $b = c$; indeed, $ab = ac$ implies $a(b - c) = 0$, and so $a \neq 0$ implies $b - c = 0$, and hence $b = c$.

RATIONAL NUMBERS

Let us consider the rational numbers $\mathbb{Q} = \{a/b : a, b \in \mathbb{Z}, b \neq 0\}$. Any rational number a/b, if we set $d := \gcd(a, b)$, and define the integers $a_0 := a/d$ and $b_0 := b/d$, then we have $a/b = a_0 / b_0$ and $\gcd(a_0, b_0) = 1$. Moreover, if $a_1 / b_1 = a_0 / b_0$, then we have $a_1 b_0 = a_0 b_1$, and so $b_0 \mid a_0 b_1$; also, since $\gcd(a_0, b_0) = 1$, we see that $b_0 \mid b_1$; writing $b_1 = b_0 c$, we see that $a_1 = a_0 c$.

Thus, we can represent every rational number as a fraction in lowest terms, which means a fraction of the form a_0 / b_0 where a_0 and b_0 are relatively prime; moreover, the values of a_0 and b_0 are uniquely determined up to sign, and every other fraction that represents the same rational number is of the form $a_0 c / b_0 c$, for some non-zero integer c.

PRIME NUMBERS

Let us consider n be a positive integer, 1 and n divide n. If $n > 1$ and no other positive integers besides 1 and n divide n, then we say n is prime. If $n > 1$ but n is not prime, then we say that n is composite. The number 1 is not considered to be either prime or composite. Evidently, n is composite if and only if $n = ab$ for some integers a, b with $1 < a < n$ and $1 < b < n$. The first few primes are

$$2, 3, 5, 7, 11, 13, 17, \ldots.$$

It is possible to extend the definition of prime and composite to negative integers, we shall not do so in this text: whenever we speak of a prime or composite number, we mean a positive integer. A basic fact is that every non-zero integer can be expressed as a signed product of primes in an essentially unique way.

Theorem. Every non-zero integer n can be expressed as

$$n = \pm p_1^{e_i} \ldots p_r^{e_r},$$

where $p_1, \ldots, p_r$ are distinct primes and $e_1, \ldots, e_r$ are positive integers. Moreover, this expression is unique, up to a reordering of the primes. If $n = \pm 1$, then $r = 0$, and the product of zero terms is interpreted (as usual) as 1.

The theorem says that the prime numbers act as the "building blocks" out of which all non-zero integers can be formed by multiplication (and negation). The reader may be so familiar with this fact that he may feel it is somehow "self evident," requiring no proof; however, this feeling is simply a delusion, and most of the rest of this section and the next are devoted to developing a proof of this theorem. We shall give a quite leisurely proof, introducing a number of other very important tools and concepts along the way that will be useful later.

To prove Theorem, we may clearly assume that n is positive, since otherwise, we may multiply n by -1 and reduce to the case where n is positive.

Every positive integer n can be expressed as a product (possibly empty) of primes. We may prove this by induction on n. If $n = 1$, the statement is true, as n is the product of zero primes. Now let $n > 1$, and assume that every positive integer smaller than n can be expressed as a product of primes. If n is a prime, then the statement is true, as n is the product of one prime. Assume, then, that n is composite, so that there exist $a, b \in \mathbb{Z}$ with $1 < a < n$, $1 < b < n$, and $n = ab$. By the induction hypothesis, both a and b can be expressed as a product of primes, and so the same holds for n.

Theorem. Let $a, b \in \mathbb{Z}$ with $b > 0$. Then there exist unique $q, r \in \mathbb{Z}$ such that $a = bq + r$ and $0 \leq r < b$.

Proof: Let us consider the set S of non-negative integers of the form $a - bt$ with $t \in \mathbb{Z}$. This set is clearly non-empty; indeed, if $a \geq 0$, set $t := 0$, and if $a < 0$, set $t := a$. Since every non-empty set of non-negative integers contains a minimum, we define r to be the smallest element of S. By definition, r is of the form $r = a - bq$ for some $q \in \mathbb{Z}$, and $r \geq 0$. Also, we must have $r < b$, since otherwise, $r - b$ would be an element of S smaller than r, contradicting the minimality of r; indeed, if $r \geq b$, then we would have $0 \leq r - b = a - b(q + 1)$. That proves the existence of r and q. For uniqueness, suppose that $a = bq + r$ and $a = bq' + r'$, where $0 \leq r < b$ and $0 \leq r' < b$. Then subtracting these two equations and rearranging terms, we obtain

$$r' - r = b(q - q').$$

Thus, $r' - r$ is a multiple of b; however, $0 \leq r < b$ and $0 \leq r' < b$ implies $|r' - r| < b$; therefore, the only possibility is $r' - r = 0$. Moreover, $0 = b(q - q')$ and $b \in 0$ implies $q - q' = 0$.

FACTORIZATION

Theorem. There are infinitely many primes.

Proof: By way of contradiction, suppose that there were only finitely many primes; call them $p_1, ..., p_k$. Then set $M := \prod_{i=1}^{k} p_i$ and $N := M + 1$. Consider a prime p that divides N. There must be at least one such prime p, since $N \geq 2$, and every positive integer can be written as a product of primes. Clearly, p cannot equal any of the p_i's, since if it did, then p would divide M, and hence also divide $N - M = 1$, which is impossible.

Therefore, the prime p is not among $p_1, ..., p_k$, which contradicts our assumption that these are the only primes. For each prime p, we may define the function v_p, mapping non-zero integers to non-negative integers, as follows: for every integer $n \neq 0$, if $n = p^e m$, where $p \mid m$, then $v_p(n) := e$. We may then write the factorization of n into primes as

$$n = \pm \prod_p p^{v_p(n)},$$

where the product is over all primes p; although syntactically this is an infinite product, all but finitely many of its terms are equal to 1, and so this expression makes sense. Observe that if a and b are non-zero integers, then

$$v_p(a \,.\, b) = v_p(a) + v_p(b) \text{ for all primes } p,$$

and

$$a \mid b \Leftrightarrow v_p(a) \leq v_p(b) \text{ for all primes } p.$$

From this, it is clear that

$$\gcd(a, b) = \prod_p p^{\min(v_p(a), v_p(b))}.$$

OPERATOR OF MOD

Let us suppose $a, b \in \mathbb{Z}$ with $b > 0$. If q and r are the unique integers from Theorem 1.4 that satisfy $a = bq + r$ and $0 \leq r < b$, we define

$$a \bmod b := r;$$

that is, $a \bmod b$ denotes the remainder in dividing a by b. It is clear that $b \mid a$ if and only if $a \bmod b = 0$. Dividing both sides of the equation $a = bq + r$ by b, we obtain $a/b = q + r/b$. Since $q \in \mathbb{Z}$ and $r/b \in [0, 1)$, we see that $q = \lfloor a/b \rfloor$. Thus,

$$(a \bmod b) = a - b\lfloor a/b \rfloor.$$

One can use this equation to extend the definition of a mod b to all integers a and b, with $b \neq 0$; that is, for $b < 0$, we simply define a mod b to be $a - b\lfloor a/b \rfloor$.

When dividing an integer a by a positive integer b, the remainder is placed in an interval other than $[0, b)$. Let x be any real number, and consider the interval $[x, x + b)$. As the reader may easily verify, this interval contains precisely b integers, namely, $\lfloor x \rfloor$, ..., $\lfloor x \rfloor + b - 1$. Applying Theorem with $a - \lfloor x \rfloor$ in place of a, we obtain:

Theorem. Let $a, b \in \mathbb{Z}$ with $b > 0$, and let $x \in \mathbb{R}$. Then there exist unique $q, r \in \mathbb{Z}$ such that $a = bq + r$ and $r \in [x, x + b)$.

FLOORS AND CEILINGS

Let us briefly recall the usual floor and ceiling functions, denoted $\lfloor . \rfloor$ and $\lceil . \rceil$, respectively. These are functions from $\mathbb{R}$ (the real numbers) to $\mathbb{Z}$. For $x \in \mathbb{R}$, $\lfloor x \rfloor$ is the greatest integer $m \leq x$; equivalently, $\lfloor x \rfloor$ is the unique integer m such that $m \leq x < m + 1$, or put another way, such that $x = m + \varepsilon$ for some $\varepsilon \in [0, 1)$. Also, $\lfloor x \rfloor$ is the smallest integer $m \geq x$; equivalently, $\lfloor x \rfloor$ is the unique integer m such that $m - 1 < x \leq m$, or put another way, such that $x = m - \varepsilon$ for some $\varepsilon \in [0, 1)$.

EXERCISE

1. Let $x \in \mathbb{R}$. Show that $\lfloor x \rfloor \in \leq \lfloor 2x \rfloor \leq \lfloor x \rfloor \in + 1$.
2. Let $x \in \mathbb{R}$ and $n \in \mathbb{Z}$ with $n > 0$. Show that $\lfloor \lfloor x \rfloor / n \rfloor = \lfloor x/n \rfloor$; in particular, $\lfloor \lfloor a/b \rfloor / c \rfloor = \lfloor a/bc \rfloor$ for all positive integers a, b, c.
3. Let $a, b, d \in \mathbb{Z}$ with $d \neq 0$. Show that $a \mid b$ if and only if $da \mid db$.
4. Let n be a composite integer. Show that there exists a prime p dividing n, with $p \geq n^{1/2}$.
5. Let $a, b \in \mathbb{Z}$ with $b < 0$. Show that $(a \bmod b) \in (b, 0]$.
6. Show that Theorem 1.5 also holds for the interval $(x, x + b]$. Does it hold in general for the intervals $[x, x + b]$ or $(x, x + b)$?
7. Let m be a positive integer. Show that for every real number $x \geq 1$, the number of multiples of m in the interval $[1, x]$ is $\lfloor x/m \rfloor$; in particular, for every integer $n \geq 1$, the number of multiples of m among 1, ..., n is $\lfloor n/m \rfloor$.

IDEALS

The notion of an ideal of $\mathbb{Z}$ is already intorduced, which is a non-empty set of integers that is closed under addition, and closed under multiplication by an arbitrary integer. That is, a non-empty set $I \subseteq \mathbb{Z}$ is an ideal if and only if for all $a, b \in I$ and all $z \in \mathbb{Z}$, we have

$$a + b \in I \text{ and } az \in I.$$

The notion of an ideal is quite useful in a number of contexts, which will be explored later. It is easy to see that every ideal I contains 0: since $a \in I$ for some integer a, we have $0 = a.0 \in I$. Also, note that if an ideal I contains an integer a, it also contains $-a$, since $-a = a.(-1) \in I$. Thus, if an ideal contains a and b, it also contains $a - b$. It is clear that $\{0\}$ and $\mathbb{Z}$ are ideals. Moreover, an ideal I is equal to Z if and only if $1 \in I$; to see this, note that $1 \in I$ implies that for every $z \in \mathbb{Z}$, we have $z = 1.z \in I$, and hence $I = \mathbb{Z}$; conversely, if $I = \mathbb{Z}$, then in particular, $1 \in I$.

For $a \in \mathbb{Z}$, define $a\mathbb{Z} := \{az : z \in \mathbb{Z}\}$; that is, $a\mathbb{Z}$ is the set of all multiples of a. If $a = 0$, then clearly $a\mathbb{Z} = \{0\}$; otherwise, $a\mathbb{Z}$ consists of the distinct integers

$$\ldots, -3a, -2a, -a, 0, a, 2a, 3a, \ldots.$$

It is easy to see that $a\mathbb{Z}$ is an ideal: for all $az, az' \in a\mathbb{Z}$ and $z'' \in \mathbb{Z}$, we have $az + az' = a(z + z') \in a\mathbb{Z}$ and $(az)z'' = a(zz'') \in a\mathbb{Z}$. The ideal $a\mathbb{Z}$ is called the ideal generated by a, and an ideal of the form $a\mathbb{Z}$ for some $a \in \mathbb{Z}$ is called a principal ideal.

Observe that for all $a, b \in \mathbb{Z}$, we have $b \in a\mathbb{Z}$ if and only if $a \mid b$. Also observe that for every ideal I, we have $b \in I$ if and only if $b\mathbb{Z} \subseteq$ I. Both of these observations are simple consequences of the definitions, as the reader may verify. Combining these two observations, we see that $b\mathbb{Z} \subseteq a\mathbb{Z}$ if and only if $a \mid b$.

Let us consider I_1 and I_2 are ideals. Then it is not hard to see that the set

$$I_1 + I_2 := \{a_1 + a_2 : a_1 \in I_1, a_2 \in I_2\}$$

is also an ideal. Indeed, suppose $a_1 + a_2 \in I_1 + I_2$ and $b_1 + b_2 \in I_1 + I_2$. Then we have $(a_1 + a_2) + (b_1 + b_2) = (a_1 + b_1) + (a_2 + b_2) \in I_1 + I_2$, and for every $z \in \mathbb{Z}$, we have $(a_1 + a_2)z = a_1z + a_2z \in I_1 + I_2$.

Example. Consider the ideal $3\mathbb{Z} + 5\mathbb{Z}$. This ideal contains $3 . \in + 5 . (-1) = 1$. Since it contains 1, it contains all integers; that is, $3\mathbb{Z} + 5\mathbb{Z} = \mathbb{Z}$.

Example. Consider the principal ideal $3\mathbb{Z}$. This consists of all multiples of 3; that is, $3\mathbb{Z} = \{\ldots, -9, -6, -3, 0, 3, 6, 9, \ldots\}$.

Example. Consider the ideal $4\mathbb{Z} + 6\mathbb{Z}$. This ideal contains $4 . (-1) + 6 . 1 = 2$, and therefore, it contains all even integers. It does not contain any odd integers, since the sum of two even integers is again even. Thus, $4\mathbb{Z} + 6\mathbb{Z} = 2\mathbb{Z}$.

In the previous two examples, we defined an ideal that turned out upon closer inspection to be a principal ideal. This was no accident: the following theorem says that all ideals of $\mathbb{Z}$ are principal.

Theorem. Let us suppose I be an ideal of $\mathbb{Z}$. Then there exists a unique non-negative integer d such that $I = d\mathbb{Z}$.

Proof. We first prove the existence part of the theorem. If $I = \{0\}$, then $d = 0$ does the job, so let us assume that $I \neq \{0\}$. Since I contains non-zero integers, it must contain positive integers, since if $a \in I$ then so is $-a$. Let d be the smallest positive integer in I.

We want to show that $I = d\mathbb{Z}$. We first show that $I \subseteq d\mathbb{Z}$. To this end, let a be any element in I. It suffices to show that $d \mid a$. Using the division with remainder property, write $a = dq + r$, where $0 \leq r < d$. Then by the closure properties of ideals, one sees that $r = a - dq$ is also an element of I, and by the minimality of the choice of d, we must have $r = 0$.

Thus, $d \mid a$. We have shown that $I \subseteq d\mathbb{Z}$. The fact that $d\mathbb{Z} \subseteq I$ follows from the fact that $d \in I$. Thus, $I = d\mathbb{Z}$. That proves the existence part of the theorem. For uniqueness, note that if $d\mathbb{Z} = e\mathbb{Z}$ for some non-negative integer e, then $d \mid e$ and $e \mid d$, from which it follows by Theorem that $d = \pm e$; since d and e are non-negative, we must have $d = e$.

GCD

For $a, b \in \mathbb{Z}$, we call $d \in \mathbb{Z}$ a common divisor of a and b if $d \mid a$ and $d \mid b$; moreover, we call such a d a greatest common divisor of a and b if d is non-negative and all other common divisors of a and b divide d.

Theorem. Let $a, b, r \in \mathbb{Z}$ and let $d := \gcd(a, b)$. Then there exist $s, t \in \mathbb{Z}$ such that $as + bt = r$ if and only if $d \mid r$. In particular, a and b are relatively prime if and only if there exist integers s and t such that $as + bt = 1$.

Proof: We have

$$as + bt = r \text{ for some } s, t \in \mathbb{Z}$$

$$\Rightarrow r \in a\mathbb{Z} + b\mathbb{Z}$$

$$\Rightarrow r \in d\mathbb{Z}$$

$$\Rightarrow d \mid r.$$

That proves the first statement. The second statement follows from the first, setting $r := 1$. Note that as we have defined it, $\gcd(0, 0) = 0$. Also note that when at least one of a or b are non-zero, $\gcd(a, b)$ may be characterized as the largest positive integer that divides both a and b, and as the smallest positive integer that can be expressed as $as + bt$ for integers s and t.

Theorem. For all $a, b \in \mathbb{Z}$, there exists a unique greatest common divisor d of a and b, and moreover, $a\mathbb{Z} + b\mathbb{Z} = d\mathbb{Z}$.

Proof. We apply the previous theorem to the ideal $I := a\mathbb{Z} + b\mathbb{Z}$. Let $d \in \mathbb{Z}$ with $I = d\mathbb{Z}$, as in that theorem. We wish to show that d is a greatest common divisor of a and b. Note that $a, b, d \in I$ and d is non-negative.

Since $a \in I = d\mathbb{Z}$, we see that $d \mid a$; similarly, $d \mid b$. So we see that d is a common divisor of a and b. Since $d \in I = a\mathbb{Z} + b\mathbb{Z}$, there exist $s, t \in \mathbb{Z}$ such that $as + bt = d$. Now suppose $a = a'd'$ and $b = b'd'$ for some $a', b', d' \in \mathbb{Z}$. Then the equation $as + bt = d$ implies that $d'(a's + b't) = d$, which says that $d' \mid d$. Thus, any common divisor d' of a and b divides d. That proves that d is a greatest common divisor of a and b. For uniqueness, note that if e is a greatest common divisor of a and b, then $d \mid e$ and $e \mid d$, and hence $d = \pm e$; since both d and e are non-negative by definition, we have $d = e$.

For $a, b \in \mathbb{Z}$, we write gcd(a, b) for the greatest common divisor of a and b. We say that $a, b \in \mathbb{Z}$ are relatively prime if gcd(a, b) = 1, which is the same as saying that the only common divisors of a and b are ± 1. The following is essentially just a restatement of Theorem, but we state it here for emphasis:

Theorem. Let $a, b, c \in \mathbb{Z}$ such that $c \mid ab$ and gcd(a, c) = 1. Then $c \mid b$.

Proof: Suppose that $c \mid ab$ and gcd(a, c) = 1. Then since gcd(a, c) = 1, by Theorem we have $as + ct = 1$ for some $s, t \in \mathbb{Z}$. Multiplying this equation by b, we obtain

$$abs + cbt = b.$$

Since c divides ab by hypothesis, and since c clearly divides cbt, it follows that c divides the left-hand side of (1.1), and hence that c divides b. Suppose that p is a prime and a is any integer. As the only divisors of p are ± 1 and $\pm p$, we have

$$p \text{ j } a \Rightarrow \gcd(a, p) = p, \text{ and}$$
$$p \mid a \Rightarrow \gcd(a, p) = 1.$$

Combining this observation with the previous theorem, we have:

Theorem. Let p be prime, and let $a, b \in \mathbb{Z}$. Then $p \mid ab$ implies that $p \mid a$ or $p \mid b$.

Proof: Assume that $p \mid ab$. If $p \mid a$, we are done, so assume that $p \mid a$. By the above observation, gcd(a, p) = 1, and so by Theorem, we have $p \mid b$. An obvious corollary to Theorem is that if a_1, ..., a_k are integers, and if p is a prime that divides the product $a_1 \ldots a_k$, then $p \mid a_i$ for some i = 1, ..., k. This is easily proved by induction on k. For $k = 1$, the statement is trivially true. Now let $k > 1$, and assume that statement holds for $k - 1$. Then by Theorem, either $p \mid a_1$ or $p \mid a_2 \ldots a_k$; if $p \mid a_1$, we are done; otherwise, by induction, p divides one of a_2, ..., a_k.

We are now in a position to prove the uniqueness part of Theorem, which we can state as follows: if p_1, ..., p_r are primes (not necessarily distinct), and q_1, ..., q_s are primes (also not necessarily distinct), such that

$$p_1 \ldots p_r = q_1 \ldots q_s,$$

then $(p_1, ..., p_r)$ is just a reordering of $(q_1, ..., q_s)$. We may prove this by induction on r. If $r = 0$, we must have $s = 0$ and we are done. Now suppose $r > 0$, and that the statement holds for $r - 1$. Since $r > 0$, we clearly must have $s > 0$. Also, as p_1 obviously divides the left-hand side of equation, it must also divide the right-hand side of equation; that is, $p_1 \mid q_1 \ldots qs$. It follows from (the corollary to) Theorem that $p_1 \mid q_j$ for some $j = 1, ..., s$, and moreover, since q_j is prime, we must have $p_1 = q_j$. Thus, we may cancel p_1 from the left-hand side of equation and q_j from the right-hand side of equation, and the statement now follows from the induction hypothesis.

EXERCISE

1. Let n be an integer. Show that if a, b are relatively prime integers, each of which divides n, then ab divides n.
2. Show that two integers are relatively prime if and only if there is no one prime that divides both of them.
3. Let I be a non-empty set of integers that is closed under addition, that is, $a + b \in I$ for all $a, b \in I$. Show that I is an ideal if and only if $-a \in I$ for all $a \in I$.
4. Show that for all integers a, b, c, we have:

   (a) $\gcd(a, b) = \gcd(b, a)$;

   (b) $\gcd(a, b) = \mid a \mid \Rightarrow a \mid b$;

   (c) $\gcd(a, 0) = \gcd(a, a) = \mid a \mid$ and $\gcd(a, 1) = 1$;

   (d) $\gcd(ca, cb) = \mid c \mid \gcd(a, b)$.
5. Show that for all integers a, b with $d := \gcd(a, b) \neq 0$, we have $\gcd(a/d, b/d) = 1$.
6. Let a, b, c be positive integers satisfying $\gcd(a, b) = 1$ and $c \geq (a - 1)(b - 1)$. Show that there exist non-negative integers s, t such that $c = as + bt$.
7. For each positive integer n, let D_n denote the set of positive divisors of n. Let n_1, n_2 be relatively prime, positive integers. Show that the sets $D_{n_1} \times D_{n_2}$ and $D_{n_1 n_2}$ are in one-to-one correspondence, via the map that sends $(d_1, d_2) \in D_{n_1} \times D_{n_2}$ to $d_1 d_2$.
8. Let $a, b_1, ..., b_k$ be integers. Show that $\gcd(a, b_1 \ldots b_k) = 1$ if and only if $\gcd(a, b_i) = 1$ for $i = 1, ..., k$.
9. An integer a is called square-free if it is not divisible by the square of any integer greater than 1. Show that:

   (a) a is square-free if and only if $a = \pm p_1 \ldots p_r$, where the p_i's are distinct primes;

   (b) Every positive integer n can be expressed uniquely as $n = ab^2$, where a and b are positive integers, and a is square-free.

10. For each positive integer m, let I_m denote $\{0, ..., m-1\}$. Let a, b be positive integers, and consider the map

    $\tau : I_b \times I_a \to I_{ab}$

    $(s, t) \to (as + bt) \bmod ab.$

    Show τ is a bijection if and only if $\gcd(a, b) = 1$.

LCM

If $a, b \in \mathbb{Z}$, a common multiple of a and b is an integer m such that $a \mid m$ and $b \mid m$; moreover, such an m is the least common multiple of a and b if m is non-negative and m divides all common multiples of a and b. It is easy to see that the least common multiple exists and is unique, and we denote the least common multiple of a and b by lcm(a, b). Indeed, for all $a, b \in \mathbb{Z}$, if either a or b are zero, the only common multiple of a and b is 0, and so lcm(a, b) = 0; otherwise, if neither a nor b are zero, we have

$$\operatorname{lcm}(a, b) = \prod_p p^{\max(v_p(a), v_p(b))},$$

or equivalently, lcm(a, b) may be characterized as the smallest positive integer divisible by both a and b.

It is convenient to extend the domain of definition of v_p to include 0, defining $v_p(0) := \infty$. If we interpret expressions involving "∞" appropriately, then for arbitrary $a, b \in \mathbb{Z}$,

$$v_p(\gcd(a, b)) = \min(v_p(a), v_p(b)) \text{ and } v_p(\operatorname{lcm}(a, b)) = \max(v_p(a), v_p(b))$$

for all primes p.

GENERALIZING INTEGERS

Let $a_1, ..., a_k$ be integers. We call $d \in \mathbb{Z}$ a common divisor of $a_1, ..., a_k$ if $d \mid a_i$ for $i = 1, ..., k$; moreover, we call such a d the greatest common divisor of $a_1, ..., a_k$ if d is non-negative and all other common divisors of $a_1, ..., a_k$ divide d. The greatest common divisor of $a_1, ..., a_k$ is denoted $\gcd(a_1, ..., a_k)$ and is the unique non-negative integer d satisfying

$$v_p(d) = \min(v_p(a_1), ..., v_p(a_k)) \text{ for all primes } p.$$

Analogously, we call $m \in \mathbb{Z}$ a common multiple of $a_1, ..., a_k$ if $a_i \mid m$ for all $i = 1, ..., k$; moreover, such an m is called the least common multiple of $a_1, ..., a_k$ if m divides all common multiples of $a_1, ..., a_k$. The least common multiple of $a_1, ..., a_k$ is denoted $\operatorname{lcm}(a_1, ..., a_k)$ and is the unique non-negative integer m satisfying

$$v_p(m) = \max(v_p(a_1), ..., v_p(ak)) \text{ for all primes } p.$$

Finally, we say that the family $\{a_i\}_{i=1}^k$ is pairwise relatively prime if for all indices i, j with $i \neq j$, we have $\gcd(a_i, a_j) = 1$. Certainly, if $\{a_i\}_{i=1}^k$ is pairwise relatively prime, and $k > 1$, then $\gcd(a_1, ..., a_k) = 1$; however, $\gcd(a_1, ..., a_k) = 1$ does not imply that $\{a_i\}_{i=1}^k$ is pairwise relatively prime.

EXERCISE

1. Show that for every positive integer k, there exist k consecutive composite integers. Thus, there are arbitrarily large gaps between primes.
2. Let p be a prime. Show that for all $a, b \in \mathbb{Z}$, we have $v_p(a + b) \geq \min\{v_p(a), v_p(b)\}$, and $v_p(a + b) = v_p(a)$ if $v_p(a) < v_p(b)$.
3. Let n be a positive integer, and let 2^k be the highest power of $\in$ in the set $S = \{1, ..., n\}$. Show that 2^k does not divide any other element in S.
4. Let $n \in \mathbb{Z}$ with $n > 1$. Show that $\sum_{i=1}^{n} 1/i$ is not an integer.
5. Show that for all integers a, b, c, we have:

   (a) $\text{lcm}(a, b) = \text{lcm}(b, a)$;

   (b) $\text{lcm}(a, b) = | a | \Leftrightarrow b \mid a$;

   (c) $\text{lcm}(a, a) = \text{lcm}(a, 1) = | a |$;

   (d) $\text{lcm}(ca, cb) = \lceil c \mid \text{lcm}(a, b)$.
6. Let n and k be positive integers, and suppose $x \in \mathbb{Q}$ such that $x^k = n$ for some $x \in \mathbb{Q}$. Show that $x \in \mathbb{Z}$. In other words, $\sqrt[k]{n}$ is either an integer or is irrational.
7. Show that $\gcd(a + b, \text{lcm}(a, b)) = \gcd(a, b)$ for all $a, b \in \mathbb{Z}$.
8. Show that for all integers a, b, we have:

   (a) $\gcd(a, b) \,.\, \text{lcm}(a, b) = | ab |$;

   (b) $\gcd(a, b) = 1 \Rightarrow \text{lcm}(a, b) = | ab |$.
9. Let $a_1, ..., a_k \in \mathbb{Z}$ with $k > 1$. Show that:

   $$\gcd(a_1, ..., a_k) = \gcd(a_1, \gcd(a_2, ..., a_k)) = \gcd(\gcd(a_1, ..., a_{k-1}), a_k);$$
   $$\text{lcm}(a_1, ..., a_k) = \text{lcm}(a_1, \text{lcm}(a_2, ..., a_k)) = \text{lcm}(\text{lcm}(a_1, ..., a_{k-1}), a_k).$$
10. Let $a_1, ..., a_k \in \mathbb{Z}$ with $d := \gcd(a_1, ..., a_k)$. Show that $d\mathbb{Z} = a_1\mathbb{Z} + ... + a_k\mathbb{Z}$; in particular, there exist integers $z_1, ..., z_k$ such that $d = a_1z_1 + ... + a_kz_k$.
11. Show that if $\{a_i\}_{i=1}^k$ is a pairwise relatively prime family of integers, then $\text{lcm}(a_1, ..., a_k) = | a_1 \dots a_k |$.
12. Let n be a positive integer, and let C_n denote the number of pairs of integers (a, b) with $a, b \in \{1, ..., n\}$ and $\gcd(a, b) = 1$, and let F_n be the number of distinct rational numbers a/b, where $0 \leq a < b \leq n$.

(a) Show that $F_n = (C_n + 1)/2$.

(b) Show that $Cn \geq n^2/4$. Hint: first show that $C_n \geq n^2(1 - \sum_{d\geq 2} 1/d^2)$, and then show that $\sum_{d\geq 2} 1/d^2 \leq 3/4$.

13. Show that every non-zero $x \in \mathbb{Q}$ can be expressed as

$$x = \pm p_1^{e_1} \ldots p_r^{e_r},$$

where the p_i's are distinct primes and the e_i's are non-zero integers, and that this expression in unique up to a reordering of the primes.

14. This exercise develops a characterization of least common multiples in terms of ideals.

(a) Arguing directly from the definition of an ideal, show that if I and J are ideals of $\mathbb{Z}$, then so is $I \cap J$.

(b) Let $a, b \in \mathbb{Z}$, and consider the ideals $I := a\mathbb{Z}$ and $J := b\mathbb{Z}$. By part

(c) We know that $I \cap J$ is an ideal. By Theorem 1.6, we know that $I \cap J = m\mathbb{Z}$ for some uniquely determined non-negative integer m. Show that $m = \text{lcm}(a, b)$.

15. For a given prime p, we may extend the domain of definition of v_p from $\mathbb{Z}$ to $\mathbb{Q}$: for non-zero integers a, b, let us define $v_p(a/b) := v_p(a) - v_p(b)$.

Show that:

(a) this definition of $v_p(a/b)$ is unambiguous, in the sense that it does not depend on the particular choice of a and b;

(b) for all $x, y \in \mathbb{Q}$, we have $v_p(xy) = v_p(x) + v_p(y)$;

(c) for all $x, y \in \mathbb{Q}$, we have $v_p(x + y) \geq \min\{v_p(x), v_p(y)\}$, and $v_p(x + y) = v_p(x)$ if $v_p(x) < v_p(y)$;

(d) for all non-zero $x \in \mathbb{Q}$, we have $x = \pm\prod_p p^{v_p(x)}$, where the product is over all primes, and all but a finite number of terms in the product are equal to 1;

(e) for all $x \in \mathbb{Q}$, we have $x \in \mathbb{Z}$ if and only if $p(x) \geq 0$ for all primes p.

Chapter 2

Arithmetic of Integers

The basic operational signs of algebra are familiar from arithmetic: addition (+), subtraction (–), multiplication (×), and division (÷). The multiplication symbol × is often omitted or replaced by a dot, as in $a \cdot b$. A group of consecutive symbols, such as *abc,* indicates the product (the result of multiplication) of *a, b,* and *c.* Division is commonly indicated by a horizontal bar (also called a vinculum), as in:

$$\frac{a}{c}$$

A virgule, or slash (/), may also be used to indicate division: a/c.

A *power* is the product of a number multiplied by itself. The notation 4^2 (read "four squared"), for example, is used as an abbreviation for $4 \cdot 4$ (4 times 4); thus $4^2 = 16$. The 4 in 4^2 is called the base, and the small raised number 2 is called the exponent. An exponent indicates how many times the number is multiplied by itself: x^3 (read "*x* cubed") means $x \cdot x \cdot x$. More generally x^n (read "*x* to the *n*th power" or "*x* to the *n*th" where *n* is any number) means the product of *x* multiplied by itself *n* times. Fractions can take exponents as well: $\left(\frac{1}{2}\right)^2 = \frac{1}{4}$.

A number whose *n*th power is equal to *x* is an *n*th root of *x*. When *n* is 2 the term "square root" is used and when *n* is 3 the term "cube root" is used. The square root of *x* is denoted like this:

The number of times the root is multiplied by itself is called the index. The index is usually omitted for square roots, but appears as a small raised number just before the root symbol for higher roots:

$$\sqrt{x}$$

The two possible values of square roots, one positive and one negative, are often written using the plus or minus symbol: ±. The equation = 2 or –2, for instance, can be abbreviated = ± 2.

$$\sqrt[3]{8} = 2$$

BASIC INTEGER ARITHMETIC

Integers will exceed the word-size of actual machines, and to satisfy the formal requirements of our random access model of computation, we shall represent large integers as vectors of digits in some base B, along with a bit indicating the sign. That is, for $a \in \mathbb{Z}$, if we write

$$a = \pm \sum_{i=0}^{k-1} a_i B^i = \pm(a_{k-1} \dots a_1 a_0)_B,$$

where $0 \leq a_i < B$ for $i = 0, \dots, k-1$, then a will be represented in memory as a data structure consisting of the vector of base-B digits $a_0, \dots, a_{k-1}$, along with a "sign bit" to indicate the sign of a. To ensure a unique representation, if a is non-zero, then the high-order digit a_{k-1} in this representation should be non-zero.For our purposes, we shall consider B to be a constant, and moreover, a power of 2. The choice of B as a power of $\in$ is convenient for a number of technical reasons. We now discuss in detail basic arithmetic algorithms for unsigned (i.e., nonnegative) integers – these algorithms work with vectors of base-B digits, and except where explicitly noted, we do not assume that the high-order digits of the input vectors are non-zero, nor do these algorithms ensure that the high-order digit of the output vector is non-zero.

These algorithms can be very easily adapted to deal with arbitrary signed integers, and to take proper care that the high-order digit of the vector representing a non-zero number is itself non-zero (the reader is asked to fill in these details in some of the exercises below). All of these algorithms can be implemented directly in a programming language that provides a "built-in" signed integer type that can represent all integers of absolute value less than B^2, and that supports the basic arithmetic operations (addition, subtraction, multiplication, integer division).

So, for example, using the C or Java programming language's int type on a typical 32-bit computer, we could take $B = 2^{15}$. The resulting software would be reasonably efficient and portable, but certainly not the fastest possible. Suppose we have the base-B representations of two unsigned integers a and b. We present algorithms to compute the base-B representation of $a + b$, $a - b$, $a\ \ b$, $\lfloor a/b \rfloor$, and a mod b. To simplify the presentation, for integers x, y with $y \neq 0$, we denote by QuoRem(x, y) the quotient/remainder pair ($\lfloor a/b \rfloor$, x mod y).

SUBTRACTION

Let us suppose $a = (a_{k-1} \dots a_0)_B$ and $b = (b_{\ell-1} \dots b_0)_B$ be unsigned integers. Assume that $k \geq \ell \geq 1$. To compute the difference $c := a - b$, we may use the same algorithm as above, but with the expression "$a_i + b_i$" replaced by "$a_i - b_i$." In every loop iteration, the value of carry is 0 or -1, and the value of tmp lies between $-B$ and $B - 1$. If $a \geq b$, then $c_k = 0$ (i.e., there is no carry out of the last loop iteration); otherwise, $c_k = -1$ (and $b - a = B^k - (c_{k-1} \dots c_0)_B$, which can be computed with another execution of the subtraction routine).

ADDITION

Let us consisder $a = (a_{k-1} \dots a_0)_B$ and $b = (b_{\ell-1} \dots b_0)_B$ be unsigned integers. Assume that $k \geq \ell \geq 1$ (if $k < \ell$, then we can just swap a and b). The sum $c := a + b$ is of the form $c = (c_k c_{k-1} \dots c_0)_B$. Using the standard "paper-and-pencil" method (adapted from base-10 to base-B, of course), we can compute the base-B representation of $a + b$ in time $O(k)$, as follows:

carry ← 0

for $i \leftarrow 0$ to $\ell - 1$ do

tmp ← $a_i + b_i$ + carry, (carry, c_i) ← QuoRem(tmp, B)

for $i \leftarrow \ell$ to $k - 1$ do

tmp ← a_i + carry, (carry, c_i) ← QuoRem(tmp, B)

c_k ← carry

Note that in every loop iteration, the value of carry is 0 or 1, and the value tmp lies between 0 and $2B - 1$.

MULTIPLICATION

Let *us suppose* $a = (a_{k-1} \dots a_0)_B$ and $b = (b_{\ell-1} \dots b_0)_B$ be unsigned integers, with $k \geq 1$ and $\ell \geq 1$. The product $c := a . b$ is of the form $(c_{k+\ell-1} \dots c_0)_B$, and may be computed in time $O(k\ell)$ as follows:

for $i \leftarrow 0$ to $k + \ell - 1$ do $c_i \leftarrow 0$

for $i \leftarrow 0$ to $k - 1$ do

carry ← 0

for $j \leftarrow 0$ to $\ell - 1$ do

tmp ← $a_i b_j + c_{i+j}$ + carry

(carry, c_{i+j}) ← QuoRem(tmp, B)

$ci+\ell$ ← carry

Note that at every step in the above algorithm, the value of carry lies between 0 and $B - 1$, and the value of tmp lies between 0 and $B^2 - 1$.

DIVISION WITH REMAINDER

Let $a = (a_{k-1} \dots a_0)_B$ and $b = (b_{\ell-1} \dots b_0)_B$ be unsigned integers, with $k \geq 1$, $\ell \geq 1$, and $b_{\ell-1} \neq 0$. We want to compute q and r such that $a = bq + r$ and $0 \leq r < b$. Assume that $k \geq \ell$; otherwise, $a < b$, and we can just set $q \leftarrow 0$ and $r \leftarrow a$. The quotient q will have at most $m := k - \ell + 1$ base-B digits. Write $q = (q_{m-1} \dots q_0)_B$.

At a high level, the strategy we shall use to compute q and r is the following:

$r \leftarrow a$

for $i \leftarrow m - 1$ down to 0 do

$q_i \leftarrow \lfloor r / B^i b \rfloor$

$r \leftarrow r - B^i \,.\, q_i b$

One easily verifies by induction that at the beginning of each loop iteration, we have $0 \leq r < B^{i+1} b$, and hence each q_i will be between 0 and $B - 1$, as required.

Turning the above strategy into a detailed algorithm takes a bit of work. In particular, we want an easy way to compute $\lfloor r / B^i b \rfloor$. Now, we could in theory just try all possible choices for *qi*—this would take time $O(B\ell)$, and viewing B as a constant, this is $O(\ell)$. However, this is not really very desirable from either a practical or theoretical point of view, and we can do much better with just a little effort.

We shall first consider a special case; namely, the case where $\ell = 1$. In this case, the computation of the quotient $\lfloor r / B^i b \rfloor$ is facilitated by the following theorem, which essentially tells us that this quotient is determined by the two high-order digits of r:

Theorem: Let x and y be integers such that

$$0 \leq x = x'2^n + s \text{ and } 0 < y = y'2^n$$

for some integers n, s, x', y', with $n \geq 0$ and $0 \leq s < 2^n$. Then $\lfloor x/y \rfloor = \lfloor x'/y' \rfloor$.

Proof: We have

$$\frac{x}{y} = \frac{x'}{y'} + \frac{s}{y'2^n} \geq \frac{x'}{y'}.$$

It follows immediately that $\lfloor x/y \rfloor \geq \lfloor x'/y' \rfloor$.

We also have

$$\frac{x}{y} = \frac{x'}{y'} + \frac{s}{y'2^n} < \frac{x'}{y'} + \frac{1}{y'} \leq \left(\left\lfloor \frac{x'}{y'} \right\rfloor + \frac{y'-1}{y'}\right) + \frac{1}{y'} \leq \left\lfloor \frac{x'}{y'} \right\rfloor + 1.$$

Thus, we have

$x/y < \lfloor x'/y' \rfloor + 1$, and hence, $\lfloor x/y \rfloor \leq \lfloor x'/y' \rfloor$.

From this theorem, one sees that the following algorithm correctly computes the quotient and remainder in time $O(k)$ (in the case $\ell = 1$):

$h_i \leftarrow 0$

for $i \leftarrow k - 1$ down to 0 do

tmp $\leftarrow$ $hi \,.\, B + a_i$

$(q_i, h_i) \leftarrow$ QuoRem(tmp, b_0)

output the quotient $q = (q_{k-1} \ldots q_0)_B$ and the remainder hi

Note that in every loop iteration, the value of hi lies between 0 and $b_0 \leq B - 1$, and the value of tmp lies between 0 and $B \,.\, b_0 + (B - 1) \leq B^2 - 1$.

That takes care of the special case where $\ell = 1$. Now we turn to the general case $\ell \geq 1$. In this case, we cannot so easily get the digits q_i of the quotient, but we can still fairly easily estimate these digits, using the following:

Theorem: Let x and y be integers such that

$$0 \leq x = x'\, 2^n + s \text{ and } 0 < y = y' 2^n + t$$

for some integers n, s, t, x', y' with $n \geq 0$, $0 \leq s < 2^n$, and $0 \leq t < 2^n$. Further, suppose that $2y' \geq x/y$. Then

$$\lfloor x/y \rfloor \leq \lfloor x'/y' \rfloor \leq \lfloor x/y \rfloor + 2.$$

Proof: We have $x/y \leq x/y'\, 2^n$, and so $\lfloor x/y \rfloor \leq \lfloor x/y'2^n \rfloor$, and by the previous theorem, $\lfloor x/y'2^n \rfloor = \lfloor x'/y' \rfloor$. That proves the first inequality.

For the second inequality, first note that from the definitions, we have $x/y \geq x'/(y' + 1)$, which implies $x'y - xy' - x \leq 0$. Further, $2y' \geq x/y$ implies $2yy' - x \geq 0$. So we have $2yy' - x \geq 0 \geq x'y - xy' - x$, which implies $x/y \geq x'/y' - 2$, and hence $\lfloor x/y \rfloor \geq \lfloor x'/y' \rfloor - 2$.

Based on this theorem, we first present an algorithm for division with remainder that works if we assume that b is appropriately "normalized," meaning that $b_{\ell-1} \geq 2^{w-1}$, where $B = 2^w$.

Some remarks are in order.

1. In line 4, we compute q_i, which by Theorem is greater than or equal to the true quotient digit, but exceeds this value by at most 2.
2. In line 5, we reduce qi if it is obviously too big.

3. In lines 6–10, we compute

$$(r_{i+\ell} \cdots r_i)B \leftarrow (r_{i+\ell} \cdots r_i)_B - q_i b.$$

In each loop iteration, the value of tmp lies between $-(B^2 - B)$ and $B - 1$, and the value carry lies between $-(B - 1)$ and 0.

1. for $i \leftarrow 0$ to $k - 1$ do $r_i \leftarrow a_i$
2. $r_k \leftarrow 0$
3. for $i \leftarrow k - \ell$ down to 0 do
4. $q_i \leftarrow \lfloor (r_{i+\ell}B + r_{i+\ell-1}) / b_{\ell-1} \rfloor$
5. if $q_i \geq B$ then $q_i \leftarrow B - 1$
6. carry $\leftarrow 0$
7. for $j \leftarrow 0$ to $\ell - 1$ do
8. tmp $\leftarrow r_{i+j} - q_i b_j$ + carry
9. (carry, r_{i+j}) $\leftarrow$ QuoRem(tmp, B)
10. $r_{i+\ell} \leftarrow r_{i+\ell}$ + carry
11. while $ri+\ell < 0$ do
12. carry $\leftarrow 0$
13. for $j \leftarrow 0$ to $\ell - 1$ do
14. tmp $\leftarrow r_{i+j} + b_i$ + carry
15. (carry, r_{i+j}) $\leftarrow$ QuoRem(tmp, B)
16. $r_{i+\ell} \leftarrow r_{i+\ell}$ + carry
17. $q_i \leftarrow q_i - 1$
18. output the quotient $q = (q_{k-\ell} \cdots q_0)_B$
    and the remainder $r = (r_{\ell-1} \cdots r_0)_B$

If the estimate q_i is too large, this is manifested by a negative value of $r_{i+\ell}$ at line 10. Lines 11–17 detect and correct this condition: the loop body here executes at most twice; in lines 12–16, we compute

$$(r_{i+\ell} \cdots r_i)_B \leftarrow (r_{i+\ell} \cdots r_i)_B + (b_{\ell-1} \cdots b_0)_B.$$

In every iteration of the loop in lines 13–15, the value of carry is 0 or 1, and the value tmp lies between 0 and $2B - 1$.

It is easily verified that the running time of the above algorithm is $O(\ell(k - \ell + 1))$.

Finally, consider the general case, where b may not be normalized. We multiply both a and b by an appropriate value $2^{w'}$, with $0 \leq w' < w$, obtaining $a' := a2^{w'}$ and $b' := b2^{w'}$, where b' is normalized; alternatively, we can use a more efficient, special-purpose "left shift" algorithm to achieve the same effect. We then compute q and r' such that $a' = b'q + r'$, using the division algorithm. Observe that $q = \lfloor a'/b' \rfloor = \lfloor a/b \rfloor$, and $r' = r2^{w'}$, where $r = a \bmod b$. To recover r, we simply divide r' by $2^{w'}$, which we can do either using the above "single precision" division algorithm, or by using a special-purpose "right shift" algorithm. All of this normalizing and denormalizing takes time $O(k + \ell)$. Thus, the total running time for division with remainder is still $O(\ell \cdot (k - \ell + 1))$.

EXERCISE

1. Work out the details for an algorithm that shifts a given unsigned integer a to the left by a specified number of bits s (i.e., computes $b := a \cdot 2^s$). The running time of your algorithm should be linear in the number of digits of the output.
2. Work out the details for an algorithm that shifts a given unsigned integer a to the right by a specified number of bits s (i.e., computes $b := \lfloor a/2^s \rfloor$). The running time of your algorithm should be linear in the number of digits of the output. Now modify your algorithm so that it correctly computes $\lfloor a/2^s \rfloor$ for signed integers a.
3. This exercise is for C/Java programmers. Evaluate the C/Java expressions

   (–17) % 4; (–17) & 3;

   and compare these values with (–17) mod 4. Also evaluate the C/Java expressions

   (–17) / 4; (–17) >> 2;

   and compare with $\lfloor -17/4 \rfloor$. Explain your findings.
4. Work out the details of algorithms for arithmetic on signed integers, using the above algorithms for unsigned integers as subroutines. You should give algorithms for addition, subtraction, multiplication, and division with remainder of arbitrary signed integers (for division with remainder, your algorithm should compute $\lfloor a/b \rfloor$ and $a \bmod b$). Make sure your algorithms correctly compute the sign bit of the results, and also strip any leading zero digits from the results.
5. Work out the details of an algorithm that compares two signed integers a and b, determining which of $a < b$, $a = b$, or $a > b$ holds.
6. Suppose that we run the division with remainder algorithm for $\ell > 1$ without normalizing b, but instead, we compute the value q_i in line 4 as follows:

$$q_i \leftarrow \left\lfloor (r_{i+\ell}B^2 + r_{i+\ell-1}B + r_{i+\ell-2})/(b_{\ell-1}B + b_{\ell-2}) \right\rfloor.$$

Show that q_i is either equal to the correct quotient digit, or the correct quotient digit plus 1. Note that a limitation of this approach is that the numbers involved in the computation are larger than B^2.

ASYMPTOTE

We review some standard notation for relating the rate of growth of functions. This notation will be useful in discussing the running times of algorithms, and in a number of other contexts as well.

Let f and g be real-valued functions. We shall assume that each is defined on the set of non-negative integers, or, alternatively, that each is defined on the set of non-negative reals. Actually, as we are only concerned about the behavior of $f(x)$ and $g(x)$ as $x \to \infty$, we only require that $f(x)$ and $g(x)$ are defined for all sufficiently large x (the phrase "for all sufficiently large x" means "for some x0 and all $x \geq x_0$"). We further assume that g is eventually positive, meaning that $g(x) > 0$ for all sufficiently large x. Then

- $f = O(g)$ means that $|f(x)| \leq cg(x)$ for some positive constant c and all sufficiently large x (read, "f is big-O of g"),
- $f = \Omega(g)$ means that $f(x) \geq cg(x)$ for some positive constant c and all sufficiently large x (read, "f is big-Omega of g"),
- $f = \Theta(g)$ means that $cg(x) \leq f(x) \leq dg(x)$ for some positive constants c and d and all sufficiently large x (read, "f is big-Theta of g"),
- $f = o(g)$ means that $f(x)/g(x) \to 0$ as $x \to \infty$ (read, "f is little-o of g"), and
- $f \sim g$ means that $f(x)/g(x) \to 1$ as $x \to \infty$ (read, "f is asymptotically equal to g").

Example. Let $f(x) := x^2$ and $g(x) := 2x^2 - 10x + 1$. Then $f = O(g)$ and $f = \Omega(g)$. Indeed, $f = \Theta(g)$.

Example. Let $f(x) := x^3 - 2x^2 + x - 3$. One could write $f(x) = x^3 + O(x^2)$. Here, the anonymous function is $g(x) := -2x^2 + x - 3$, and clearly $g(x) = O(x^2)$. One could also write $f(x) = x^3 - (2 + o(1))x^2$. Here, the anonymous function is $g(x) := -1/x + 3/x^2$. While $g = o(1)$, it is only defined for $x > 0$. This is acceptable, since we will only regard statements such as this asymptotically, as $x \to \infty$.

As an even further use (abuse?) of the notation, one may use the big-O, big- Omega, and big-Theta notation for functions on an arbitrary domain, in which case the relevant inequalities should hold throughout the entire domain. This usage includes functions of several independent variables, as well as functions defined on sets with no natural ordering.

Example. Let $f(x) := x^2$ and $g(x) := x^2 - 10x + 1$. Then $f \sim g$.

Example. Let $f(x) := 100x^2$ and $g(x) := x3$. Then $f = o(g)$.

Note that by definition, if we write $f = \Omega(g)$, $f = \Theta(g)$, or $f \sim g$, it must be the case that f (in addition to g) is eventually positive; however, if we write $f = O(g)$ or $f = o(g)$, then f need not be eventually positive.

When one writes "$f = O(g)$," one should interpret ". = O(.)" as a binary relation between f with g. Analogously for "$f = \Omega(g)$," "$f = \Omega(g)$," and "$f = o(g)$."

One may also write "$O(g)$" in an expression to denote an anonymous function f such that $f = O(g)$. Analogously, $\Omega(g)$, $\Theta(g)$, and $o(g)$ may denote anonymous functions. The expression $O(1)$ denotes a function bounded in absolute value by a constant, while the expression $o(1)$ denotes a function that tends to zero in the limit.

EXERCISE

1. Suppose that f and g are functions defined on the integers k, $k + 1$, ..., and that both f and g are eventually positive. For $n \geq k$, define $F(n) := \sum_{i=k}^{n} f(i)$ and $G(n) := \sum_{i=k}^{n} g(i)$. Show that if $f \sim g$ and $G(n)$ ® ¥ as n ® ¥, then $F \sim G$.
2. Suppose that f and g are piece-wise continuous on $[a, \infty)$ and that both f and g are eventually positive. For $x \geq a$, define $F(x) := \int_a^x f(t)$ and $G(x) := \int_a^x g(t)\,dt$. Show that if $f \sim g$ and $G(x) \to \infty$ as $x \to \infty$, then $F \sim G$.
3. Show that:
   (a) $f = o(g)$ implies $f = O(g)$ and $g \neq O(f)$;
   (b) $f = O(g)$ and $g = O(h)$ implies $f = O(h)$;
   (c) $f = O(g)$ and $g = o(h)$ implies $f = o(h)$;
   (d) $f = o(g)$ and $g = O(h)$ implies $f = o(h)$.
4. Suppose f and g are defined on the integers $i \geq k$, and that $g(i) > 0$ for all $i \geq k$. Show that if $f = O(g)$, then there exists a positive constant c such that $|f(i)| \leq cg(i)$ for all $i \geq k$.
5. Give an example of two non-decreasing functions f and g, each mapping positive integers to positive integers, such that $f \neq O(g)$ and $g \neq O(f)$.
6. Show that all of the claims in the previous exercise also hold when the relation "$\sim$" is replaced with the relation ". = Θ(.)."
7. Let f and g be eventually positive functions, and assume that $f(x)/g(x)$ tends to a limit L (possibly $L = \infty$) as $x \to \infty$. Show that:

(a) if $L = 0$, then $f = o(g)$;

(b) if $0 < L < \infty$, then $f = \Theta(g)$;

(c) if $L = \infty$, then $g = o(f)$.

8. Suppose that f and g are piece-wise continuous on $[a, \infty)$ and that g is eventually positive. For $x \geq a$, define $F(x) := \int_a^x f(t)dt$ and $G(x) := \int_a^x g(t)dt$. Show that if $f = O(g)$ and G is eventually positive, then $F = O(G)$.

9. Suppose that $f(x) \leq c + dg(x)$ for some positive constants c and d, and for all sufficiently large x. Show that if $g = \Omega(1)$, then $f = O(g)$.

10. Let f and g be eventually positive functions. Show that:

    (a) $f \sim g$ if and only if $f = (1 + o(1))g$;

    (b) $f \sim g$ implies $f = \Theta(g)$;

    (c) $f = \Theta(g)$ if and only if $f = O(g)$ and $f = \Omega(g)$;

    (d) $f = \Omega(g)$ if and only if $g = O(f)$.

11. Suppose $f_1 = O(g_1)$ and $f_2 = O(g_2)$. Show that $f_1 + f_2 = O(\max(g_1, g_2))$, $f_1 f_2 = O(g_1 g_2)$, and that for every constant c, $cf_1 = O(g_1)$.

12. Let f, g be eventually positive functions. Show that:

    (a) $f = \Theta(g)$ if and only if $\log f = \log g + O(1)$;

    (b) $f \sim g$ if and only if $\log f = \log g + o(1)$.

13. Let $f(x) := x^\alpha (\log x)^\beta$ and $g(x) := x^\gamma (\log x)^\delta$, where, α, β, γ, δ are non-negative constants. Show that if $\alpha < \gamma$, or if $\alpha = \gamma$ and $\beta < \delta$, then $f = o(g)$.

14. Order the following functions in x so that for each adjacent pair f, g in the ordering, we have $f = O(g)$, and indicate if $f = o(g)$, $f \sim g$, or $g = O(f)$:

    x^3, $e^x x^2$, $1/x$, $x^2(x + 100) + 1/x$, $x + \sqrt{x}$, $\log_2 x$, $\log_3 x$, $2x^2$, x,

    e^{-x}, $2x^2 - 10x + 4$, $e^{x+\sqrt{x}}$, 2^x, 3^x, x^{-2}, $x^2(\log x)^{1000}$.

15. Show that:

    (a) the relation "$\sim$" is an equivalence relation on the set of eventually positive functions;

    (b) for all eventually positive functions f_1, f_2, g_1, g_2, if $f_1 \sim g_1$ and $f_2 \sim g_2$, then $f_1 * f_2 \sim g_1 * g_2$, where "*" denotes addition, multiplication, or division;

    (c) for all eventually positive functions f, g, and every $\alpha > 0$, if $f \sim g$, then $f^\alpha \sim g^\alpha$;

(d) for all eventually positive functions f, g, and every function h such that $h(x) \to \infty$ as $x \to 1$, if $f \sim g$, then $f \circ h \sim g \circ h$, where "$\circ$" denotes function composition.

16. Suppose that f and g are functions defined on the integers $k, k+1, \ldots$, and that g is eventually positive. For $n \geq k$, define $F(n) = \sum_{i=k}^{n} f(i)$ and $G(n) = \sum_{i=k}^{n} g(i)$. Show that if $f = O(g)$ and G is eventually positive, then $F = O(G)$.

COMPLEX THEORY

When presenting an algorithm, we shall always use a high-level, and somewhat informal, notation. However, all of our high-level descriptions can be routinely translated into the machine-language of an actual computer.

So that our theorems on the running times of algorithms have a precise mathematical meaning, we formally define an "idealized" computer: the random access machine or RAM. A RAM consists of an unbounded sequence of memory cells

$$m[0], m[1], m[2], \ldots,$$

each of which can store an arbitrary integer, together with a program. A program consists of a finite sequence of instructions $I_0, I_1, \ldots$, where each instruction is of one of the following types:

- **Branching**: Instruction is of the form IF a ◊ b GOTO i, where i is the index of an instruction, and where ◊ is one of the comparison operations =, ¹, <, >, $\leq$, $\geq$, and a and b are as above. Execution of this type of instruction causes the "flow of control" to pass conditionally to instruction I_i.
- **Arithmetic:** Instruction is of the form g ¬ a * b, where * represents one of the operations addition, subtraction, multiplication, or integer division (i.e., ./.). The values a and b are of the form c, $m[a]$, or $m[m[a]]$, and g is of the form $m[a]$ or $m[m[a]]$, where c is an integer constant and a is a non-negative integer constant. Execution of this type of instruction causes the value a * b to be evaluated and then stored in g.
- **Halt:** The HALT instruction stops the execution of the program.

A RAM works by executing instruction *I*0, and continues to execute instructions, following branching instructions as appropriate, until a HALT instruction is reached.

We do not specify input or output instructions, and instead assume that the input and output are to be found in memory cells at some prescribed locations, in some standardized format. To determine the running time of a program on a given input, we charge 1 unit of time to each instruction executed. This model of computation closely resembles a typical modern-day computer, except that we have abstracted away many annoying details. However, there are two details of real machines that cannot be ignored; namely, any real machine has a finite number of memory cells, and each cell can store numbers only in some fixed range. The first limitation must be dealt with by either purchasing sufficient memory or designing more space-efficient algorithms. The second limitation is especially annoying, as we will want to perform computations with quite large integers – much larger than will fit into any single memory cell of an actual machine.

To deal with this limitation, we shall represent such large integers as vectors of digits in some fixed base, so that each digit is bounded in order to fit into a memory cell. This is discussed in more detail in the next section. The only other numbers we actually need to store in memory cells are "small" numbers representing array indices, counters, and the like, which we hope will fit into the memory cells of actual machines. Below, we shall make a more precise, formal restriction on the magnitude of numbers that may be stored in memory cells. Even with these caveats and restrictions, the running time as we have defined it for a RAM is still only a rough predictor of performance on an actual machine.

On a real machine, different instructions may take significantly different amounts of time to execute; for example, a division instruction may take much longer than an addition instruction. Also, on a real machine, the behavior of the cache may significantly affect the time it takes to load or store the operands of an instruction. Finally, the precise running time of an algorithm given by a high-level description will depend on the quality of the translation of this algorithm into "machine code." However, despite all of these problems, it still turns out that measuring the running time on a RAM as we propose here is a good "first order" predictor of performance on real machines in many cases. Also, we shall only state the running time of an algorithm using a big-O estimate, so that implementation-specific constant factors are anyway "swept under the rug." If we have an algorithm for solving a certain problem, we expect that "larger" instances of the problem will require more time to solve than "smaller" instances, and a general goal in the analysis of any algorithm is to estimate the rate of growth of the running time of the algorithm as a function of the size of its input.

For this purpose, we shall simply measure the size of an input as the number of memory cells used to represent it. Theoretical computer scientists sometimes equate the notion of "efficient" with "polynomial time" (although not everyone takes theoretical computer scientists very seriously, especially on this point): a polynomialtime algorithm is one whose running time on inputs of size n is at most $an^b + c$, for some constants a, b, and c (a "real" theoretical computer scientist will write this as $n^{O(1)}$).

Furthermore, we also require that for a polynomial-time algorithm, all numbers stored in memory are at most $a'n^{b'} + c'$ in absolute value, for some constants a', b', and c'. Even for algorithms that are not polynomial time, we shall insist that after executing t instructions, all numbers stored in memory are at most $a'(n + t)^{b'} + c'$ in absolute value, for some constants a', b', and c'. Note that in defining the notion of polynomial time on a RAM, it is essential that we restrict the magnitude of numbers that may be stored in the machine's memory cells, as we have done above. Without this restriction, a program could perform arithmetic on huge numbers, being charged just one unit of time for each arithmetic operation — not only is this intuitively "wrong," it is possible to come up with programs that solve some problems using a polynomial number of arithmetic operations on huge numbers, and these problems cannot otherwise be solved in polynomial time.

EXERCISE

1. Show that given integers $a, n_1, ..., n_k$, with each $n_i > 1$, where $0 \leq a < n := \prod_i n_i$, we can compute ($a$ mod n_1, ..., a mod n_k) in time $O(\text{len}(n)^2)$.
2. Show that given integers $n_1, ..., n_k$, with each $n_i > 1$, we can compute $(n/n_1, ..., n/n_k)$, where $n = \prod_i n_i$, in time $O(\text{len}(n)^2)$.
3. Let $a, b \neq \mathbb{Z}$ with $a \geq b > 0$, and let $q := \lfloor a/b \rfloor$. Show that len(a) – len (b) – 1 $\leq$ len (q) $\leq$ len(a) – len(b) + 1.
7. Modify the algorithm in the previous exercise so that given positive integers n and e, with $n \geq 2^e$, it computes $\lfloor n^{1/e} \rfloor$ in time $O(\text{len}(n)^3/e)$.
4. An integer $n > 1$ is called a perfect power if $n = a^b$ for some integers $a > 1$ and $b > 1$. Using the algorithm from the previous exercise, design an efficient algorithm that determines if a given n is a perfect power, and if it is, also computes a and b such that $n = a^b$, where $a > 1$, $b > 1$, and a is as small as possible. Your algorithm should run in time $O(\ell^3 \text{len}(\ell))$, where $\ell := \text{len}(n)$.
5. Show how to convert (in both directions) in time $O(\text{len}(n)^2)$ between the base-10 representation and our implementation's internal representation of an integer n.
6. Let $n_1, ..., n_k$ be positive integers. Show that

$$\sum_{i=1}^{k} len(n_i) - k \leq len\left(\sum_{i=1}^{k} n_i\right) \leq \sum_{i=1}^{k} len(n_i).$$

7. Show that given integers $n_1, ..., n_k$, with each $n_i > 1$, we can compute the product $n = \prod_i n_i$ in time $O(\text{len}(n)^2)$.

FASTER INTEGER ARITHMETIC

The quadratic-time algorithms presented for integer multiplication and division are by no means the fastest possible. The next exercise develops a faster multiplication algorithm.

EXERCISE

1. Give an algorithm that runs in time $O(M(\text{len}(n)) \text{len}(k))$.
2. Give an algorithm whose running time is $O(M(\text{len}(n)))$.
3. Suppose we have an algorithm that computes the square of an ℓ-bit integer in time at most $S(\ell)$, where S is a well-behaved complexity function.

   Show how to use this algorithm to compute the product of two arbitrary integers of length at most ' in time $O(S('))$.

4. Suppose we have two positive integers a and b, each of length at most ℓ, such that $a = a_1 2^k + a_0$ and $b = b_1 2^k + b_0$, where $0 \leq a_0 < 2^k$ and $0 \leq b_0 < 2^k$. Then

   $ab = a_1 b_1 2^{2k} + (a_0 b_1 + a_1 b_0) 2^k + a_0 b_0.$

   Show how to compute the product ab in time $O(\ell)$, given the products $a_0 b_0$, $a_1 b_1$, and $(a_0 - a_1)(b_0 - b_1)$. From this, design a recursive algorithm that computes ab in time $O(\ell^{\log_2 3})$.

The algorithm in the previous exercise is also not the best possible. In fact, it is possible to multiply two integers of length at most ℓ on a RAM in time $O(\ell)$, but we do not explore this any further for the moment.

The following exercises explore the relationship between integer multiplication and related problems. We assume that we have an algorithm that multiplies two integers of length at most ℓ in time at most $M(\ell)$. It is convenient (and reasonable) to assume that M is a well-behaved complexity function. By this, we mean that M maps positive integers to positive real numbers, such that for some constant $\gamma \geq 1$, and all positive integers a and b, we have

$$1 \leq \frac{M(a+b)}{M(a)+M(b)} \leq \gamma.$$

5. Show that if M is a well-behaved complexity function, then it is strictly increasing.
6. In the previous exercise, suppose all the inputs n_i have the same length, and that $M(\ell) = \alpha \ell \beta$, where α and β are constants with $\alpha > 0$ and $\beta > 1$.

   Show that your algorithm runs in time $O(M(\text{len}(n)))$.
7. We can represent a "floating point" number $\hat{z}$ as a pair (a, e), where a and e are integers – the value of $\hat{z}$ is the rational number $a2^e$, and we call len(a) the precision of $\hat{z}$. We say that $\hat{z}$ is a k-bit approximation of a real number z if $\hat{z}$ has precision k and $\hat{z} = (1 + \varepsilon)z$ for some $|\varepsilon| \leq 2^{-k+1}$. Show that given positive integers b and k, we can compute a k-bit approximation of $1/b$ in time $O(M(k))$.
8. Using the result of the previous exercise, show that, given positive integers a and b of bit length at most ℓ, we can compute $\lfloor a/b \rfloor$ and $a \bmod b$ in time $O(M(\ell))$. From this we see that, up to a constant factor, division with remainder is no harder than multiplication.
9. Give algorithms whose running times are $O(M(\ell)\ \text{len}(\ell))$, where $\ell := \text{len}(n)$.
10. Show that if $N(\ell) := M(\ell)/\ell$ is a non-decreasing function, and $M(2\ell)/M(\ell) = O(1)$, then M is a well-behaved complexity function.

11. Let $\alpha > 0, \beta \geq 1, \gamma \geq 0, \delta \geq 0$ be real constants. Show that

$M(\ell) := \alpha \ell^{\beta} \text{ len } (\ell)^{\gamma} \text{ len } (\text{len}(\ell))^{\delta}$

is a well-behaved complexity function.

COMPUTING WITH $\mathbb{Z}N$

Let us subppose that n be a positive integer. For every $\alpha \in \mathbb{Z}_n$, there exists a unique integer $a \in \{0, ..., n-1\}$ such that $\alpha = [a]_n$; we call this integer a the canonical representative of α, and denote it by rep(α). For computational purposes, we represent elements of $\mathbb{Z}_n$ by their canonical representatives. Addition and subtraction in $\mathbb{Z}_n$ can be performed in time $O(\text{len}(n))$: given , $\alpha, \beta \in \mathbb{Z}_n$, to compute rep($\alpha + \beta$), we first compute the integer sum rep(α) + rep(β), and then subtract n if the result is greater than or equal to n; similarly, to compute rep($\alpha - \beta$), we compute the integer difference rep(α) – rep(β), adding n if the result is negative.

Multiplication in $\mathbb{Z}_n$ can be performed in time $O(\text{len}(n)^2)$: given , $\alpha, \beta \in \mathbb{Z}_n$, we compute rep($\alpha.\beta$) as rep($\alpha$) rep($\beta$) mod n, using one integer multiplication and one division with remainder.A note on notation: "rep," "mod," and "$[.]_n$." In describing algorithms, as well as in other contexts, if α, β are elements of $\mathbb{Z}_n$, we may write, for example, $\gamma \leftarrow \alpha + \beta$ or $\gamma \leftarrow \alpha\beta$, and it is understood that elements of $\mathbb{Z}_n$ are represented by their canonical representatives as discussed above, and arithmetic on canonical representatives is done modulo n. Thus, we have in mind a "strongly typed" language for our pseudo-code that makes a clear distinction between integers in the set $\{0, ..., n-1\}$ and elements of $\mathbb{Z}_n$.

If $\alpha \in \mathbb{Z}$, we can convert a to an object $\alpha \in \mathbb{Z}_n$ by writing $\alpha \leftarrow [a]_n$, and if $a \in \{0, ..., n-1\}$, this type conversion is purely conceptual, involving no actual computation. Conversely, if $\alpha \in \mathbb{Z}_n$, we can convert α to an object $a \in \{0, ..., n-1\}$, by writing $a \leftarrow$ rep(α); again, this type conversion is purely conceptual, and involves no actual computation. It is perhaps also worthwhile to stress the distinction between a mod n and $[a]_n$–the former denotes an element of the set $\{0, ..., n-1\}$, while the latter denotes an element of $\mathbb{Z}_n$. Another interesting problem is exponentiation in $\mathbb{Z}_n$: given $\alpha \in \mathbb{Z}_n$ and a nonnegative integer e, compute $\alpha^e \in \mathbb{Z}_n$. Perhaps the most obvious way to do this is to iteratively multiply by α a total of e times, requiring time $O(e \text{ len}(n)^2)$.

For small values of e, this is fine; however, a much faster algorithm, the repeated-squaring algorithm, computes α^e using just $O(\text{len}(e))$ multiplications in $\mathbb{Z}_n$, thus taking time $O(\text{len}(e) \text{ len}(n)^2)$. This method is based on the following observation. Let $e = (b_{\ell-1} \cdots b_0)_2$ be the binary expansion of e (where b_0 is the low-order bit). For $i = 0, ..., \ell$, define $e_i := \left\lfloor e/2^i \right\rfloor$; the binary expansion of e_i is $e_i = (b_{\ell-1} \cdots b_i)_2$. Also define $\beta_i := \alpha^{e_i}$ for $i = 0, ..., \ell$, so $\beta_\ell = 1$ and $\beta_0 = \alpha^e$. Then we have

$$e_i = 2e_{i+1} + b_i \text{ and } \beta_i = \beta_{i+1}^2 . \beta^{b_i} \text{ for } i = 0, ..., \ell - 1.$$

THIS OBSERVATION YIELDS THE FOLLOWING ALGORITHM FOR COMPUTING A^E

MULTIPLICATIVE INVERSES

Suppose we are given a prime p and an element $\alpha \in \mathbb{Z}^*_p$, and we want to compute α^{-1}. By Euler's theorem, we have $\alpha^{p-1} = 1$, and multiplying this equation by α^{-1}, we obtain $\alpha^{p-2} = \alpha^{-1}$. Thus, we can use the repeated-squaring algorithm to compute α^{-1} by raising α to the power $p - 2$. This algorithm runs in time $O(\text{len}(p)^3)$.

TESTING FOR PRIMALITY

Suppose we are given an integer $n > 1$, and we want to determine whether n is prime or composite. For large n, searching for prime factors of n is hopelessly impractical. A better idea is to use Euler's theorem, combined with the repeated-squaring algorithm: we know that if n is prime, then every non-zero $\alpha \in \mathbb{Z}_n$ satisfies $\alpha^{n-1} = 1$. Conversely, if n is composite, there exists a non-zero $\alpha \in \mathbb{Z}_n$ such that $\alpha^{n-1} \neq 1$. This suggests the following "trial and error" strategy for testing if n is prime:

repeat k times

choose $\alpha \in \mathbb{Z}_n \setminus \{[0]\}$

compute $\beta \leftarrow \alpha^{n-1}$

if $\beta \neq 1$ output "composite" and halt

output "maybe prime"

As stated, this is not a fully specified algorithm: we have to specify the loopiteration parameter k, and more importantly, we have to specify a procedure for choosing α in each loop iteration. One approach might be to just try $\alpha = [1], [2], [3], \ldots$. Another might be to choose at random in each loop iteration: this would be an example of a probabilistic algorithm. In any case, if the algorithm outputs "composite," we may conclude that n is composite (even though the algorithm does not find a non-trivial factor of n).

However, if the algorithm completes all k loop iterations and outputs "maybe prime," it is not clear what we should conclude: certainly, we have some reason to suspect that n is prime, but not really a proof; indeed, it may be the case that n is composite, but we were just unlucky in all of our choices for α. Thus, while this rough idea does not quite give us an effective primality test, it is not a bad start, and is the basis for several effective primality tests.

THE REPEATED-SQUARING ALGORITHM

On input α, e, where $\alpha \in \mathbb{Z}_n$ and e is a nonnegative

integer, do the following, where $e = (b_{\ell-1} \ldots b_0)_2$ is the binary expansion of e:

$\beta \leftarrow [1]n$

for $i \leftarrow \ell - 1$ down to 0 do

$\beta \leftarrow \beta^2$

if $b_i = 1$ then $\beta \leftarrow \beta \, . \, \alpha$

output β

It is clear that when this algorithm terminates, we have $\beta = \alpha^e$, and that the running-time estimate is as claimed above. Indeed, the algorithm uses ℓ squarings in $\mathbb{Z}_n$, and at most ℓ additional multiplications in $\mathbb{Z}_n$.

TESTING QUADRATIC RESIDUOSITY

Suppose we are given an odd prime p and an element $\alpha \in \mathbb{Z}^*_p$, and we want to test whether $\alpha \in (\mathbb{Z}^*_p)^2$. By Euler's criterion, we have $\alpha \in (\mathbb{Z}^*_p)^2$ if and only if $\alpha^{(p-1)/2} = 1$. Thus, we can use the repeated-squaring algorithm to test if $\alpha \in (\mathbb{Z}^*_p)^2$ by raising to the power $(p - 1)/2$. This algorithm runs in time $O(\text{len}(p)^3)$. While this is also reasonably efficient, we will develop an even more efficient method later in the text.

Example. Suppose $e = 37 = (100101)_2$. The above algorithm performs the following operations in this case:

	// computed exponent (in binary)
$\beta \leftarrow [1]$	// 0
$\beta \leftarrow \beta^2, \beta \leftarrow \beta \, . \, \alpha$	// 1
$\beta \leftarrow \beta^2$	// 10
$\beta \leftarrow \beta^2$	// 100
$\beta \leftarrow \beta^2, \beta \leftarrow \beta \, . \, \alpha$	// 1001
$\beta \leftarrow \beta^2$	// 10010
$\beta \leftarrow \beta^2, \beta \leftarrow \beta.\alpha$	// 100101.

The repeated-squaring algorithm has numerous applications. We mention a few here, but we will see many more later on.

EXERCISE

1. Suppose we are given $\alpha \in \mathbb{Z}_n$, along with non-negative integers $e_1, ..., e_r$, where $\text{len}(e_i) \leq \ell$ for $i = 1, ..., r$, and $r = O(\text{len}(\ell))$. Using the previous exercise, show how to compute $(\alpha^{e_1}, ..., \alpha^{e_r})$ using $O(\ell)$ multiplications in $\mathbb{Z}_n$.

2. Suppose we are given $\alpha \in \mathbb{Z}_n$, along with integers $m_1, ..., m_r$, with each $m_i > 1$. Let $m := \prod_i m_i$. Also, for $i = 1, ..., r$, let $m_i^* := m/m_i$.

   Show how to compute $(\alpha^{m_1^*},...\alpha^{m_r^*})$ using $O(\text{len}(r)\,\ell)$ multiplications in $\mathbb{Z}_n$, where $\ell := \text{len}(m)$.

3. The repeated-squaring algorithm we have presented here processes the bits of the exponent from left to right (i.e., from high order to low order). Develop an algorithm for exponentiation in $\mathbb{Z}_n$ with similar complexity that processes the bits of the exponent from right to left.

4. Suppose we are given $\alpha_1, ..., \alpha_k \in \mathbb{Z}_n$, along with non-negative integers $e_1, ..., e_k$, where $\text{len}(e_i) \leq \ell$ for $i = 1, ..., k$. Show how to compute b := $\beta := \alpha_1^{e_1}...\alpha_k^{e_k}$, using at most ℓ squarings and $\ell + 2^k$ additional multiplications in $\mathbb{Z}_n$. Your algorithm should work in two phases: the first phase uses only the values $\alpha_1, ..., \alpha_k$, and performs at most 2^k multiplications in $\mathbb{Z}_n$; in the second phase, the algorithm computes β, using the exponents $e_1, ..., e_k$, along with the data computed in the first phase, and performs at most ℓ squarings and ℓ additional multiplications in $\mathbb{Z}_n$.

5. Suppose that we are to compute a^e, where $\alpha \in \mathbb{Z}_n$, for many exponents e of length at most ℓ, but with fixed. Show that for every positive integer parameter k, we can make a pre-computation (depending on α, ℓ, and k) that uses at most ℓ squarings and $2k$ additional multiplications in $\mathbb{Z}_n$, so that after the pre-computation, we can compute e for every exponent e of length at most ℓ using at most $\ell/k + O(1)$ squarings and $\ell/k + O(1)$ additional multiplications in $\mathbb{Z}_n$.

6. Show that given a prime p, $\alpha \in \mathbb{Z}_p$, and an integer $e \geq p$, we can compute e in time $O(\text{len}(e)\,\text{len}(p) + \text{len}(p)^3)$. The following exercises develop some important efficiency improvements to the basic repeated-squaring algorithm.

7. Let k be a constant, positive integer. Suppose we are given $\alpha_1, ..., \alpha_k \in \mathbb{Z}_n$, along with non-negative integers $e_1, ..., e_k$, where $\text{len}(e_i) \leq \ell$ for $i = 1, ..., k$. Show how to compute the value $\alpha_1^{e_1}...\alpha_k^{e_k}$, using at most ℓ squarings and $O(\ell/\text{len}(\ell))$ additional multiplications in $\mathbb{Z}_n$.

8. The goal of this exercise is to develop a "2^t-ary" variant of the above repeated-squaring algorithm, in which the exponent is effectively treated as a number in base 2^t, for some parameter t, rather than in base 2. Let $\alpha \in .\mathbb{Z}_n$ and let e be a positive integer of length ℓ. Let us write e in base 2^t as $e = (e_k ... e_0)2^t$, where $e_k \neq 0$. Consider the following algorithm:

compute a table of values $T[0 \ldots 2^t - 1]$,

where $T[j] := \alpha^j$ for $j = 0, \ldots, 2^t - 1$

$\beta \leftarrow T[e_k]$

for $i \leftarrow k - 1$ down to 0 do

$\beta \leftarrow \beta^{2^t} . T[e_i]$

(a) Show that this algorithm correctly computes α^e, and work out the implementation details; in particular, show that it may be implemented in such a way that it uses at most ℓ squarings and $2^t + \ell/t + O(1)$ additional multiplications in $\mathbb{Z}_n$.

(b) Improve the algorithm so that it only uses no more than ℓ squarings and $2^{t-1} + \ell/t + O(1)$ additional multiplications in $\mathbb{Z}_n$.

(c) Show that, by appropriately choosing the parameter t, we can bound the number of multiplications in $\mathbb{Z}_n$ (besides the squarings) by $O(\ell/\text{len}(\ell))$. Thus, from an asymptotic point of view, the cost of exponentiation is essentially the cost of about ℓ squarings in $\mathbb{Z}_n$.

Chapter 3

Prime Numbers

In mathematics, a prime number (or a prime) is a natural number greater than 1 which has exactly two distinct natural number divisors: 1 and itself. An infinitude of prime numbers exists, as demonstrated by Euclid around 300 BC. The first thirty prime numbers are:

2, 3, 5, 7, 11, 13, 17, 19, 23, 29, 31, 37, 41, 43, 47, 53, 59, 61, 67, 71, 73, 79, 83, 89, 97, 101, 103, 107, 109, 113.

See the list of prime numbers for a longer list. The number one is by definition not a prime number; see the discussion below under *Primality of one*.

The property of being a prime is called primality, and the word prime is also used as an adjective. Since two is the only even prime number, the term odd prime refers to any prime number greater than two.

The study of prime numbers is part of number theory, the branch of mathematics which encompasses the study of natural numbers. Prime numbers have been the subject of intense research, yet some fundamental questions, such as the Riemann hypothesis and the Goldbach conjecture, have been unresolved for more than a century. The problem of modelling the distribution of prime numbers is a popular subject of investigation for number theorists: when looking at individual numbers, the primes seem to be randomly distributed, but the "global" distribution of primes follows well-defined laws.

The notion of prime number has been generalized in many different branches of mathematics.

- In ring theory, a branch of abstract algebra, the term "prime element" has a specific meaning. Here, a non-zero, non-unit ring element a is defined to be prime if whenever a divides bc for ring elements b and c, then a divides at least one of b or c. With this meaning, the additive inverse of any prime number is also prime. In other words, when considering the set of integers as a ring, –7 is a prime element. Without further specification, however, "prime number" always means a positive integer prime. Among rings of complex algebraic integers, Eisenstein primes and Gaussian primes may also be of interest.

- In knot theory, a prime knot is a knot which can not be written as the knot sum of two lesser nontrivial knots.

PRIME DIVISORS

The fundamental theorem of arithmetic states that every positive integer larger than 1 can be written as a product of one or more primes in a way which is unique except possibly for the order of the prime factors. The same prime factor may occur multiple times. Primes can thus be considered the "basic building blocks" of the natural numbers. For example, we can write

$$23244 = 22 \times 3 \times 13 \times 149$$

and any other factorization of 23244 as the product of primes will be identical except for the order of the factors. There are many prime factorization algorithms to do this in practice for larger numbers.

The importance of this theorem is one of the reasons for the exclusion of 1 from the set of prime numbers. If 1 were admitted as a prime, the precise statement of the theorem would require additional qualifications.

PROPERTIES OF PRIMES

- When written in base 10, all prime numbers except 2 and 5 end in 1, 3, 7 or 9. (Numbers ending in 0, 2, 4, 6 or 8 represent multiples of 2 and numbers ending in 0 or 5 represent multiples of 5.)
- If p is a prime number and p divides a product ab of integers, then p divides a or p divides b. This proposition was proved by Euclid and is known as Euclid's lemma. It is used in some proofs of the uniqueness of prime factorizations.
- The ring Z/nZ is a field if and only if n is a prime. Put another way: n is prime if and only if $\varphi(n) = n - 1$.
- If p is prime and a is any integer, then $a^p - a$ is divisible by p (Fermat's little theorem).
- If p is a prime number other than 2 and 5, $^1/_p$ is always a recurring decimal, whose period is $p - 1$ or a divisor of $p - 1$. This can be deduced directly from Fermat's little theorem. $^1/_p$ expressed likewise in base q (other than base 10) has similar effect, provided that p is not a prime factor of q. The article on recurring decimals shows some of the interesting properties.
- An integer $p > 1$ is prime if and only if the factorial $(p - 1)! + 1$ is divisible by p (Wilson's theorem). Conversely, an integer $n > 4$ is composite if and only if $(n - 1)!$ is divisible by n.
- If n is a positive integer greater than 1, then there is always a prime number p with $n < p < 2n$ (Bertrand's postulate).

- Adding the reciprocals of all primes together results in a divergent infinite series (proof). More precisely, if $S(x)$ denotes the sum of the reciprocals of all prime numbers p with p £ x, then $S(x)$ = ln ln x + O(1) for x ® ¥.
- In every arithmetic progression a, $a + q$, $a + 2q$, $a + 3q$,... where the positive integers a and q are coprime, there are infinitely many primes (Dirichlet's theorem on arithmetic progressions).
- The characteristic of every field is either zero or a prime number.
- If G is a finite group and p^n is the highest power of the prime p which divides the order of G, then G has a subgroup of order p^n. (Sylow theorems.)
- If G is a finite group and p is a prime number dividing the order of G, then G contains an element of order p. (Cauchy Theorem)
- The prime number theorem says that the proportion of primes less than x is asymptotic to $^1/_{\ln x}$ (in other words, as x gets very large, the likelihood that a number less than x is prime is inversely proportional to the number of digits in x).
- The Copeland-Erdυs constant 0.23571113171923293137 4143..., obtained by concatenating the prime numbers in base ten, is known to be an irrational number.
- The value of the Riemann zeta function at each point in the complex plane is given as a meromorphic continuation of a function, defined by a product over the set of all primes for Re(s) > 1:

$$\varsigma(s) = \sum_{n=1}^{\infty} \frac{1}{n^s} = \prod_p \frac{1}{1 - p^{-s}}$$

Evaluating this identity at different integers provides an infinite number of products over the primes whose values can be calculated, the first two being

$$\prod_p \frac{1}{1 - p^{-s}} = \infty$$

$$\prod_p \frac{1}{1 - p^{-s}} = \frac{\pi^2}{6}$$

- If $p > 1$, the polynomial is irreducible over Z/pZ if and only if p is prime.
- All prime numbers above 3 are of form $6n - 1$ or $6n + 1$, because all other numbers are divisible by 2 or 3. Generalizing this, all prime numbers above q are of form q # $\cdot n + m$, where $0 < m < q$, and m has no prime factor £ q.

CLASSIFICATION OF PRIME NUMBERS

Two ways of classifying prime numbers, class n+ and class n", were studied by Paul Erdõs and John Selfridge.

Determining the class $n+$ of a prime number p involves looking at the largest prime factor of $p + 1$. If that largest prime factor is 2 or 3, then p is class 1+. But if that largest prime factor is another prime q, then the class $n+$ of p is one more than the class $n+$ of q. The class n^- is almost the same as class n^+, except that the factorization of $p - 1$ is looked at instead.

THE NUMBER OF PRIME NUMBERS

THERE ARE INFINITELY MANY PRIME NUMBERS

The oldest known proof for the statement that there are infinitely many prime numbers is given by the Greek mathematician Euclid in his *Elements*. Euclid states the result as "there are more than any given [finite] number of primes", and his proof is essentially the following:

Consider any finite set of primes. Multiply all of them together and add one. The resulting number is not divisible by any of the primes in the finite set we considered, because dividing by any of these would give a remainder of one. Because all non-prime numbers can be decomposed into a product of underlying primes, then either this resultant number is prime itself, or there is a prime number or prime numbers which the resultant number could be decomposed into but are not in the original finite set of primes. Either way, there is at least one more prime that was not in the finite set we started with. This argument applies no matter what finite set we began with. So there are more primes than any given finite number.

This previous argument explains why the product P of finitely many primes plus 1 must be divisible by some prime not among those finitely many primes (possibly itself).

The proof is sometimes phrased in a way that leads the student to conclude that $P + 1$ must itself be prime, and think that Euclid's proof says the prime product plus 1 is always prime. In fact, any set of primes which does not include 2 will have an odd product. Adding 1 to this product will always produce an even number, which will be divisible by 2 (and therefore not be prime).

Even though the total number of primes is infinite, one could still ask "Approximately how many primes are there below 100,000?", or "How likely is a random 20-digit number to be prime?".

The prime-counting function ð(x) is defined as the number of primes up to x. There are known algorithms to compute exact values of ð(x) faster than it would be possible to compute each prime up to x. Values as large as ð(10^{20}) can be calculated quickly and accurately with modern computers. Thus, e.g., ð(100000) = 9592, and ð(10^{20}) = 2,220,819,602,560,918,840.

For larger values of x, beyond the reach of modern equipment, the prime number theorem provides a good estimate: ð(x) is approximately $x/\ln(x)$. Even better estimates are known.

THE PRIME NUMBER THEOREM

In this section, we survey a number of theorems and conjectures related to the distribution of primes. This is a vast area of mathematical research, with a number of very deep results. We shall be stating a number of theorems from the literature in this section without proof; while our intent is to keep the text as self contained as possible, and to avoid degenerating into "mathematical tourism," it nevertheless is a good idea to occasionally have a somewhat broader perspective.

The main theorem in the theory of the density of primes is the following.

Theorem: We have

$$\pi(x) \sim x/\log x.$$

EXERCISE

1. Using the prime number theorem, show that Bertrand's postulate can be strengthened (asymptotically) as follows: for every $\varepsilon > 0$, there exist positive constants c and x_0, such that for all $x \geq x_0$, we have

$$\pi((1+\varepsilon)x) - \pi(x) \geq c\frac{x}{\log x}.$$

2. Using the prime number theorem, show that $\vartheta(x) \sim x$.
3. Using the prime number theorem, show that $p_n \sim n \log n$, where p_n denotes the nth prime.

ERRORSIN THE PRIME NUMBER THEOREM

The prime number theorem says that

$$|\,\pi(x) - x/\log x\,| \leq \delta(x),$$

where $\delta(x) = o(x/\log x)$. A natural question is: how small is the "error term" $\delta(x)$? It can be shown that

$$\pi(x) = x/\log x + O(x/(\log x)^2).$$

This bound on the error term is not very impressive, but unfortunately, cannot be improved upon. The problem is that $x/\log x$ is not really the best "simple" function that approximates $\pi(x)$. It turns out that a better approximation to $\pi(x)$ is the logarithmic integral, defined for all real numbers $x \geq \in$ as

$$li(x) := \int_2^x \frac{dt}{\log t}.$$

It is not hard to show that

$$\mathrm{li}(x) = x/\log x + O(x/(\log x)^2).$$

Thus, $\mathrm{li}(x) \sim x/\log x \sim \pi(x)$. However, the error term in the approximation of $\pi(x)$ by $\mathrm{li}(x)$

is much better. This is illustrated numerically in Table for example, at $x = 1018$, li(x) approximates $\pi(x)$ with a relative error just under 10^{-9}, while $x/\log x$ approximates $\pi(x)$ with a relative error of about 0.025.

The sharpest proven result on the error in approximating $\pi(x)$ by li(x) is the following:

Theorem: Let $\kappa(x) := (\log x)^{3/5} (\log \log x)^{-1/5}$. Then for some $c > 0$, we have

$$\pi(x) = \text{li}(x) + O(xe^{-c\kappa(x)}).$$

Proof:

Table: Values of $\pi(x)$, li(x), and $x/\log x$

x	$\pi(x)$	$li(x)$	$x/\log x$
10^3	168	176.6	144.8
10^6	78498	78626.5	72382.4
10^9	50847534	50849233.9	48254942.4
10^{12}	37607912018	37607950279.8	36191206825.3
10^{15}	29844570422669	29844571475286.5	28952965460216.8
10^{18}	24739954287740860	24739954309690414.0	24127471216847323.8

Note that the error term $xe^{-c\kappa(x)}$ is $o(x/(\log x)^k)$ for every fixed $k \geq 0$.

Although the above estimate on the error term in the approximation of $\pi(x)$ by li(x) is pretty good, it is conjectured that the actual error term is much smaller:

Conjecture: For all $x \geq 2.01$, we have

$$| \pi(x) - \text{li}(x) | < x^{1/2} \log x.$$

This conjecture is equivalent to the famous Riemann hypothesis, which is a conjecture about the location of the zeros of a certain function, called Riemann's zeta function. We give a very brief, high-level account of this conjecture, and its connection to the theory of the distribution of primes.

For all real numbers $s > 1$, the zeta function is defined as

$$\zeta(s) := \sum_{n=1}^{\infty} \frac{1}{n^s}.$$

Euler's identity

Note that because $s > 1$, the infinite series defining $\zeta(s)$ converges. A simple, but important, connection between the zeta function and the theory of prime numbers is the following:

Theorem: For every real number $s > 1$, we have

$$\zeta(s) = \prod_p (1 - p^{-s})^{-1},$$

where the product is over all primes p.

Proof. The rigorous interpretation of the infinite product on the right-hand side of is as a limit of finite products. Thus, if p_i denotes the ith prime, for $i = 1, 2, \ldots$, then we are really proving that

$$\zeta(s) = \lim_{r\to\infty} \prod_{i=1}^{r} (1 - p_i^{-s})^{-1}.$$

Now, from the identity

$$(1 - p_i^{-s})^{-1} = \prod_{e=0}^{\infty} p_i^{-es},$$

we have

$$\prod_{i=1}^{r} (1 - p_i^{-s})^{-1} = (1 + p_1^{-s} + p_1^{-2s} + \ldots)(1 + p_r^{-s} + p_r^{-2s} + \ldots)$$

$$= \sum_{n=1}^{\infty} \frac{h_r(n)}{n^s}.$$

where

$$h_r(n) := \begin{cases} 1 & \text{if } n \text{ is divisible only by the primes } p_1, \ldots, p_r; \\ 0 & \text{otherwise.} \end{cases}$$

Here, we have made use of the fact that we can multiply term-wise infinite series with non-negative terms. Now, for every $\varepsilon > 0$, there exists n_0 such that $\sum_{n=n_0}^{\infty} n^{-s} < \varepsilon$ (because the series defining $\zeta(s)$ converges). Moreover, there exists an r_0 such that $h_r(n) = 1$ for all $n < n_0$ and $r \geq r_0$. Therefore, for all $r \geq r_0$, we have

$$\left| \sum_{n=1}^{\infty} \frac{h_r(n)}{n^s} - \zeta(s) \right| \leq \sum_{n=n_0}^{\infty} n^{-s} < \varepsilon.$$

It follows that

$$\lim_{r\to\infty}\sum_{n=1}^{\infty}\frac{h_r(n)}{n^s}=\zeta(s),$$

which proves the theorem.

The domain of definition of the zeta function to the complex plane. For the reader who is familiar with just a little complex analysis, it is easy to see that the infinite series defining the zeta function in converges absolutely for all complex numbers s whose real part is greater than 1, and that holds as well for such s.

However, it is possible to extend the domain of definition of $\zeta(s)$ even further – in fact, one can extend the definition of $\zeta(s)$ in a "nice way" (in the language of complex analysis, analytically continue) to the entire complex plane (except the point $s = 1$, where there is a simple pole). Exactly how this is done is beyond the scope of this text, but assuming this extended definition of $\zeta(s)$.

Riemann hypothesis

Suppose s is a complex number with $s = x + yi$, where $x, y \in \mathbb{R}$, such that $\zeta(s) = 0$ and $0 < x < 1/2$. Then $x = 1/2$.

A lot is known about the zeros of the zeta function in the "critical strip," which consists of those points s whose real part is greater than 0 and less than 1: it is known that there are infinitely many such zeros, and there are even good estimates about their density. It turns out that one can apply standard tools in complex analysis, like contour integration, to the zeta function (and functions derived from it) to answer various questions about the distribution of primes. Indeed, such techniques may be used to prove the prime number theorem. However, if one assumes the Riemann hypothesis, then these techniques yield much sharper results.

EXERCISE

1. For any arithmetic function a (mapping positive integers to reals), we can form the Dirichlet series

$$F_a(s) := \sum_{n=1}^{\infty}\frac{a(n)}{n^s}.$$

For simplicity we assume that s takes only real values, even though such series are usually studied for complex values of s.

(a) Show that if the Dirichlet series $F_a(s)$ converges absolutely for some real s, then it converges absolutely for all real $s' \geq s$.

(b) From part (a), conclude that for any given arithmetic function a, there is an interval of absolute convergence of the form (s_0, ∞), where we allow $s_0 = -\infty$ and $s_0 = \infty$, such that $F_a(s)$ converges absolutely for $s > s_0$, and does not converge absolutely for $s < s_0$.

(c) Let a and b be arithmetic functions such that $F_a(s)$ has an interval of absolute convergence (s_0, ∞) and $F_b(s)$ has an interval of absolute convergence (s_0', ∞), and assume that $s_0 < \infty$ and $s_0' < \infty$. Let $c := a * b$ be the Dirichlet product of a and b. Show that for all $s \in (\max(s_0, s_0'), \infty)$, the series $F_c(s)$ converges absolutely and, moreover, that $F_a(s) F_b(s) = F_c(s)$.

PRIMES IN ARITHMETIC PROGRESSIONS

Earlier we proved that there are infinitely many primes $p \equiv 1 \pmod 4$ and infinitely many primes $p \equiv 3 \pmod 4$. These results are actually special cases of a much more general result.

Let d be a positive integer, and let a be any integer. An arithmetic progression with first term a and common difference d consists of all integers of the form

$$a + dm, m = 0, 1, 2, ...$$

The question is: under what conditions does such an arithmetic progression contain infinitely many primes? An equivalent formulation is: under what conditions are there infinitely many primes $p \equiv a \pmod d$? If a and d have a common factor $c > 1$, then every term in the progression is divisible by c, and so there can be at most one prime in the progression. So a necessary condition for the existence of infinitely many primes $p \equiv a \pmod d$ is that $\gcd(a, d) = 1$. A famous theorem due to Dirichlet states that this is a sufficient condition as well.

Dirichlet's theorem

Theorem : Let $a, d \in \mathbb{Z}$ with $d > 0$ and $\gcd(a, d) = 1$. Then there are infinitely many primes $p \in a \pmod d$.

We can also ask about the density of primes in arithmetic progressions. One might expect that for a fixed value of d, the primes are distributed in roughly equal measure among the $\varphi(d)$ different residue classes $[a]_d$ with $\gcd(a, d) = 1$ (here, φ is Euler's phi function). This is in fact the case. To formulate such assertions, we define $\pi(x; d, a)$ to be the number of primes p up to x with $p \equiv a \pmod d$.

Theorem: Let $a, d \in \mathbb{Z}$ with $d > 0$ and $\gcd(a, d) = 1$. Then

$$\pi(x; d, a) \sim \frac{1}{\varphi(d) \log x}.$$

Proof: Literature

The above theorem is only applicable in the case where d and a are fixed as $x \to \infty$. For example, it says that roughly half the primes up to x are congruent to 1 modulo 4, and roughly half the primes up to x are congruent to 3 modulo 4. However, suppose $d \to \infty$, and we want to estimate, say, the number of primes $p \equiv 1 \pmod d$ up to d^3. The following conjecture does, however:

Conjecture: Let $x \in \mathbb{R}$, $a, d \in \mathbb{Z}$ with $x \geq 2$, $d \geq 2$, and $\gcd(a, d) = 1$. Then

$$\left|\pi(x;d,a) - \frac{li(x)}{\varphi(d)}\right| \leq x^{1/2}(\log x + \in \log d).$$

The above conjecture is in fact a consequence of a generalization of the Riemann hypothesis. This conjecture implies that for every constant $\alpha < 1/2$, if $\in \leq d \leq x^{\alpha}$, then $\pi(x; d, a)$ is closely approximated by $\mathrm{li}(x)/\varphi(d)$. It can also be used to get an upper bound on the least prime $p \equiv a \pmod{d}$. The following theorem is the best rigorously proven upper bound on the smallest prime in an arithmetic progression:

Theorem: There exists a constant c such that for all $a, d \in \mathbb{Z}$ with $d \geq \in$ and $\gcd(a, d) = 1$, the least prime $p \equiv a \pmod{d}$ is at most $cd^{11/2}$.

Chebyshev's theorem

The natural way of measuring the density of primes is to count the number of primes up to a bound x, where x is a real number. To this end, we introduce the function $\pi(x)$, whose value at each real number $x \geq 0$ is defined to be the number of primes up to (and including) x. It might seem more natural to define $\pi(x)$ only on the integers, but it is the tradition to define it over the real numbers (and there are some technical benefits in doing so).

Let us first take a look at some values of $\pi(x)$. Table shows values of $\pi(x)$ for $x = 10^{3i}$ and $i = 1, \ldots, 6$. The third column of this table shows the value of $x/\pi(x)$ (to five decimal places). One can see that the differences between successive rows of this third column are roughly the same – about 6.9 – which suggests that the function $x/\pi(x)$ grows logarithmically in x. Indeed, as $\log(10^3) \approx 6.9$, it would not be unreasonable to guess that $x/\pi(x) \approx \log x$, or equivalently, $\pi(x) \approx x/\log x$ (as discussed in the Preliminaries, $\log x$ denotes the natural logarithm of x).

The following theorem is a first – and important – step towards making the above guesswork more rigorous:

Table : Some values of $\pi(x)$

x	$\pi(x)$	$x/\pi(x)$
10^3	168	5.95238
10^6	78498	12.73918
10^9	50847534	19.66664
10^{12}	37607912018	26.59015
10^{15}	29844570422669	33.50693
10^{18}	24739954287740860	40.42045

Theorem: We have

$$\pi(x) = \Theta(x/\log x).$$

It is not too diffcult to prove this theorem, which we now proceed to do in several steps. We begin with some elementary bounds on binomial coe—cients:

Lemma: If m is a positive integer, then

$$\binom{2m}{m} \geq 2^{2m}/2m \text{ and } \binom{2m+1}{m} < 2^{2m}.$$

Proof: As $\binom{2m}{m}$ is the largest binomial coeffcient in the binomial expansion of $(1 + 1)^{2m}$, we have

$$2^{2m} = \sum_{i=0}^{2m}\binom{2m}{i} = 1 + \sum_{i=1}^{2m-1}\binom{2m}{i} + 1 \leq 2 + (2m-1)\binom{2m}{m} \leq 2m\binom{2m}{m}.$$

The proves the first inequality. For the second, observe that the binomial coeffcient $\binom{2m+1}{m}$ occurs twice in the binomial expansion of $(1 + 1)^{2m+1}$, and is therefore less than $2^{2m+1}/2 = 2^{2m}$.

Next, recalling that $v_p(n)$ denotes the power to which a prime p divides an integer n, we continue with the following observation:

Lemma: Let n be a positive integer. For every prime p, we have

$$v_p(n!) = \sum_{k\geq 1}\lfloor n/p^k \rfloor.$$

Proof. For all positive integers j, k, define $d_{jk} := 1$ if $p^k \mid j$, and $d_{jk} := 0$, otherwise. Observe that $v_p(j) = \sum_{k\geq 1} d_{jk}$ (this sum is actually finite, since $d_{jk} = 0$ for all suffciently large k). So we have

$$v_p(n!) = \sum_{j=1}^{n} v_p(j) = \sum_{j=1}^{n}\sum_{k\geq 1} d_{jk} = \sum_{k\geq 1}\sum_{j=1}^{n} d_{jk}.$$

Finally, note that $\sum_{j=1}^{n} d_{jk}$ is equal to the number of multiples of p^k among the integers 1, ..., n, is equal to $\lfloor n/p^k \rfloor$.

The following theorem gives a lower bound on $\pi(x)$.

Theorem: $\pi(n) \geq \frac{1}{2}$ $(\log 2)n/ \log n$ for every integer $n \geq 2$.

Proof: Let m be a positive integer, and consider the binomial coefficient

$$N := \binom{2m}{m} = \frac{(2m)!}{(m!)^2}$$

It is clear that N is divisible only by primes p up to $2m$. Applying Lemma 5.3 to the identity $N = (2m)!/(m!)^2$, we have

$$v_p(N) = \sum_{k \geq 1} \left(\lfloor 2m/p^k \rfloor - 2\lfloor m/p^k \rfloor \right).$$

Each term in this sum is either 0 or 1 and for $k > \log(2m)/\log p$, each term is zero. Thus, $v_p(N) \leq \log(2m)/\log p$. So we have

$$\pi(2m)\log(2m) = \sum_{p \leq 2m} \frac{\log(2m)}{\log p} \log p$$

$$\geq \sum_{p \leq 2m} v_p(N) \log p = \log N,$$

where the summations are over the primes p up to $2m$. By Lemma, we have $N \geq 2^{2m}/2m \geq 2^m$, and hence

$$\pi(2m)\log(2m) \geq m \log \in = \frac{1}{2}(\log 2)(2m).$$

That proves the theorem for even n. Now consider odd $n \geq 3$, so $n = 2m - 1$ for some $m \geq 2$. It is easily verified that the function $x/\log x$ is increasing for $x \geq 3$; therefore,

$$\pi(2m - 1) = \pi(2m)$$

$$\geq \frac{1}{2}(\log 2)(2m)/\log(2m)$$

$$\geq \frac{1}{2}(\log 2)(2m - 1)/\log(2m - 1).$$

That proves the theorem for odd n.

As a consequence of the above theorem, we have $\pi(x) = \Omega(x/\log x)$. Indeed, for every real number $x \geq 2$, setting $c := \frac{1}{2}(\log 2)$, we have

$$\pi(x) = \pi(\lfloor x \rfloor) \geq c\lfloor x \rfloor / \log \lfloor x \rfloor \geq c(x - 1) = \log x = \Omega(x/\log x).$$

To obtain a corresponding upper bound for $\pi(x)$, we introduce an auxiliary function, called Chebyshev's theta function:

$$\vartheta(x) := \sum_{p \le x} \log p,$$

where the sum is over all primes p up to x.

Chebyshev's theta function is an example of a summation over primes, and in this chapter, we will be considering a number of functions that are defined in terms of sums or products over primes. To avoid excessive tedium, we adopt the usual convention used by number theorists: if not explicitly stated, summations and products over the variable p are always understood to be over primes. For example, we may write $\pi(x) = \sum_{p \le x} 1$.

Theorem: We have

$$\vartheta(x) = \Theta(\pi(x) \log x).$$

Proof: On the one hand, we have

$$\vartheta(x) := \sum_{p \le x} \log p \le \log x \sum_{p \le x} 1 = \pi(x) \log x.$$

On the other hand, we have

$$\vartheta(x) = \sum_{p \le x} \log p \ge \sum_{x^{1/2} < p \le x} \log p \ge \frac{1}{2} \log x \sum_{x^{1/2} < p \le x} 1$$

$$= \frac{1}{2} \log(\pi(x) - \pi(x^{1/2})) = \frac{1}{2} (1 - \pi(x^{1/2}) / \pi(x) \pi(x) \log x.$$

It will therefore suffice to show that $\pi(x^{1/2})/\pi(x) = o(1)$. Clearly, $\pi(x^{1/2}) \le x^{1/2}$. Moreover, by the previous theorem, $\pi(x) = \Omega(x/ \log x)$. Therefore,

$$\pi(x^{1/2})/\pi(x) = O(\log x/x^{1/2}) = o(1),$$

and the theorem follows.

Theorem: $\vartheta(x) < 2(\log 2)x$ for every real number $x \ge 1$.

Proof: It suffices to prove that $\vartheta(n) < 2(\log 2)n$ for every positive integer n, since then $\vartheta(x) = \vartheta(\lfloor x \rfloor) < 2(\log 2)\lfloor x \rfloor \le 2(\log 2)x$. We prove this by induction on n.

For $n = 1$ and $n = 2$, this is clear, so assume $n > 2$. If n is even, then using the induction hypothesis for $n - 1$, we have

$$\vartheta(n) = \vartheta(n-1) < 2(\log 2)(n-1) < 2(\log 2)n.$$

Now consider the case where n is odd. Write $n = 2m + 1$, where m is a positive integer, and consider the binomial coefficient

$$M := \binom{2m+1}{m} = \frac{(2m+1)\ldots(m+2)}{m!}.$$

Observe that M is divisible by all primes p with $m + 1 < p \leq 2m + 1$. Moreover, we have $M < 2^{2m}$. It follows that

$$\vartheta(2m + 1) - \vartheta(m + 1) = \sum_{m+1<p\leq 2m+1} \log p \leq \log M < 2(\log 2)m.$$

Using this, and the induction hypothesis for $m + 1$, we obtain

$$\vartheta(n) = \vartheta(2m + 1) - \vartheta(m + 1) + \vartheta(m + 1)$$
$$< 2(\log 2)m + 2(\log 2)(m + 1) = 2(\log 2)n.$$

Another way of stating the above theorem is:

$$\prod_{p\leq x} p < 4^x.$$

Note that we have also proved:

Theorem: We have

$$\vartheta(x) = \Theta(x).$$

EXERCISE

1. Show that $\sum_{p\leq x} 1/\log p = \Theta(x/(\log x)^2)$.
2. For each positive integer n, let p_n denote the nth prime. Show that $p_n = \Theta(n \log n)$.
3. For each positive integer n, let $\omega(n)$ denote the number of distinct primes dividing n. Show that $\omega(n) = O(\log n/ \log\log n)$.

Mertens' theorem

Our next goal is to prove the following theorem, which turns out to have a number of applications.

Theorem: We have

$$\sum_{p\leq x} \frac{1}{p} = \log\log x + O(1).$$

The proof of this theorem, while not difficult, is a bit technical, and we proceed in several steps.

Theorem: We have

$$\sum_{p \le x} \frac{\log p}{p} = \log x + O(1).$$

Proof: Let $n := \lfloor x \rfloor$. The idea of the proof is to estimate $\log(n!)$ in two different ways. By Lemma, we have

$$\log(n!) = \sum_{p \le n} \sum_{k \ge 1} \lfloor n / p^k \rfloor \log p = \sum_{p \le n} \lfloor n / p \rfloor \log p + \sum_{k \ge 2} \sum_{p \le n} \lfloor n / p^k \rfloor \log p.$$

We next show that the last sum is $O(n)$. We have

$$\sum_{p \le n} \log p \sum_{k \ge 2} \lfloor n / p^k \rfloor \le n \sum_{p \le n} \log p \sum_{k \ge 2} p^{-k}$$

$$= n \sum_{p \le n} \frac{\log p}{p^2} \cdot \frac{1}{1 - 1/p} = n \sum_{p \ge n} \frac{\log p}{p(p-1)}$$

$$\le n \sum_{k \ge 2} \frac{\log k}{k(k-1)} = O(n).$$

Thus, we have shown that

$$\log(n!) = \sum_{p \ge n} \lfloor n / p \rfloor \log p + O(n).$$

Since $\lfloor n / p \rfloor = n/p + O(1)$, applying Theorem , we obtain

$$\log(n!) = \sum_{p \ge n} (n/p) \log p + O\left(\sum_{p \le n} \log p \right) + O(n) = n \sum_{p \le n} \frac{\log p}{p} + O(n).$$

We can also estimate $\log(n!)$ by estimating a sum by an integral:

$$\log(n!) = \sum_{k=1}^{n} \log k = \int_1^n \log t dt + O(\log n) = n \log n - n + O(\log n).$$

$\log x - \log n = o(1)$, we obtain

$$\sum_{p \le x} \frac{\log p}{p} = \log n + O(1) = \log x + O(1),$$

which proves the theorem.

We shall also need the following theorem, which is a very useful tool in its own right; it is essentially a discrete variant of "integration by parts."

Abel's Identity

Theorem: Let $\{c_i\}_{i=k}^{\infty}$ be a sequence of real numbers, and for each real number t, define

$$C(t) := \sum_{k \le i \le t} c_i.$$

Further, suppose that $f(t)$ is a function with a continuous derivative $f'(t)$ on the interval $[k, x]$, where x is a real number, with $x \ge k$. Then

$$\sum_{k \le i \le x} c_i f(i) = C(x) f(x) - \int_k^x C(t) f'(t) dt.$$

Note that since $C(t)$ is a step function, the integrand $C(t)f'(t)$ is piece-wise continuous on $[k, x]$, and hence the integral is well defined.

Proof: Let $n := \lfloor x \rfloor$. We have

$$\sum_{i=k}^{n} c_i f(i) = C(k)f(k) + \sum_{i=k+1}^{n} [C(i) - C(i-1)] f(i)$$

$$= \sum_{i=k}^{n-1} C(i)[f(i) - f(i+1)] + C(n)f(n)$$

$$= \sum_{i=k}^{n-1} C(i)[f(i) - f(i+1)] + C(n)[f(n) - f(x)] + C(x)f(x).$$

Observe that for $i = k, \ldots, n-1$, we have $C(t) = C(i)$ for all $t \in [i, i+1)$, and so

$$C(i)[f(i) - f(i+1)] = -C(i) \int_i^{i+1} f'(t) dt = -\int_i^{i+1} C(t) f'(t) dt;$$

likewise,

$$C(n)[f(n) - f(x)] = -\int_n^x C(t) f'(t) dt,$$

from which the theorem directly follows.

Proof of For $i \ge 2$, set

$$c_i := \begin{cases} (\log i)/i & \text{if } i \text{ is prime,} \\ 0 & \text{otherwise.} \end{cases}$$

By Theorem, we have

$$C(t) := \sum_{2 \le i \le t} c_i = \sum_{p \le t} \frac{\log p}{p} = \log t + R(t),$$

where $R(t) = O(1)$. Applying Theorem with $f(t) := 1/\log t$ (and using Exercise 3.13), we obtain

$$\sum_{p\le x}\frac{1}{p} = \sum_{2\le i\le x} c_i f(i) = \frac{C(x)}{\log x} + \int_2^x \frac{C(t)}{t(\log t)^2}dt$$

$$= 1 + \frac{R(x)}{\log x} + \int_2^x \frac{dt}{t\log t} + \int_2^x \frac{R(t)}{t(\log t)^2}dt$$

$$= 1 + O(1/\log x) + (\log\log x - \log\log 2) + O(1)$$

$$= \log\log x + O(1).$$

Mertens' Theorem

Theorem: We have

$$\prod_{p\le x}(1 - 1/p) = \Theta(1/\log x).$$

Proof: Using parts (i) and (iii) of §A1, for any fixed prime p, we have

$$-\frac{1}{p^2} \le \frac{1}{p} + \log(1 - 1/p) \le 0.$$

Moreover, since

$$\sum_{p\le x}\frac{1}{p^2} \le \sum_{i\ge 2}\frac{1}{i^2} < \infty,$$

summing the inequality over all primes $p \le x$ yields

$$-C \le \sum_{p\le x}\frac{1}{p} + \log g(x) \le 0,$$

where C is a positive constant, and $g(x) := \prod_{p\le x}(1 - 1/p)$. From this, and from Theorem, we obtain $\log g(x) = -\log\log x + O(1)$, which implies that $g(x) = \Theta(1/\log x)$

EXERCISE

1. Use Euler's summation formula (previous exercise) to show that

$$\log(n!) = n\log n - n + \frac{1}{2}\log n + O(1),$$

and from this, conclude that $n! = \Theta((n/e)^n\sqrt{n})$. This is a weak form of Stirling's approximation; a sharper form states that $n! \sim (n/e)^n\sqrt{2\pi n}$.

2. Show that for some constant B_2, we have

$$\prod_{2<p\leq x} (1-2/p) \sim B_2/(\log x)^2.$$

You do not need to estimate B_2, but in fact $B_2 \approx 0.832429065662$.

3. Use Stirling's approximation (previous exercise) to show that

$$\binom{2m}{m} = \Theta\left(2^{2m}/\sqrt{m}\right)$$

4. For each positive integer k, let P_k denote the product of the first k primes. Show that $\varphi(P_k) = \Theta(P_k/\log\log P_k)$ (here, φ is Euler's phi function).

5. Show that for some constant A, we have $\sum_{p\leq x} 1/p = \log\log x + A + o(1)$. You do not need to estimate A, but in fact $A \approx 0.261497212847643$.

6. The previous exercise showed that $\varphi(n)$ could be as small as (about) $n/\log\log n$ for infinitely many n. Show that this is the "worst case," in the sense that $\varphi(n) = \Omega(n/\log\log n)$.

7. Show that

$$\prod_{2<p\leq x} (1-2/p) = \Theta(1/(\log x)^2).$$

8. Show that for every positive integer constant k,

$$\int_2^x \frac{dt}{(\log t)^k} = \frac{x}{(\log x)^k} + O\left(\frac{x}{(\log x)^{k+1}}\right).$$

This fact may be useful in some of the following exercises.

9. Use the result from the previous exercise to strengthen Mertens' theorem: show that for some constant B_1, we have $\prod_{p\leq x}(1-1/p) \sim B_1/(\log x)$.

You do not need to estimate B_1, but in fact $B_1 \approx 0.561459483566885$.

10. Use Chebyshev's theorem and Abel's identity to show that

$$\sum_{p\leq x} \frac{1}{\log p} = \frac{\pi(x)}{\log x} + O(x/\log x)^3).$$

11. Use Abel's identity to derive Euler's summation formula: if $f(t)$ has a continuous derivative $f'(t)$ on the interval $[a, b]$, where a and b are integers, then

$$\sum_{i=a}^{b} f(i) - \int_a^b f(t)dt = f(a) + \int_a^b (t - \lfloor t \rfloor f'(t)dt.$$

12. Show that if $\pi(x) \sim cx/\log x$ for some constant c, then we must have $c = 1$.

ERATOSTHENES ALGORITHM

This is an algorithm that generates all the primes up to a given bound n. It uses an array $A[2 \ldots n]$, and runs as follows.

for $k \leftarrow \in$ to n do $A[k] \leftarrow 1$

for $k \leftarrow \in$ to $\lfloor \sqrt{n} \rfloor$ do

if $A[k] = 1$ then

$i \leftarrow 2k$

while $i \leq n$ do

$A[i] \leftarrow 0,\ i \leftarrow i + k$

When the algorithm finishes, we have $A[k] = 1$ if and only if k is prime, for $k = 2, \ldots, n$. This can easily be proven using the fact that a composite number k between $\in$ and n must be divisible by a prime that is at most $\sqrt{n}$, and by proving by induction on k that at the beginning of each iteration of the main loop, $A[i] = 0$ if and only if i is divisible by a prime less than k, for $i = k, \ldots, n$. We leave the details of this to the reader.

We are more interested in the running time of the algorithm. To analyze the running time, we assume that all arithmetic operations take constant time; this is reasonable, since all the numbers computed are used as array indices and thus should fit in single machine words. Therefore, we can assume that built-in arithmetic instructions are used for operating on such numbers.

Every time we execute the inner loop of the algorithm, we perform $O(n/k)$ steps to clear the entries of A indexed by multiples of k. Pessimistically, then, we could bound the total running time by $O(n\,T(n))$, where

$$T(n) := \sum_{k \leq \sqrt{n}} 1/k.$$

Estimating the sum by an integral, we have

$$T(n) = \sum_{k=1}^{\lfloor \sqrt{n} \rfloor} 1/k = \int_1^{\lfloor \sqrt{n} \rfloor} \frac{dy}{y} + O(1) \sim \frac{1}{2}\log n.$$

This implies a $O(n \text{ len}(n))$ bound on the running time of the algorithm. However, this rather crude analysis ignores the fact that the inner loop is executed only for prime values of k; taking this fact into account, we see that the running time is $O(n\ T_1(n))$, where

$$T_1(n) := \sum_{p \le \sqrt{n}} 1/p.$$

By Theorem, $T_1(n) = \log \log n + O(1)$, which implies a $O(n \text{ len}(\text{len}(n)))$ bound on the running time of the algorithm. This is a substantial improvement over the above, rather crude analysis.

EXERCISE

1. Design and analyze an algorithm that on input n outputs the table of values $\tau(k)$ for $k = 1, ..., n$, where $\tau(k)$ is the number of positive divisors of k. Your algorithm should run in time $O(n \text{ len}(n))$.
2. Give a detailed proof of the correctness of the above algorithm.
3. One drawback of the above algorithm is its use of space: it requires an array of size n. Show how to modify the algorithm, without substantially increasing its running time, so that one can enumerate all the primes up to n, using an auxiliary array of size just $O(\sqrt{n})$.

BERTRAND'S POSTULATE

Suppose we want to know how many primes there are of a given bit length, or more generally, how many primes there are between m and $2m$ for a given positive integer m. Neither the statement, nor our proof, of Chebyshev's theorem imply that there are any primes between m and $2m$, let alone a useful density estimate of such primes. Bertrand's postulate is the assertion that for every positive integer m, there exists a prime between m and $2m$.

We shall in fact prove a stronger result: there is at least one prime between m and $2m$, and moreover, the number of such primes is $\Omega(m/\log m)$.

Theorem: For every positive integer m, we have

$$\pi(2m) - \pi(m) > \frac{m}{3\log(2m)}.$$

The theorem is clearly true for $m \le 2$, so we may assume that $m \ge 3$. $N := \binom{2m}{m}$, and recall that N is divisible only by primes less than $2m$, and that we have the identity

$$v_p(N) = \sum_{k \ge 1} \left(\lfloor 2m/p^k \rfloor - 2\lfloor m/p^k \rfloor \right),$$

where each term in the sum is either 0 or 1. We can characterize the values $v_p(N)$ a bit more precisely, as follows:

Lemma: Let $m \geq 3$ and $N := \binom{2m}{m}$. For all primes p, we have

$$p^{v_p(N)} \leq 2m;$$

if $p > \sqrt{2m}$, then $v_p(N) \leq 1$;

if $2m/3 < p \leq m$, then $v_p(N) = 0$;

if $m < p < 2m$, then $v_p(N) = 1$.

Proof: For all terms with $k > \log(2m)/\log p$ in vanish, and hence $v_p(N) \leq \log(2m)/\log p$, from which it follows that $p^{v_p(N)} \leq 2m$.

For, if $2m/3 < p \leq m$, then $2m/p < 3$, and we must also have $p \geq 3$, since $p = \in$ implies $m < 3$. We have $p^2 > p(2m/3) = 2m(p/3) \geq 2m$, and hence all terms with $k > 1$ in vanish. The term with $k = 1$ also vanishes, since $1 \leq m/p < 3/2$, from which it follows that $\in \leq 2m/p < 3$, and hence $\lfloor 2m/p \rfloor = 1$ and $\lfloor 2m/p \rfloor = 2$.

If $m < p < 2m$, it follows that $1 < 2m/p < 2$, so $\lfloor 2m/p \rfloor = 1$. Also, $m/p < 1$, so $\lfloor m/p \rfloor = 0$. It follows that the term with $k = 1$ in is 1, and it is clear that $2m/p^k < 1$ for all $k > 1$, and so all the other terms vanish.

We now have the necessary technical ingredients to prove Theorem. Define

$$P_m := \prod_{m<p<2m} p,$$

and define Q_m so that

$$N = Q_m P_m.$$

we see that

$$Q_m = \prod_{p \leq 2m/3} p^{v_p(N)}$$

Moreover, $v_p(N) > 1$ for at most those $p \leq \sqrt{2m}$, so there are at most $\sqrt{2m}$ such primes, the contribution of each such prime to the above product is at most $2m$. Combining this with Theorem we obtain

$$Q_m < (2m)^{\sqrt{2m}} . 4^{2m/3}.$$

We now apply Lemma, obtaining

$$P_m = NQ_m^{-1} \geq 2^{2m}(2m)^{-1}Q_m^{-1} > 4^{m/3}(2m)^{-(1+\sqrt{2m})}.$$

It follows that

$$\pi(2m) - \pi(m) \geq \log P_m / \log(2m) > \frac{m \log 4}{3\log(2m)} - (1+\sqrt{2m})$$

$$= \frac{m}{3\log(2m)} + \frac{m\log(4-1)}{3\log(2m)} - (1+\sqrt{2m})$$

Clearly, for all sufficiently large *m*, we have

$$\frac{m(\log(4-1))}{3\log(2m)} > 1+\sqrt{2m}.$$

That proves Theorem for all sufficiently large *m*. Moreover, a simple calculation shows that holds for all $m \geq 13{,}000$, and one can verify by brute force (with the aid of a computer) that the theorem holds for $m < 13{,}000$.

SOPHIE GERMAIN PRIMES

A Sophie Germain prime is a prime *p* such that $2p + 1$ is also prime. Such primes are actually useful in a number of practical applications, and so we discuss them briefly here. It is an open problem to prove (or disprove) that there are infinitely many Sophie Germain primes. However, numerical evidence, and heuristic arguments, strongly suggest not only that there are infinitely many such primes, but also a fairly precise estimate on the density of such primes.

Let $\pi^*(x)$ denote the number of Sophie Germain primes up to *x*.

Conjecture: We have

$$\pi^*(x) \sim C\frac{x}{(\log x)^2},$$

where *C* is the constant

$$C := 2\prod_{p>2} \frac{p(p-2)}{(p-1)^2} \approx 1.32032,$$

and the product is over all primes $p > 2$.

The above conjecture is a special case of the following, more general conjecture.

Dickson's Conjecture

Conjecture: Let $(a_1, b_1), \ldots, (a_k, b_k)$ be distinct pairs of integers, where each a_i is positive. Let $P(x)$ be the number of positive integers m up to x such that $a_im + b_i$ are simultaneously prime for $i = 1, \ldots, k$. For each prime p, let $\omega(p)$ be the number of integers $m \in \{0, \ldots, p-1\}$ that satisfy

$$\prod_{i=1}^{k}(a_im+b_i) \equiv 0 \pmod{p}.$$

If $\omega(p) < p$ for each prime p, then

$$p(x) \sim D\frac{x}{(\log x)^k},$$

where

$$D := \prod_p \frac{1-\omega(p)/p}{(1-1/p)^k},$$

the product being over all primes p.

In Exercise 5.26 below, you are asked to verify that the quantity D appearing in Conjecture satisfies $0 < D < \infty$. Conjecture is implied with $k := 2$, $(a_1, b_1) := (1, 0)$, and $(a_2, b_2) := (2, 1)$; in this case, $\omega(2) = 1$ and $\omega(p) = \in$ for all $p > 2$. The above conjecture also includes (a strong version of) the famous twin primes conjecture as a special case: the number of primes p up to x such that $p + \in$ is also prime is $\sim Cx/(\log x)^2$, where C is the same constant. A heuristic argument in favor runs as follows. In some sense, the chance that a large positive integer m is prime is about $1/\log m$.

Since $\log(a_im + b_i) \sim \log m$, the chance that $a_1m + b_1, \ldots, a_km + b_k$ are all prime should be about $1/(\log m)^k$. But this ignores the fact that $a_1m + b_1, \ldots, a_km + b_k$ are not quite random integers. For each prime p, we must apply a "correction factor" r_p/s_p, where r_p is the chance that for random m, none of $a_1m + b_1, \ldots, a_km + b_k$ is divisible by p, and s_p is the chance that for k truly random, large integers, none of them is divisible by p. One sees that $r_p = 1 - \omega(p)/p$ and $s_p = (1 - 1/p)^k$. This implies that $P(x)$ should be about

$$D\sum_{m\leq x} 1/(\log m)^k \sim D\int_2^x dt/(\log t)^k \sim Dx/(\log x)^k.$$

EXPLICIT ESTIMATES

Sometimes, it is useful to have explicit estimates for $\pi(x)$, as well as related functions, like $\vartheta(x)$ and the nth prime function p_n. The following theorem presents a number of bounds that have been proved without relying on any unproved conjectures.

1. $\log\log x + A - \frac{1}{2(\log x)^2} < \sum_{p \le x} 1/p < \log\log x + A + \frac{1}{2(\log x)^2},$

   for $x \geq 286$, where $A \approx 0.261497212847643$;

2. $\frac{x}{\log x}\left(1+\frac{1}{2\log x}\right) < \pi(x) < \frac{x}{\log x}\left(1+\frac{3}{2\log x}\right)$, for $x \geq 59$;

3. $\frac{B}{\log x}\left(1-\frac{1}{2(\log x)^2}\right) < \prod_{p \le x}\left(1-\frac{1}{p}\right) < \frac{B_1}{\log x}\left(1+\frac{1}{2(\log x)^2}\right),$

   for $x \geq 285$, where $B_1 \approx 0.561459483566885$.

4. $x\left(1-\frac{1}{2\log x}\right) < \vartheta(x) < x\left(1+\frac{1}{2\log x}\right)$, for $x \geq 563$;

5. $n(\log n + \log\log n - 3/2) < p_n < n(\log n + \log\log n - 1/2)$, for $n \geq 20$;

Chapter 4

Modular Square Roots

In mathematics, a square root of a number x is a number r such that $r^2 = x$, or in words, a number r whose *square* (the result of multiplying the number by itself) is x. Every non-negative real number x has a unique non-negative square root, called the principal square root and denoted with a radical symbol as $\sqrt{x}$. For example, the principal square root of 9 is 3, denoted $\sqrt{9} = 3$, because $3^2 = 3 \times 3 = 9$. If otherwise unqualified, "*the* square root" of a number refers to the principal square root: the square root of 2 is approximately 1.4142.

Square roots often arise when solving quadratic equations, or equations of the form $ax^2 + bx + c = 0$, due to the variable x being squared.

Every positive number x has two square roots. One of them is $\sqrt{x}$, which is positive, and the other $\sqrt{x}$, which is negative. Together, these two roots are denoted $\pm\sqrt{x}$. Square roots of negative numbers can be discussed within the framework of complex numbers. Square roots of objects other than numbers can also be defined.

Square roots of integers that are not perfect squares are always *irrational numbers*: numbers not expressible as a ratio of two integers. For example, $\sqrt{2}$ cannot be written exactly as m/n, where n and m are integers. Nonetheless, it is exactly the length of the diagonal of a square with side length 1. This has been known since ancient times, with the discovery that $\sqrt{2}$ is irrational attributed to Hipparchus, a disciple of Pythagoras.

PROPERTIES

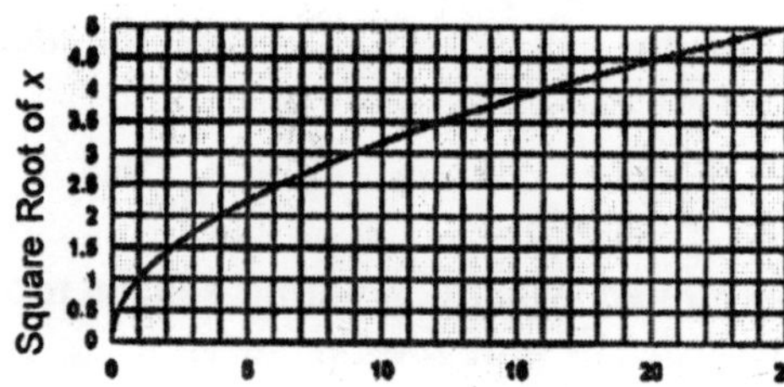

Fig. The Graph of the Function $f(x) = \sqrt{x}$, Made up of half a Parabola with a Vertical Directrix.

The principal square root function $f(x) = \sqrt{x}$ (usually just referred to as the "square root function") is a function which maps the set of non-negative real numbers $R^+ \cup \{0\}$ onto itself, and, like all functions, always returns a unique value. The square root function also maps rational numbers into algebraic numbers (a superset of the rational numbers); $\sqrt{x}$ is rational if and only if x is a rational number which can be represented as a ratio of two perfect squares. In geometrical terms, the square root function maps the area of a square to its side length.

- For all real numbers x,

$$\sqrt{x^2} = |x| = \begin{cases} x, & \text{if } x \geq 0 \\ -x, & \text{if } x \leq 0 \end{cases}$$

- For all non-negative real numbers x and y,

$$\sqrt{xy} = \sqrt{x}\sqrt{y}$$

and

$$\sqrt{x} = x^{\frac{1}{2}}.$$

- The square root function is continuous for all non-negative x and differentiable for all positive x. Its derivative is given by

$$f'(x) = \frac{1}{2\sqrt{x}}$$

- The Taylor series of $\sqrt{1+x}$ about $x = 0$ converges for $|x| < 1$ and is given by

$$\sqrt{1+x} = 1 + \frac{1}{2}x - \frac{1}{8}x^2 + \frac{1}{16}x^3 - \frac{5}{128}x^4 + \ldots$$

COMPUTATION

Many methods of calculating square roots exist today, some meant to be done by hand and some meant to be done by machine.

Many, but not all pocket calculators have a square root key. Computer spreadsheets and other software are also frequently used to calculate square roots. Computer software programs typically implement good routines to compute the exponential function and the natural logarithm or logarithm, and then compute the square root of x using the identity

$$\sqrt{x} = e^{\frac{1}{2}\ln x} \text{ or } \sqrt{x} = 10^{\frac{1}{2}\log x}$$

The same identity is exploited when computing square roots with logarithm tables or slide rules.

The most common method of square root calculation by hand is known as the "Babylonian method". It involves a simple algorithm, which results in a number closer to the actual square root each time it is repeated. To find r, the square root of a real number x:

- Start with an arbitrary positive start value r (the closer to the square root of x, the better).
- Replace r by the average between r and x/r. (It is sufficient to take an approximate value of the average, not too close to the previous value of r and x/r in order to ensure convergence.)
- Repeat step 2 until r and x/r are as close as desired.

The best known time complexity for computing a square root with n digits of precision is the same as that for multiplying two n-digit numbers.

SQUARE ROOTS OF NEGATIVE AND COMPLEX NUMBERS

The square of any positive or negative number is positive, and the square of 0 is 0. Therefore, no negative number can have a real square root. However, it is possible to work with a larger set of numbers, called the complex numbers, that does contain solutions to the square root of a negative number. This is done by introducing a new number, denoted by i (sometimes j, especially in the context of electricity) and called the imaginary unit, which is *defined* such that $i^2 = -1$. Using this notation, we can think of i as the square root of -1, but notice that we also have $(-i)^2 = i^2 = -1$ and so $-i$ is also a square root of -1. Similarly to the real numbers, we say the principal square root of -1 is i, or more generally, if x is any positive number, then the principal square root of $-x$ is

$$\sqrt{-x} = i\sqrt{x}$$

because

$$\left(i\sqrt{x}\right)^2 = i^2\left(\sqrt{x}\right)^2 = (-1)x = -x.$$

By the argument given above, i can be neither positive nor negative. This creates a problem: for the complex number z, we cannot define $\sqrt{z}$ to be the "positive" square root of z.

For every non-zero complex number z there exist precisely two numbers w such that $w^2 = z$. For example, the square roots of i are:

$$\sqrt{i} = \frac{1}{\sqrt{2}}(1+i)$$

and

$$-\sqrt{i} = -\frac{1}{\sqrt{2}}(1+i)$$

The usual definition of $\sqrt{z}$ is by introducing the following branch cut: if $z = re^{iö}$ is represented in polar coordinates with $-ð < ö \le ð$, then we set the principal value to

$$\sqrt{z} = \sqrt{x}\, e^{\frac{i\phi}{2}}.$$

Thus defined, the square root function is holomorphic everywhere except on the non-positive real numbers (where it isn't even continuous). The above Taylor series for $\sqrt{1+x}$ remains valid for complex numbers x with $|x| < 1$.

When the number is in rectangular form the following formula can be used for the principal value:

$$\sqrt{x+iy} = \sqrt{\frac{r+x}{2}} + i\frac{y}{\sqrt{2(r+x)}}$$

where

$$r = |x+iy| = \sqrt{x^2+y^2}$$

is the absolute value or modulus of the complex number, unless $x = -r$ and $y = 0$. Notice that the sign of the imaginary part of the root is the same as the sign of the imaginary part of the original number. The real part of the principal value is always non-negative.

Note that because of the discontinuous nature of the square root function in the complex plane, the law $\sqrt{zw} = \sqrt{z}\sqrt{w}$ is in general not true. (Equivalently, the problem occurs because of the freedom in the choice of branch. The chosen branch may or may not yield the equality; in fact, the choice of branch for the square root need not contain the value of $\sqrt{z}\sqrt{w}$ at all, leading to the equality's failure. A similar problem appears with the complex logarithm and the relation log z + log ŒEw = log(zw).) Wrongly assuming this law underlies several faulty "proofs", for instance the following one showing that $-1 = 1$:

$$-1 = i \,.\, i = \sqrt{-1}\cdot\sqrt{-1} \qquad = \sqrt{-1\cdot -1} = \sqrt{1} = 1$$

The third equality cannot be justified (see invalid proof). It can be made to hold by changing the meaning of $\sqrt{\ }$ so that this no longer represents the principal square root but selects a branch for the square root that contains $\sqrt{-1}\cdot\sqrt{-1}$. The left hand side becomes either

$$\sqrt{-1}\cdot\sqrt{-1} \qquad = i\cdot i = 1$$

if the branch includes $+i$ or

$$\sqrt{-1}\cdot\sqrt{-1} \qquad = \sqrt{1} = 1,$$

if the branch includes $\sqrt{1} = -i$, while the right hand side becomes

where the last equality,, is a consequence of the choice of branch in the redefinition of $\sqrt{\ }$.

Square Roots of Matrices and Operators

If A is a positive-definite matrix or operator, then there exists precisely one positive definite matrix or operator B with $B^2 = A$; we then define $\sqrt{A} = B$.

More generally, to every normal matrix or operator A there exist normal operators B such that $B^2 = A$. In general, there are several such operators B for every A and the square root function cannot be defined for normal operators in a satisfactory manner. Positive definite operators are akin to positive real numbers, and normal operators are akin to complex numbers.

MODULAR SQUARE ROOTS

We will discuss the problem of computing a square root of a modulo n, given integers a and n, where a is a quadratic residue modulo n.

PRIME MODULUS

Let p be an odd prime, and let a be an integer such that $0 < a < p$ and $(a \mid p) = 1$. We would like to compute a square root of a modulo p. Let $\alpha := [a]_p \in \mathbb{Z}_n^*$, so that we can restate our problem of that of finding $\beta \in \mathbb{Z}_n^*$ such that $\beta^2 = \alpha$, given $\alpha \in (\mathbb{Z}_n^*)^2$.

We first consider the special case where $p \equiv 3 \pmod 4$, in which it turns out that this problem can be solved very easily. Indeed, we claim that in this case

$$\beta := \alpha^{(p+1)/4}$$

is a square root of α — note that since $p \equiv 3 \pmod 4$, the number $(p + 1)/4$ is an integer. To show that $\beta^2 = \alpha$, suppose $\alpha = \tilde{\beta}^2$ for some $\tilde{\beta} \in \mathbb{Z}_p^*$. We know that there is such a $\tilde{\beta}$, since we are assuming that $\alpha \in (\mathbb{Z}_p^*)^2$. Then we have

$$\beta^2 = \alpha^{(p+1)/2} = \tilde{\beta}^{p+1} = \tilde{\beta}^2 = \alpha,$$

where we used Fermat's little theorem for the third equality. Using a repeatedsquaring algorithm, we can compute β in time $O(\text{len}(p)^3)$.

Now we consider the general case, where we may have $p \not\equiv 3 \pmod 4$. Here is one way to efficiently compute a square root of α, assuming we are given, in addition to α, an auxiliary input $\gamma \in \mathbb{Z}_p^* \setminus (\mathbb{Z}_p^*)^2$ (how one obtains such a γ is discussed below).

Let us write $p - 1 = 2^h m$, where m is odd. For every $\gamma \in \mathbb{Z}_p^*$, δ^m has multiplicative order dividing $2h$. Since $2^{h-1} m = 1$, α^m has multiplicative order dividing 2^{h-1}. Since $\gamma^{2h-1}\, m = -1$, γ^m has multiplicative order precisely 2^h. Since there is only one subgroup of $\mathbb{Z}_p^*$ of order 2^h, it follows that m generates this subgroup, and that $\alpha^m = \gamma^{mx}$ for $0 \leq x < 2^h$ and x is even. We can find x by computing the discrete logarithm of α^m to the base γ^m. Setting $\kappa = \gamma^{mx/2}$, we have

$$\kappa^2 = \alpha^m.$$

We are not quite done, since we now have a square root of α^m, and not of α.

Since m is odd, we may write $m = 2t + 1$ for some non-negative integer t. It then follows that

$$(\kappa\alpha^{-t})^2 = \kappa^2 \alpha^{-2t} = \alpha^m\alpha^{-2t} = \alpha^{m-2t} = \alpha.$$

Thus, $\kappa\alpha^{-t}$ is a square root of α.

Let us summarize the above algorithm for computing a square root of $\alpha \in (\mathbb{Z}_p^*)^2$, assuming we are given $\gamma \in \mathbb{Z}_p^* \setminus (\mathbb{Z}_p^*)^2$, in addition to α:

compute positive integers m, h such that $p - 1 = 2^h m$ with m odd

$\gamma' \leftarrow \gamma^m$, $\alpha' \leftarrow \alpha^m$

compute $x \leftarrow \log_{\gamma'} \alpha'$ // note that $0 \leq x < 2^h$ and x is even

$$\beta \leftarrow (\gamma')^{x/2}\alpha^{-\lfloor m/2 \rfloor}$$

output β

The work done outside the discrete logarithm calculation amounts to just a handful of exponentiations modulo p, and so takes time $O(\text{len}(p)^3)$. The time to compute the discrete logarithm is $O(h\ \text{len}(h)\ \text{len}(p)^2)$. So the total running time of this procedure is

$$O(\text{len}(p)^3 + h\ \text{len}(h)\ \text{len}(p)^2).$$

The above procedure assumed we had at hand a non-square γ. If $h = 1$, which means that $p \equiv 3 \pmod 4$, then $(-1 \mid p) = -1$, and so we are done. However, we have already seen how to efficiently compute a square root in this case.

If $h > 1$, we can find a non-square γ using a probabilistic search algorithm. Simply choose γ at random, test if it is a square, and if so, repeat. The probability that a random element of $\mathbb{Z}_p^*$ is a square is 1/2; thus, the expected number of trials until we find a non-square is 2; moreover, the running time per trial is $O(\text{len}(p)^2)$, and hence the expected running time of this probabilistic search algorithm is $O(\text{len}(p)^2)$.

PRIME-POWER MODULUS

Let p be an odd prime, let a be an integer relatively prime to p, and let $e > 1$ be an integer. We know that a is a quadratic residue modulo p^e if and only if a is a quadratic residue modulo p. Suppose that a is a quadratic residue modulo p, and that we have found an integer b such that $b^2 \equiv a \pmod p$. From this, we can easily compute a square root of a modulo p^e using the following technique, which is known as Hensel lifting.

More generally, suppose that for some $f \geq 1$, we have computed an integer b satisfying the congruence $b^2 \equiv a \pmod{pf}$, and we want to find an integer c satisfying the congruence $c^2 \equiv a \pmod{p^{f+1}}$. Clearly, if $c^2 \equiv a \pmod{p^{f+1}}$, then $c^2 \equiv a \pmod{p^f}$, and so $c \equiv \pm b \pmod{p^f}$. So let us set $c = b + p^f h$, and solve for h. We have

$$c^2 \equiv (b + p^f h)^2 \equiv b^2 + 2bp^f h + p^{2f} h^2 \equiv b^2 + 2bp^f h \pmod{p^{f+1}}.$$

So we want to find an integer h satisfying the linear congruence

$$2bp^f h \equiv a - b^2 \pmod{p^{f+1}}.$$

Since $p \mid 2b$, we have $\gcd(2bp^f, p^{f+1}) = p^f$. Furthermore, since $b^2 \equiv a \pmod{p^f}$, we have $p^f \mid (a - b^2)$.

By iterating the above procedure, starting with a square root of a modulo p, we can quickly find a square root of a modulo p^e. We leave a detailed analysis of the running time of this procedure to the reader.

COMPOSITE MODULUS

To find square roots modulo n, where n is an odd composite modulus, if we know the prime factorization of n, then we can use the above procedures for finding square roots modulo primes and prime powers, and then use the algorithm of the Chinese remainder theorem to get a square root modulo n.

However, if the factorization of n is not known, then there is no efficient algorithm known for computing square roots modulo n. In fact, one can show that the problem of finding square roots modulo n is at least as hard as the problem of factoring n, in the sense that if there is an efficient algorithm for computing square roots modulo n, then there is an efficient (probabilistic) algorithm for factoring n.

We now present an algorithm to factor n, using a modular square-root algorithm A as a subroutine. For simplicity, we assume that A is deterministic, and that for all n and for all $\alpha \in (\mathbb{Z}_n^*)^2$, $A(n, \alpha)$ outputs a square root of α. Also for simplicity, we shall assume that n is of the form $n = pq$, where p and q are distinct, odd primes. algorithm runs as follows:

$\beta \xleftarrow{\$} \mathbb{Z}_n^+, d \leftarrow \gcd \text{rep}(\beta), n)$

if $d > 1$ then

output d

else

$\alpha \leftarrow \beta^2$, $\beta' \leftarrow A(n, \alpha)$

if $\beta = \pm\beta'$

then output "failure"

else output gcd (rep($\beta - \beta'$), n)

Here, $\mathbb{Z}_n^+$ denotes the set of non-zero elements of $\mathbb{Z}_n$. Also, recall that rep(β) denotes the canonical representative of β.

First, we argue that the algorithm outputs either "failure" or a non-trivial factor of n. Clearly, if $\beta \notin \mathbb{Z}_n^*$, then the value d computed by the algorithm is a non-trivial factor. So suppose $\beta \in \mathbb{Z}_n^*$. In this case, the algorithm invokes A on inputs n and $\alpha := \beta^2$, obtaining a square root β' of α. Suppose that $\beta \neq \pm\beta'$, and set $\gamma := \beta - \beta'$.

What we need to show is that gcd(rep(γ), n) is a non-trivial factor of n. To see this, consider the ring isomorphism of the Chinese remainder theorem

$$\theta : \mathbb{Z}_n \to \mathbb{Z}_p \times \mathbb{Z}_q$$
$$[a]_n \mapsto ([a]_p, [a]_q).$$

Suppose $\theta(\beta') = (\beta_1', \beta_2')$. Then the four square roots of α are

$$\alpha' = \theta^{-1}(\beta_1', \beta_2'), -\beta' = \theta^{-1}(-\beta_1', -\beta_2'), \theta^{-1}(-\beta_1', \beta_2'), \theta^{-1}(\beta_1', -\beta_2').$$

The assumption that $\beta \neq \pm\beta'$ implies that $\theta(\beta) = (-\beta_1', \beta_2')$ or $\theta(\beta) = (\beta_1', -\beta_2')$. In the first case, $\theta(\gamma) = (-2\beta_1', 0)$, which implies gcd(rep($\gamma$), n) $= q$. In the second case, $\theta(\gamma) = (0, -2\beta_2')$, which implies gcd(rep($\gamma$), n) $= p$.

Second, we argue that $P[F] \leq 1/2$, where F is the event that the algorithm outputs "failure." Viewed as a random variable, is uniformly distributed over $\mathbb{Z}_n^+$. Clearly, $P[F \mid \notin \mathbb{Z}_n^+] = 0$. Now consider any fixed $\alpha' \in (\mathbb{Z}_n^*)^2$. Observe that the conditional distribution of β given that $\beta^2 = \alpha'$ is the uniform distribution on the set of four square roots of α'. Also observe that the output of A depends only on n and 2, and so with respect to the conditional distribution given that $\beta^2 = \alpha'$, the output β' of A is fixed. Thus,

$$P[F \mid \beta^2 = \alpha'] = P[\beta = \pm \beta' \mid \beta^2 = \alpha'] = 1/2.$$

Putting everything together, using total probability, we have

$$P[F] = P[F \mid \beta \notin \mathbb{Z}_n^*]\, P[\beta \notin \mathbb{Z}_n^*] +$$
$$\sum_{\alpha' \in (\mathbb{Z}_n^*)^2} P[F \mid \beta^2 = \alpha']P[\beta^2 = \alpha']P[\beta^2 = \alpha']$$
$$= 0 \,.\, P[\beta \notin \mathbb{Z}_n^*] + \sum_{\alpha' \in (\mathbb{Z}_n^*)^2} \frac{1}{2} . P[\beta^2 = \alpha'] \leq \frac{1}{2}.$$

Thus, the above algorithm fails to split n with probability at most 1/2. If we like, we can repeat the algorithm until it succeeds. The expected number of iterations performed will be at most 2.

EXERCISE

1. Show that the following two problems are deterministic, polytime equivalent :

   (a) Given an odd prime p and $(\mathbb{Z}_p^*)^2$, find $\beta \in \mathbb{Z}_p^*$ such that $\beta^2 = \alpha$.

   (b) Given an odd prime p, find an element of $\mathbb{Z}_p^* \setminus (\mathbb{Z}_p^*)^2$.

2. This exercise develops an alternative algorithm for computing square roots modulo a prime. Let p be an odd prime, let $\beta \in \mathbb{Z}_p^*$, and set $\alpha := \beta^2$.

   Define $B_\alpha := \{\gamma \in \mathbb{Z}_p : \gamma^2 - \alpha \in (\mathbb{Z}_p^*)^2\}$.

   (a) Show that $B = \{\gamma \in \mathbb{Z}_p : g(\gamma) = 0\}$, where
   $g := (X - \beta)^{(p-1)/2} - (X + \beta)^{(p-1)/2} \in \mathbb{Z}_p[X]$.

   (b) Let $\gamma \in \mathbb{Z}_p \setminus B_\alpha$, and suppose $\gamma^2 \neq \alpha$. Let μ, v, be the uniquely determined elements of $\mathbb{Z}_p$ satisfying the polynomial congruence
   $\mu + vX \equiv (\gamma - X)^{(p-1)/2} \pmod{X^2 - \alpha}$.
   Show that $\mu = 0$ and $v^{-2} = \alpha$.

   (c) Using parts (a) and (b), design and analyze a probabilistic algorithm that computes a square root of a given $\alpha \in (\mathbb{Z}_p^*)^2$ in expected time $O(\text{len}(p)^3)$.

   Note that when $p - 1 = 2^h m$ (m odd), and h is large (e.g., $h \approx \text{len}(p)/2$), the algorithm in the previous exercise is asymptotically faster than the one in §12.5.1; however, the latter algorithm is likely to be faster in practice for the typical case where h is small.

3 Let p be an odd prime, and let $f \in \mathbb{Z}_p[X]$ be a polynomial with $0 \leq \deg(f) \leq 2$. Design and analyze an efficient, deterministic algorithm that takes as input p, f, and an element of $\mathbb{Z}_p^* \setminus (\mathbb{Z}_p^*)^2$, and which determines if f has any roots in $\mathbb{Z}_p$, and if so, finds all of the roots.

4. Design and analyze an efficient, deterministic algorithm that takes as input primes p and q, such that $q \mid (p - 1)$, along with an element $\alpha \in \mathbb{Z}_p^*$, and determines whether or not $\alpha \in (\mathbb{Z}_p^*)^q$.

5. Suppose there is a probabilistic algorithm A that takes as input positive integers n

and m, and an element $\alpha \in (\mathbb{Z}_n^*)^m$. It outputs either "failure," or an mth root of α. Furthermore, assume that A runs in expected polynomial time, and that for all n and m, and for randomly chosen $\alpha \in (\mathbb{Z}_n^*)^m$, A succeeds in computing an mth root of with probability $\varepsilon(n, m)$. Here, the probability is taken over the random choice of α, as well as the random choices made during the execution of A. Show how to use A to construct another probabilistic algorithm A' that takes as input n, m, and $\alpha \in (\mathbb{Z}_n^*)^m$, runs in expected polynomial time, and that satisfies the following property:

if $\varepsilon(n, m) \geq 0.001$, then for all $\alpha \in (\mathbb{Z}_n^*)^m$, A' computes an mth root of with probability at least 0.999.

6. Design and analyze an efficient, deterministic algorithm that takes as input primes p and q, such that $q \mid (p-1)$ but $q^2 \mid (p-1)$, along with an element $\alpha \in (\mathbb{Z}_p^*)^q$, and computes a qth root of , that is, an element $\beta \in \mathbb{Z}_p^*$ such that $\beta^q = \alpha$.
7. Suppose there is a probabilistic algorithm A that takes as input a positive integer n, and an element $\alpha \in (\mathbb{Z}_n^*)^2$. Assume that for all n, and for a randomly chosen $\alpha \in (\mathbb{Z}_n^*)^2$, A computes a square root of with probability at least 0.001. Here, the probability is taken over the random choice of α and the random choices of A. Show how to use A to construct another probabilistic algorithm A' that takes n as input, runs in expected polynomial time, and that satisfies the following property:

   for all n, A' outputs the complete factorization of n into primes with probability at least 0.999.
8. Show how to deterministically compute square roots modulo primes $p \equiv 5 \pmod 8$ in time $O(\text{len}(p)^3)$.
9. Let p be an odd prime, γ be a generator for $\mathbb{Z}_p^*$, and α be any element of $\mathbb{Z}_p^*$. Define

$$B(p,\gamma,\alpha) := \begin{cases} 1 & \text{if } \log_\gamma \alpha \geq (p-1)/2; \\ 0 & \text{if } \log_\gamma \alpha < (p-1)/2. \end{cases}$$

   Suppose that there is an algorithm that efficiently computes $B(p,\gamma,\alpha)$ for all p, γ, α as above. Show how to use this algorithm as a subroutine in an efficient, probabilistic algorithm that computes $\log_\gamma \alpha$ for all p, γ, α as above.

THE LEGENDRE SYMBOL

For an odd prime p and an integer a with $\gcd(a, p) = 1$, the Legendre symbol $(a \mid p)$ is defined to be 1 if a is a quadratic residue modulo p, and -1 otherwise. For completeness, one defines $(a \mid p) = 0$ if $p \mid a$. The following theorem summarizes the essential properties of the Legendre symbol.

Theorem: Let p be an odd prime, and let $a, b \in Z$. Then we have

(i) $(a \mid p) \equiv a^{(p-1)/2} \pmod p$; in particular, $(-1 \mid p) = (-1)^{(p-1)/2}$;

(ii) $(a \mid p)(b \mid p) = (ab \mid p)$;

(iii) $a \equiv b \pmod p$ implies $(a \mid p) = (b \mid p)$;

(iv) $(2 \mid p) = (-1)^{(p^2-1)/8}$;

(v) if q is an odd prime, then $(p \mid q) = (-1)^{\frac{p-1}{2}\frac{q-1}{2}}(q \mid p)$.

Part (i) of the theorem is just a restatement of Euler's criterion.

As was observed in previous Theorem, this implies that -1 is a quadratic residue modulo p if and only if $p \equiv 1 \pmod 4$. Thus, the quadratic residuosity of -1 modulo p is determined by the residue class of p modulo 4.

Part (ii) of the theorem follows immediately from part (i), and part (iii) is an immediate consequence of the definition of the Legendre symbol.

Part (iv), which we will prove below, can also be recast as saying that 2 is a quadratic residue modulo p if and only if $p \equiv \pm 1 \pmod 8$. Thus, the quadratic residuosity of 2 modulo p is determined by the residue class of p modulo 8.

Part (v), which we will also prove below, is the law of quadratic reciprocity.

Note that when $p = q$, both $(p \mid q)$ and $(q \mid p)$ are zero, and so the statement of part (v) is trivially true – the interesting case is when $p \neq q$, and in this case, part (v) is equivalent to saying that

$$(p \mid q)(q \mid p) = (-1)^{\frac{p-1}{2}\frac{q-1}{2}}.$$

Thus, the Legendre symbols $(p \mid q)$ and $(q \mid p)$ have the same values if and only if either $p \equiv 1 \pmod 4$ or $q \equiv 1 \pmod 4$. As the following examples illustrate, this result also shows that for a given odd prime q, the quadratic residuosity of q modulo another odd prime p is determined by the residue class of p modulo either q or $4q$.

Example. Let us characterize those primes p modulo which 5 is a quadratic residue. Since $5 \equiv 1 \pmod 4$, the law of quadratic reciprocity tells us that $(5 \mid p) = (p \mid 5)$. Now, among the numbers $\pm 1, \pm 2$, the quadratic residues modulo 5 are 1. It follows that 5 is a quadratic residue modulo p if and only if $p \equiv \pm 1 \pmod 5$. This example obviously generalizes, replacing 5 by any prime $q \equiv 1 \pmod 4$, and replacing the above congruences modulo 5 by appropriate congruences modulo q.

Example. Let us characterize those primes p modulo which 3 is a quadratic residue. Since $3 \not\equiv 1 \pmod 4$, we must be careful in our application of the law of quadratic reciprocity. First, suppose that $p \equiv 1 \pmod 4$. Then $(3 \mid p) = (p \mid 3)$, and so 3 is a quadratic residue modulo p if and only if $p \equiv 1 \pmod 3$. Second, suppose that $p \not\equiv 1 \pmod 4$. Then $(3 \mid p) = -(p \mid 3)$, and so 3 is a quadratic residue modulo p if and only if $p \equiv -1 \pmod 3$. Putting this all together, we see that 3 is quadratic residue modulo p if and only if

$$p \equiv 1 \pmod 4 \text{ and } p \equiv 1 \pmod 3$$

$$\text{or} \quad p \equiv -1 \pmod 4 \text{ and } p \equiv -1 \pmod 3.$$

Using the Chinese remainder theorem, we can restate this criterion in terms of residue classes modulo 12 : 3 is quadratic residue modulo p if and only if $p \equiv 1 \pmod{12}$. This example obviously generalizes, replacing 3 by any prime $q \equiv -1 \pmod 4$, and replacing the above congruences modulo 12 by appropriate congruences modulo $4q$.

Gauss' Lemma

Theorem: Let p be an odd prime and let a be an integer not divisible by p. Define $\alpha_j := ja \bmod p$ for $j = 1, \ldots, (p-1)/2$, and let n be the number of indices j for which $j > p/2$. Then $(a \mid p) = (-1)^n$.

Proof: Let $r_1, \ldots, r_n$ denote the values α_j that exceed $p/2$, and let $s_1, \ldots, s_k$ denote the remaining values α_j. The r_i's and s_i's are all distinct and non-zero. We have $0 < p - r_i < p/2$ for $i = 1, \ldots, n$, and no $p - r_i$ is an s_j; indeed, if $p - r_i = s_j$, then $s_j \equiv -r_i \pmod p$, and writing $s_j = ua \bmod p$ and $r_i = va \bmod p$, for some $u, v = 1, \ldots, (p-1)/2$, we have $ua \equiv -va \pmod p$, which implies $u \equiv -v \pmod p$, which is impossible.

It follows that the sequence of numbers $s_1, \ldots, s_k, p - r_1, \ldots, p - r_n$ is just a reordering of $1, \ldots, (p-1)/2$. Then we have

$$\begin{aligned}((p-1)/2)! &\equiv s_1 \ldots s_k(-r_1) \ldots (-r_n)\\ &\equiv (-1)^n s_1 \ldots s_k r_1 \ldots r_n\\ &\equiv (-1)^n((p-1)/2)!\, a^{(p-1)/2} \pmod p,\end{aligned}$$

and canceling the factor $((p-1)/2)!$, we obtain $a^{(p-1)/2} \equiv (-1)^n \pmod p$, and the result follows from the fact that $(a \mid p) \equiv a^{(p-1)/2} \pmod p$.

Theorem: If p is an odd prime and $\gcd(a, 2p) = 1$, then $(a \mid p) = (-1)^t$ where $t = \sum_{j=1}^{(p-1)/2} \lfloor ja/p \rfloor$. Also, $(2 \mid p) = (-1)^{(p^2-1)/8}$.

Proof: Let a be an integer not divisible by p, but which may be even, and let us adopt the same notation as in the statement and proof of Theorem in particular, $\alpha_1, ..., \alpha_{(p-1)/2}, r_1, ..., r_n$, and $s_1, ..., s_k$ are as defined there. Note that $ja = p\lfloor ja/p \rfloor + \alpha_j$, for $j = 1, ..., (p-1)/2$, so we have

$$\sum_{j=1}^{(p-1)/2} ja = \sum_{j=1}^{(p-1)/2} p\lfloor ja/p \rfloor + \sum_{j=1}^{n} r_j + \sum_{j=1}^{k} s_j.$$

The sequence of numbers $s_1, ..., s_k, p - r_1, ..., p - r_n$ is a reordering of $1, ..., (p-1)/2$, and hence

$$\sum_{j=1}^{(p-1)/2} j = \sum_{j=1}^{n} (p - r_j) + \sum_{j=1}^{k} s_j = np - \sum_{j=1}^{n} r_j + \sum_{j=1}^{k} s_j.$$

Subtracting we get

$$(a-1)\sum_{j=1}^{(p-1)/2} j = p\left(\sum_{j=1}^{(p-1)/2} \lfloor ja/p \rfloor - n\right) + 2\sum_{j=1}^{n} r_j.$$

Note that

$$\sum_{j=1}^{(p-1)/2} j = \frac{p^2-1}{8},$$

which together with implies

$$(a-1)\frac{p^2-1}{8} \equiv \sum_{j=1}^{(p-1)/2} \lfloor ja/p \rfloor - n \pmod 2.$$

If a is odd, implies

$$n \equiv \sum_{j=1}^{(p-1)/2} \lfloor ja/p \rfloor \pmod 2.$$

If $a = 2$, then $\lfloor 2j/p \rfloor = 0$ for $j = 1, ..., (p-1)/2$, and implies

$$n \equiv \frac{p^2-1}{8} \pmod 2.$$

Theorem: If p and q are distinct odd primes, then

$$(p \mid q)(q \mid p) = (-1)^{\frac{p-1}{2}\frac{q-1}{2}}.$$

Proof: Let S be the set of pairs of integers (x, y) with $1 \leq x \leq (p-1)/2$ and $1 \leq y \leq (q-1)/2$. Note that S contains no pair (x, y) with $qx = py$, so let us partition S into two subsets: S_1 contains all pairs (x, y) with $qx > py$, and S_2 contains all pairs (x, y) with $qx < py$. Note that $(x, y) \in S_1$ if and only if $1 \leq x \leq (p-1)/2$ and $1 \leq y \leq \lfloor qx/p \rfloor$. So $|S_1| = \sum_{x=1}^{(p-1)/2} \lfloor qx/p \rfloor$.

Similarly, $|S_2| = \sum_{y=1}^{(q-1)/2} \lfloor px/q \rfloor$. So we have

$$\frac{p-1}{2}\frac{q-1}{2} = |S| = |S_1| + |S_2|$$

$$= \sum_{x=1}^{(p-1)/2} \lfloor qx/p \rfloor + \sum_{y=1}^{(q-1)/2} \lfloor py/q \rfloor,$$

and Theorem implies

$$(p \mid q)(q \mid p) = (-1)^{\frac{p-1}{2}\frac{q-1}{2}}.$$

EXERCISE

1. Let p be an odd prime. Show that the following are equivalent:
   (a) $p = r^2 + 2t^2$ for some $r, t \in \mathbb{Z}$.
   (b) $(-2 \mid p) = 1$;
   (c) $p \equiv 1$ or $3 \pmod 8$;
2. Characterize those odd primes p for which $(15 \mid p) = 1$, in terms of the residue class of p modulo 60.

THE QUADRATIC RESIDUOSITY ASSUMPTION

QUDRATIC RECIPROCAL

Also called the aureum theorema (golden theorem) by Gauss. If and are distinct odd primes, then the congruences

$x^2 \equiv q \pmod p$

$x^2 \equiv p \pmod q$

are both solvable or both unsolvable unless both and leave the remainder 3 when divided by 4 (in which case one of the congruences is solvable and the other is not). Written symbolically,

$$\left(\frac{p}{q}\right)\left(\frac{q}{p}\right) = (-1)(p^{-1})(q^{-1})^{/4},$$

where

$$\left(\frac{p}{q}\right) = \begin{cases} 1 & \text{for } x^2 \equiv p (\bmod q) \text{ solvable for } x \\ -1 & \text{for } x^2 \equiv p (\bmod q) \text{ solvable for } x \end{cases}$$

is known as a Legendre symbol.

Euler stated the theorem in 1783 without proof. Legendre was the first to publish a proof, but it was fallacious. In 1796, Gauss became the first to publish a correct proof. The quadratic reciprocity theorem was Gauss's favorite theorem from number theory, and he devised no fewer than eight different proofs of it over his lifetime.

The genus theorem states that the Diophantine equation

$$x^2 + y^2 = p$$

can be solved for a prime iff $p \equiv 1 \pmod 4$ or $p = 2$.

The elementary properties of quadratic congruences and a method for their solution were studied in a previous chapter. Now, we focus our attention on some deeper properties of numbers that were discovered by Euler, Legendre, and Gauss. The simplest of these are the following.

- The odd prime divisors of numbers of the form $n^2 + 1$ are of the form $4k + 1$.
- The odd prime divisors of numbers of the form $n^2 - 1$ are of the form $8k + 1$ or $8k - 1$.
- The odd prime divisors (also not 3) of numbers of the form $n^2 - 3$ are of the form $12k + 1$ or $12k - 1$.
- The prime divisors (not equal to 2 or 5) of numbers of the form $n^2 - 5$ are of the form $20k + 1$, $20k - 1$, $20k + 9$ or $20k - 9$.

Generalizing these results, Euler conjectured that the prime divisors p of numbers of the form $n^2 - q$ are of the form $4qk + b^2$ or $4qk - b^2$, for some odd b. This is the Quadratic Reciprocity Law. The first complete proof of this law was given by Gauss in 1796. Gauss gave eight different proofs of the law and we discuss a proof that Gauss gave in 1808. The chapter concludes with a discussion of the Jacobi symbol and its applications.

Loosely speaking, the quadratic residuosity (QR) assumption is the assumption that it is hard to distinguish squares from non-squares in $\mathbb{Z}_n^*$, where n is of the form $n = pq$, and p and q are distinct primes. This assumption plays an important role in cryptography. Of course, since the Jacobi symbol is easy to compute, for this assumption to make sense, we have to restrict our attention to elements of Ker J_n, where $J_n : \mathbb{Z}_n^* \rightarrow \{\pm 1\}$ is the Jacobi map. We know that $(\mathbb{Z}_n^*)^2 \subseteq \text{Ker } J_n$. Somewhat more precisely, the QR assumption is the assumption that it is hard to distinguish a random element in Ker $J_n \setminus (\mathbb{Z}_n^*)^2$ from a random element in $(\mathbb{Z}_n^*)^2$, given n (but not its factorization!).

To give a rough idea as to how this assumption may be used in cryptography, assume that $p \equiv q \equiv 3 \pmod 4$, so that $[-1]_n \in \text{Ker } J_n \setminus (\mathbb{Z}_n^*)^2$, and moreover, $\text{Ker } J_n \setminus (\mathbb{Z}_n^*)^2 = [-1]_n(\mathbb{Z}_n^*)^2$. The value n can be used as a public key in a public-key cryptosystem. Alice, knowing the public key, can encrypt a single bit $b \in \{0, 1\}$ as $\beta := (-1)^b \alpha^2$, where Alice chooses $\alpha \in \mathbb{Z}_n^*$ at random. The point is, if $b = 0$, then β is uniformly distributed over $(\mathbb{Z}_n^*)^2$, and if $b = 1$, then β is uniformly distributed over Ker $J_n \setminus (\mathbb{Z}_n^*)^2$. Now Bob, knowing the secret key, which is the factorization of n, can easily determine if $\beta \in (\mathbb{Z}_n^*)^2$ or not, and hence deduce the value of the encrypted bit b. However, under the QR assumption, an eavesdropper, seeing just n and β, cannot effectively figure out what b is.

Of course, the above scheme is much less efficient than the RSA cryptosystem presented, but nevertheless, has attractive properties; in particular, its security is very closely tied to the QR assumption, whereas the security of RSA is a bit less well understood.

EXERCISE

1. Assume the same notation as in the previous exercise. Define the random variable X_n' to be the output of A on input n and a value α chosen at random from Ker J_n. Show that $|P[X_n' = 1] - P[Y_n = 1]| = \varepsilon(n)/2$. Thus, the problem of distinguishing Ker J_n from $(\mathbb{Z}_n^*)^2$ is essentially equivalent to the problem of distinguishing Ker $J_n \setminus (\mathbb{Z}_n^*)^2$ from $(\mathbb{Z}_n^*)^2$.

2. Suppose that A is a probabilistic algorithm that takes as input n of the form $n = pq$, where p and q are distinct primes such that $p \equiv q \equiv 3 \pmod 4$. The algorithm also takes as input $\alpha \in \text{Ker } J_n$, and outputs either 0 or 1. Furthermore, assume that A runs in expected polynomial time. Define two random variables, X_n and Y_n, as follows: X_n is defined to be the output of A on input n and a value chosen at random from Ker $J_n \setminus (\mathbb{Z}_n^*)^2$, and Y_n is defined to be the output of A on input n and a value chosen at random from $(\mathbb{Z}_n^*)^2$. In both cases, the

value of the random variable is determined by the random choice of , as well as the random choices made by the algorithm. Define $\varepsilon(n) := | P[X_n = 1] - P[Y_n = 1]|$.

Show how to use A to design a probabilistic, expected polynomial time algorithm A' that takes as input n as above and $\alpha \in \operatorname{Ker} J_n$, and outputs either "square" or "non-square," with the following property:

if $\varepsilon(n) \geq 0.001$, then for all $\alpha \in \operatorname{Ker} J_n$, the probability that A' correctly identifies whether $\alpha \in (\mathbb{Z}_n^*)^2$ is at least 0.999.

THE JACOBI SYMBOL

Let a, n be integers, where n is positive and odd, so that $n = q_1 \ldots q_k$, where the q_i's are odd primes, not necessarily distinct. Then the Jacobi symbol $(a \mid n)$ is defined as

$$(a \mid n) := (a \mid q_1) \ldots (a \mid q_k),$$

where $(a \mid q_i)$ is the Legendre symbol. By definition, $(a \mid 1) = 1$ for all $a \in \mathbb{Z}$. Thus, the Jacobi symbol essentially extends the domain of definition of the Legendre symbol. Note that $(a \mid n) \in \{0,1\}$, and that $(a \mid n) = 0$ if and only if $\gcd(a, n) > 1$. The following theorem summarizes the essential properties of the Jacobi symbol.

Theorem: Let m, n be odd, positive integers, and let $a, b \in \mathbb{Z}$. Then

(i) $(ab \mid n) = (a \mid n)(b \mid n)$;

(ii) $(a \mid mn) = (a \mid m)(a \mid n)$;

(iii) $a \equiv b \pmod{n}$ implies $(a \mid n) = (b \mid n)$;

(iv) $(-1 \mid n) = (-1)^{(n-1)/2}$;

(v) $(2 \mid n) = (-1)^{(n^2-1)/8}$;

(vi) $(m \mid n) = (-1)^{\frac{m-1}{2}\frac{n-1}{2}} (n \mid m)$

Proof: Parts (i) – (iii) follow directly from the definition (exercise).

For parts (iv) and (vi), one can easily verify (exercise) that for all odd integers $n_1, \ldots, n_k$,

$$\sum_{i=1}^{k} (n_1 - 1)/2 \equiv (n_1 \ldots n_k - 1)/2 \pmod{2}.$$

Part (iv) easily follows from this fact, along with part (ii) of this theorem and part (i) of Theorem (exercise). Part (vi) easily follows from this fact, along with parts (i) and (ii) of this theorem, and part (v) of Theorem.

For part (v), one can easily verify (exercise) that for odd integers $n_1, \ldots, n_k$,

$$\sum_{i=1}^{k}(n_i^2-1)/8 \equiv \left(n_1^2 \dots n_k^2 - 1\right)/8 \pmod 2.$$

Part (v) easily follows from this fact, along with part (ii) of this theorem, and part (iv) of Theorem.

As we shall see later, this theorem is extremely useful from a computational point of view – with it, one can efficiently compute $(a \mid n)$, without having to know the prime factorization of either a or n. Also, in applying this theorem it is useful to observe that for all odd integers m, n,

- $(-1)^{(n-1)/2} = 1 \Leftrightarrow n \equiv 1 \pmod 4$;
- $(-1)^{(n^2-1)/8} = 1 \Leftrightarrow n \equiv \pm 1 \pmod 8$;
- $(-1)^{((m-1)/2)((n-1)/2)} = 1 \Leftrightarrow m \equiv 1 \pmod 4$ or $n \equiv 1 \pmod 4$.

Suppose a is a quadratic residue modulo n, so that $a \equiv b^2 \pmod n$, where $\gcd(a, n) = 1 = \gcd(b, n)$. Then by parts (iii) and (i) of Theorem, we have $(a \mid n) = (b^2 \mid n) = (b \mid n)^2 = 1$. Thus, if a is a quadratic residue modulo n, then $(a \mid n) = 1$. The converse, however, does not hold: $(a \mid n) = 1$ does not imply that a is a quadratic residue modulo n.

It is sometimes useful to view the Jacobi symbol as a group homomorphism. Let n be an odd, positive integer. Define the Jacobi map

$$J_n : \mathbb{Z}_n^* \to \{\pm 1\}$$

$$[a]_n \mapsto (a \mid n).$$

First, we note that by part (iii) of Theorem, this definition is unambiguous. Second, we note that since $\gcd(a, n) = 1$ implies $(a \mid n) = \pm 1$, the image of J_n is indeed contained in $\{\pm 1\}$. Third, we note that by part (i) of Theorem, J_n is a group homomorphism. Since J_n is a group homomorphism, it follows that its kernel, Ker J_n, is a subgroup of $\mathbb{Z}_n^*$.

EXERCISE

1. Let p and q be distinct primes, with p q 3 (mod 4), and let

   $n := pq$.

   (a) Show that $[-1]_n \in \text{Ker } J_n \setminus (\mathbb{Z}_n^*)2$, and from this, conclude that the cosets of $(\mathbb{Z}_n^*)^2$ in Ker J_n are the two distinct cosets $(\mathbb{Z}_n^*)^2$ and $[-1]_n(\mathbb{Z}_n^*)^2$.

   (b) Let $\delta \in \mathbb{Z}_n^* \setminus \text{Ker } J_n$. Show that the map from $\{0, 1\} \times \{0, 1\} \times (\mathbb{Z}_n^*)^2$ to $\mathbb{Z}_n^*$ that sends (a, b, γ) to $\delta^a(-1)^b \gamma$ is a bijection.

2. Let n be an odd, positive integer, and consider the Jacobi map J_n.

   (a) Show that $(\mathbb{Z}_n^*)^2 \subseteq \text{Ker } J_n$.

   (b) Show that if n is the square of an integer, then $\text{Ker } J_n = \mathbb{Z}_n^*$.

   (c) Show that if n is not the square of an integer, then $[\mathbb{Z}_n^* : \text{Ker } J_n] = 2$ and $[\text{Ker } Jn : (\mathbb{Z}_n^*)_2] = 2^{r-1}$, where r is the number of distinct prime divisors of n.

TESTING QUADRATIC RESIDUOSITY

In this section, we consider the problem of testing whether a is a quadratic residue modulo n, for given integers a and n, from a computational perspective.

PRIME MODULUS

For an odd prime p, we can test if an integer a is a quadratic residue modulo p by either performing the exponentiation $a^{(p-1)/2} \bmod p$ or by computing the Legendre symbol $(a \mid p)$. Assume that $0 \leq a < p$. Using a standard repeated squaring algorithm, the former method takes time $O(\text{len}(p)^3)$, while using the Euclideanlike algorithm of the previous section, the latter method takes time $O(\text{len}(p)^2)$. So clearly, the latter method is to be preferred.

PRIME-POWER MODULUS

For an odd prime p, we know that a is a quadratic residue modulo p^e if and only if a is a quadratic residue modulo p. So this case immediately reduces to the previous case.

COMPOSITE MODULUS

For odd, composite n, if we know the factorization of n, then we can also determine if a is a quadratic residue modulo n by determining if it is a quadratic residue modulo each prime divisor p of n. However, without knowledge of this factorization (which is in general believed to be hard to compute), there is no efficient algorithm known. We can compute the Jacobi symbol $(a \mid n)$; if this is -1 or 0, we can conclude that a is not a quadratic residue; otherwise, we cannot conclude much of anything.

COMPUTING THE JACOBI SYMBOL

Suppose we are given an odd, positive integer n, along with an integer a, and we want to compute the Jacobi symbol $(a \mid n)$. Theorem suggests the following algorithm:

$\sigma \leftarrow 1$

repeat

// loop invariant: n is odd and positive

$a \leftarrow a \bmod n$

if $a = 0$ then

if $n = 1$ then return σ else return 0

compute a', h such that $a = 2^h a'$ and a' is odd

if $h \not\equiv 0 \pmod 2$ and $n \not\equiv \pm 1 \pmod 8$ then $\sigma \leftarrow -\sigma$

if $a' \not\equiv 1 \pmod 4$ and $n \not\equiv 1 \pmod 4$ then $\sigma \leftarrow -\sigma$

$(a, n) \leftarrow (n, a')$

forever

That this algorithm correctly computes the Jacobi symbol $(a \mid n)$ follows directly from Theorem. Using an analysis similar to that of Euclid's algorithm, one easily sees that the running time of this algorithm is $O(\text{len}(a)\ \text{len}(n))$.

EXERCISE

1. This exercise develops a probabilistic primality test based on the Jacobi symbol. For odd integer $n > 1$, define

   $G_n := \{\alpha \in \mathbb{Z}_n^* : \alpha^{(n-1)/2} = J_n(\alpha)\}$,

   where $J_n : \mathbb{Z}_n^* \to \{\pm 1\}$ is the Jacobi map.

   (a) Show that if n is prime, then $Gn = \mathbb{Z}_n^*$.

   (b) Show that if n is composite, then $Gn \subseteq \mathbb{Z}_n^*$.

   (c) Show that G_n is a subgroup of $\mathbb{Z}_n^*$.

   (d) Based on parts (a) – (c), design and analyze an efficient probabilistic primality test that works by choosing a random, non-zero element $\alpha \in \mathbb{Z}_n$, and testing if $\alpha \in G_n$.

Chapter 5

Arithmetic of Polynomials

Polynomial arithmetic includes basic mathematical operations such as addition, subtraction, and multiplication. These operations are defined naturally as if the variable x was an element of S. Division is defined similarly, but requires that S be a field. Examples of fields include rational numbers, Z_p for p prime, and real numbers. The set of all integers is not a field and does not support polynomial division.

ADDITION AND SUBTRACTION

Addition and subtraction are performed by adding or subtracting corresponding coefficients. If

$$f(x) = \sum_{i=0}^{n} a_i x^i ; g(x) = \sum_{i=0}^{m} b_i x^i$$

then addition is defined as

$$f(x) + g(x) = \sum_{i=0}^{m} (a_i + b_i) x^i + \sum_{i=0}^{n} a_i x^i$$

MULTIPLICATION

Multiplication is performed much the same way as addition and subtraction, but instead by multiplying the corresponding coefficients. If $f(x) = \sum_{i=0}^{n} a_i x^i ; g(x) = \sum_{i=0}^{m} b_i x^i$ then multiplication is defined as $f(x) \times g(x) = \sum_{i=0}^{n+m} c_i x^i$ where $c_k = a_0 b_k + a_1 b_{k+1} + \ldots + a_{k-1} b_1 + a_k b_0$. Note that we treat a_i as zero for $i > m$ and that the degree of the product is equal to the sum of the degrees to the two polynomials.

BASIC ARITHMETIC

Non-trivial ring is denoted by R. For computational purposes, we shall assume that a polynomial $g = \sum_{i=0}^{k-1} a_i X^i \in R[X]$ is represented as a coefficient vector $(a_0, a_1, ..., a_{k-1})$. Further, when g is non-zero, the coefficient a_{k-1} should be non-zero.

The basic algorithms for addition, subtraction, multiplication, and division of polynomials are quite straightforward adaptations of the corresponding algorithms for integers. In fact, because of the lack of "carries," these algorithms are actually much simpler in the polynomial case. We briefly discuss these algorithms here—analogous to our treatment of integer arithmetic, we do not discuss the details of "stripping" leading zero coefficients.

For addition and subtraction, all we need to do is to add or subtract coefficient vectors.

For multiplication, let $g = \sum_{i=0}^{k-1} a_i X^i \in R[X]$ and $h = \sum_{i=0}^{\ell-1} a_i X^i \in R[X]$, where $k \geq 1$ and $\ell \geq 1$. The product $f := g.h$ is of the form $f = \sum_{i=0}^{k+\ell-2} c_i X^i$, the coefficients of which can be computed using $O(k\ell)$ operations in R as follows:

for $i \leftarrow 0$ to $k + \ell - 2$ do $c_i \leftarrow 0$

for $i \leftarrow 0$ to $k - 1$ do

for $j \leftarrow 0$ to $\ell - 1$ do

$c_{i+j} \leftarrow c_{i+j} + a_i \,.\, b_j$

For division, let $g = \sum_{i=0}^{k-1} a_i X^i \in R[X]$ and $h = \sum_{i=0}^{\ell-1} b_i X^i \in R[X]$, where $b_{\ell-1} \in R^*$. We want to compute polynomials $q, r \in R[X]$ such that $g = hq + r$, where $\deg(r) < \ell - 1$. If $k < \ell$, we can simply set $q \leftarrow 0$ and $r \leftarrow g$; otherwise, we can compute q and r using $O(\ell .(k - \ell + 1))$ operations in R using the following algorithm:

$t \leftarrow b_{\ell-1}^{-1} \in R$

for $i \leftarrow 0$ to $k - 1$ do $r_i \leftarrow a_i$

for $i \leftarrow k - \ell$ down to 0 do

$q_i \leftarrow t \,.\, r_{i+\ell-1}$

for $j \leftarrow 0$ to $\ell - 1$ do

$r_{i+j} \leftarrow r_{i+j} - q_i \,.\, b_j$

$q \leftarrow \sum_{i=0}^{k-\ell} q_i X^i \quad , r \leftarrow \sum_{i=0}^{\ell-2} r_i X^i$

With these simple algorithms, we obtain the polynomial analog of Theorem. Let us define the length of $g \in R[X]$, denoted len(g), to be the length of its coeefficient vector; more precisely, we define

$$\text{len}(g) := \begin{cases} \deg(g)+1 & \text{if } g \neq 0, \\ 1 & \text{if } g = 0. \end{cases}$$

Sometimes (but not always) it is clearer and more convenient to state the running times of algorithms in terms of the length, rather than the degree, of a polynomial (the latter has the inconvenient habit of taking on the value 0, or worse, $-\infty$).

Theorem: Let g and h be arbitrary polynomials in R[X].

(i) We can compute $g \pm h$ with $O(\text{len}(g) + \text{len}(h))$ operations in R.

(ii) We can compute $g \,.\, h$ with $O(\text{len}(g)\ \text{len}(h))$ operations in R.

(iii) If $\text{lc}(h) \in R^*$, we can compute $q, r \in R[X]$ such that $g = hq + r$ and $\deg(r) < \deg(h)$ with $O(\text{len}(h)\ \text{len}(q))$ operations in R.

Analogous to algorithms for modular integer arithmetic, we can also do arithmetic in the residue class ring $R[X]/(f)$, where $f \in R[X]$ is a polynomial with $\text{lc}(f) \in R^*$. For each $\alpha \in R[X]/(f)$, there exists a unique polynomial $g \in R[X]$ with $\deg(g) < \deg(f)$ and $= [g]_f$; we call this polynomial g the canonical representative of α, and denote it by rep(α). For computational purposes, we represent elements of $R[X]/(f)$ by their canonical representatives.

With this representation, addition and subtraction in $R[X]/(f)$ can be performed using $O(\text{len}(f))$ operations in R, while multiplication takes $O(\text{len}(f)^2)$ operations in R.

The repeated-squaring algorithm for computing powers works equally well in this setting: given $\alpha \in R[X]/(f)$ and a non-negative exponent e, we can compute α^e using $O(\text{len}(e))$ multiplications in $R[X]/(f)$, for a total of $O(\text{len}(e)\ \text{len}(f)^2)$ operations in R.

EXERCISE

1. Given polynomials $g, h \in R[X]$, show how to compute their composition $g(h) \in R[X]$ using $O(\text{len}(g)^2\ \text{len}(h)^2)$ operations in R.

2. Let $x, a_0, ..., a_{\ell-1} \in R$, and let k be an integer with $0 < k — \ell$. For $i = 0, ..., \ell - k$, define $g_i := \sum_{j=1}^{i+k-1} a_j X^j \in R[X]$. Show how to compute the $\ell - k + 1$ values $g_0(x), ..., g_{\ell-k}(x)$ using $O(\ell)$ operations in R..

3. Let $f \in R[X]$ be a polynomial of degree $\ell > 0$ with $\text{lc}(f) \in R^*$, and let $E := R[X]/(f)$. Suppose that in addition to f, we are given a polynomial $g \in R[X]$ of degree less than k and an element $\alpha \in E$, and we want to compute $g(\alpha) \in E$.

   (a) Show that a straightforward application of Horner's rule yields an algorithm that uses $O(k\ell^2)$ operations in R, and requires space for storing $O(\ell)$ elements of R.

(b) Show how to compute $g(\alpha)$ using just $O(k\ell + k^{1/2}\ell^2)$ operations in R, at the expense of requiring space for storing $O(k^{1/2}\ell)$ elements of R.

4. Suppose you are given three polynomials $f, g, h \in \mathbb{Z}_p[X]$, where p is a large prime, in particular, $p \geq 2 \deg(g) \deg(h)$. Design an efficient probabilistic algorithm that tests if $f = g(h)$ (i.e., if f equals g composed with h). Your algorithm should have the following properties: if $f = g(h)$, it should always output "true," and otherwise, it should output "false" with probability at least 0.999. The expected running time of your algorithm should be $O((\text{len}(f) + \text{len}(g) + \text{len}(h)) \text{len}(p)^2)$.

5. Given a polynomial $g \in R[X]$ and an element $x \in R$, a particularly elegant and efficient way of computing $g(x)$ is called Horner's rule. Suppose $g = \sum_{i=0}^{k-1} a_i X^i$, where $k \geq 0$ and $a_i \in R$ for $i = 0, ..., k - 1$. Horner's rule computes $g(x)$ as follows:

   $y \leftarrow 0_R$
   for $i \leftarrow k - 1$ down to 0 do
   $y \leftarrow yx + a_i$
   output y

   Show that this algorithm correctly computes $g(x)$ using k multiplications in R and k additions in R.

RATIONAL FUNCTION RECONSTRUCTION

Throughout this section, F denotes a field.

We next state and prove the polynomial analog of Theorem. As we are now "reconstituting" a rational function, rather than a rational number, we call this procedure rational function reconstruction. Because of the relative simplicity of polynomials compared to integers, the rational reconstruction theorem for polynomials is a bit "sharper" than the rational reconstruction theorem for integers, and much simpler to prove.

To state the result precisely, let us introduce some notation. For polynomials $g, h \in F[X]$ with $\deg(g) \geq \deg(h)$ and $g \neq 0$, let us define

$$EEA(g, h) := \{(r_i, s_i, t_i)\}_{i=0}^{\lambda+1},$$

where r_i, s_i, and t_i, for $i = 0, ..., \lambda + 1$.

Theorem: Let $f, h \in F[X]$ be polynomials, and let r^*, t^* be non-negative integers, such that

$$\deg(h) < \deg(f) \text{ and } r^* + t^* \leq \deg(f).$$

Further, let $\text{EEA}(f, h) = \{(r_i, s_i, t_i)\}_{i=0}^{\lambda+1}$, and let j be the smallest index (among 0, ..., λ + 1) such that $\deg(r_j) < r^*$, and set

$$r' := r_j,\ s' := s_j, \text{ and } t' := t_j.$$

Finally, suppose that there exist polynomials $r, s, t \in F[X]$ such that

$$r = fs + ht,\ \deg(r) < r^*,\ \text{and}\ 0 \leq \deg(t) \leq t^*.$$

Then for some non-zero polynomial $q \in F[X]$, we have

$$r = r'q,\ s = s'q,\ t = t'q.$$

Proof. Since $\deg(r_0) = \deg(f) \geq r^* > -\infty = \deg(r_{\lambda+1})$, the value of j is well defined, and moreover, $j \geq 1$, $\deg(r_{j-1}) \geq r^*$, and $t_j \neq 0$.

From the equalities $rj = fs_j + ht_j$ and $r = f_s + ht$, we have the two congruences:

$$r_j \equiv ht_j \pmod f,$$

$$r \equiv ht \pmod f.$$

Subtracting t times the first from tj times the second, we obtain

$$rt_j \equiv r_j t \pmod f.$$

This says that f divides $rt_j - r_j t$.

We want to show that, in fact, $rt_j - r_j t = 0$. The inequality $\deg(r_{j-1}) \geq r^*$, we have

$$\deg(t_j) = \deg(f) - \deg(r_{j-1}) \leq \deg(f) - r^*.$$

Combining this with the inequality $\deg(r) < r^*$, we see that

$$\deg(rt_j) = \deg(r) + \deg(t_j) < \deg(f).$$

Furthermore, using the inequalities

$$\deg(r_j) < r^*,\ \deg(t) \leq t^*,\ \text{and}\ r^* + t^* \leq \deg(f),$$

we see that

$$\deg(r_j t) = \deg(r_j) + \deg(t) < \deg(f),$$

and it immediately follows that

$$\deg(rt_j - r_j t) < \deg(f).$$

Since f divides $rt_j - r_j t$ and $\deg(rt_j - r_j t) < \deg(f)$, the only possibility is that

$$rt_j - r_j t = 0.$$

RECOVERING RATIONAL FUNCTIONS

This is an entirely straightforward translation of the results in §4.6.1, but we shall see in the next chapter that this problem has its own interesting applications.

Suppose Alice knows a rational function $z = s/t \in F(X)$, where s and t are polynomials with $\deg(s) < \deg(t)$, and tells Bob some of the high-order coefficients of the reversed Laurent series representing z in $F((X^{-1}))$. We shall show that if $\deg(t) \leq \ell$ and Bob is given the bound ℓ on $\deg(t)$, along with the high-order 2ℓ coefficients of z, then Bob can determine z, expressed as a rational function in lowest terms.

So suppose that $z = s/t = \sum_{i=1}^{\infty} z_i X^{-i}$, and that Alice tells Bob the coefficients $z_1, ..., z_{2\ell}$. Equivalently, Alice gives Bob the polynomial

$$h := z_1 X_{2\ell-1} + ... + z_{2\ell-1} X + z_{2\ell}.$$

Also, let us define $f := X^{2\ell}$. Here is Bob's algorithm for recovering z:

1. Run the extended Euclidean algorithm on input f, h to obtain EEA(f, h), and apply with f, h, $r^* := \ell$, and $t^* := \ell$, to obtain the polynomials r', s', t'.
2. Output s', t'.

   We claim that $z = -s'/t'$. To prove this, first observe that $h = \lfloor fz \rfloor = \lfloor fs/t \rfloor$. So if we set $r := fs \bmod t$, then we have

   $r = fs - ht$, $\deg(r) < r^*$, $0 \leq \deg(t) \leq t^*$, and $r + t \leq \deg(f)$.

   It follows that the polynomials s', t' from Theorem satisfy $s = s'q$ and $-t = t'q$ for some non-zero polynomial q, and thus, $s' / t' = -s/t$, which proves the claim.

   We may further observe that since the extended Euclidean algorithm guarantees that $\gcd(s', t') = 1$, not only do we obtain z, but we obtain z expressed as a fraction in lowest terms.

   It is clear that this algorithm takes $O(\ell^2)$ operations in F.

POLYNOMIAL INTERPOLATION WITH ERRORS

If we "encode" a polynomial $g \in F[X]$, with $\deg(g) < k$, as the sequence $(y_1, ..., y_k) \in F_{\times k}$, where $y_i = g(x_i)$, then we can efficiently recover g from this encoding, using an algorithm for polynomial interpolation. Here, of course, the x_i's are distinct elements of F.

Now suppose that Alice encodes g as $(y_1, ..., y_k)$, and sends this encoding to Bob, but that some, say at most ℓ, of the yi's may be corrupted during transmission. Let $(z_1, ..., z_k)$ denote the vector actually received by Bob.

Here is how we can use Theorem to recover the original value of g from $(z_1, ..., z_k)$, assuming:

- The original polynomial g has degree less than m,
- At most ℓ errors occur in transmission, and
- $k \geq 2\ell + m$.

Let us set $f_i := X - x_i$ for $i = 1, ..., k$, and $f := f_1 ... f_k$. Now, suppose Bob obtains the corrupted encoding $(z_1, ..., z_k)$. Here is what Bob does to recover g:

1. Interpolate, obtaining a polynomial h, with $\deg(h) < k$ and $h(x_i) = z_i$ for $i = 1, ..., k$.
2. Run the extended Euclidean algorithm on input f, h to obtain EEA(f, h), and apply Theorem 17.8 with f, h, $r^* := m + \ell$ and $t^* := \ell$, to obtain the polynomials r', s', t'.
3. If $t' \mid r'$, output r'/t'; otherwise, output "error."

We claim that the above procedure outputs g, under the assumptions listed above. To see this, let t be the product of the fi's for those values of i where an error occurred. Now, assuming at most ℓ errors occurred, we have $\deg(t) \leq \ell$. Also, let $r := gt$, and note that $\deg(r) < m + \ell$. We claim that

$$r \equiv ht \pmod{f}.$$

To show that holds, it suffices to show that

$$gt \equiv ht \pmod{f_i}$$

for all $i = 1, ..., k$. To show this, consider first an index i at which no error occurred, so that $y_i = z_i$. Then $g_t \equiv y_i t \pmod{f_i}$ and $h_t \equiv z_i t \equiv y_i t \pmod{f_i}$, and so holds for this i. Next, consider an index i for which an error occurred. Then by construction, $gt \equiv 0 \pmod{f_i}$ and $ht \equiv 0 \pmod{f_i}$, and so holds for this i. Thus, holds, from which it follows that the values r', t' obtained from Theorem 17.8 satisfy

$$\frac{r'}{t'} = \frac{r}{t} = \frac{gt}{t} = g.$$

One easily checks that both the procedures to encode and decode a value g run in time $O(k^2)$. The above scheme is an example of an error correcting code called a Reed–Solomon code.

APPLICATIONS TO SYMBOLIC ALGEBRA

We have already discussed the application of solving systems of linear equations over the integers using rational reconstruction. In exactly the same way, one can use rational function reconstruction to solve systems of linear equations over $F[X]$ – the solution to such a system of equations will be a vector whose entries are elements of $F(X)$, the field of rational functions.

EXERCISE

1. Let F be a field. Show that given polynomials $s, t \in F[X]$ and integer k, with $\deg(s) < \deg(t)$ and $k > 0$, we can compute the kth coefficient in the reversed Laurent series representing s/t using $O(\text{len}(k)\,\text{len}(t)^2)$ operations in F.
2. Let F be a field. Let $z = s/t$, where $s, t \in F[X]$, $\deg(s) < \deg(t)$, and $\gcd(s, t) = 1$.

(a) Show that if F is finite, there exist integers k, k' such that $0 \leq k < k'$ and $sX^k \equiv sX^{k'} \pmod{t}$.

(b) Show that for integers k, k' with $0 \leq k < k'$, the sequence of coefficients of the reversed Laurent series representing z is $(k, k'-k)$-periodic if and only if $sX^k \equiv sX^{k'} \pmod{t}$.

(c) Show that if F is finite and $X \nmid t$, then the reversed Laurent series representing z is purely periodic with period equal to the multiplicative order of $[X]_t \in (F[X]/(t))$.

(d) More generally, show that if F is finite and $t = X^k t'$, with $X \nmid t'$, then the reversed Laurent series representing z is ultimately periodic with pre-period k and period equal to the multiplicative order of $[X]_{t'} \in (F[X]/(t'))^*$.

3. Let F be a field. Let $z \in F((X^{-1}))$ be a reversed Laurent series whose coefficient sequence is ultimately periodic. Show that $z \in F(X)$.

COMPUTING MINIMAL POLYNOMIALS IN $F[X]/(F)$ (I)

In this section, we shall examine a computational problem to which we shall return on several occasions, as it will serve to illustrate a number of interesting algebraic and algorithmic concepts.

Let F be a field, and let $f \in F[X]$ be a monic polynomial of degree $\ell > 0$. Also, let $E := F[X]/(f)$, which is an F-algebra, and in particular, an F-vector space. As an F-vector space, E has dimension ℓ. Suppose we are given an element $\alpha \in E$, and want to efficiently compute the minimal polynomial of α over F, that is, the monic polynomial $\phi \in F[X]$ of least degree such that $\phi(\alpha) = 0$, which we know has degree at most ℓ.

We can solve this problem using polynomial arithmetic and Gaussian elimination, as follows. Consider the F-linear map $\rho : F[X]_{\leq \ell} \to E$ that sends a polynomial $g \in F[X]$ of degree at most ℓ to g(). To perform the linear algebra, we need to specify bases for $F[X]_{\leq \ell}$ and E. For $F[X]_{\leq \ell}$, let us work with the basis $S := \{X^{\ell+1-i}\}_{i=1}^{\ell+1}$. With this choice of basis, for $g = \sum_{i=0}^{\ell} a_i X^i \in F[X]_{\leq \ell}$, the coordinate vector of g is $\mathrm{Vec}_S(g) = (a_\ell, \ldots, a_0) \in F^{1\times(\ell+1)}$. For E, let us work with the basis $T := \{\xi^{i-1}\}_{i=1}^{\ell}$, where $\xi := [X]_f \in E$. Let

$$A := \mathrm{Mat}_{S,T}(\rho) \in F^{(\ell+1)\times \ell};$$

that is, A is the matrix of ρ relative to S and T (see §14.2). For $i = 1, \ldots, \ell + 1$, the ith row of A is the coordinate vector $\mathrm{Vec}_T(\alpha^{\ell+1-i}) \in F^{1\times \ell}$.

We compute the matrix A by computing the powers $1, \alpha, \ldots, \alpha^{\ell}$; reading off the ith row of A directly from the canonical representative of the $\alpha^{\ell+1-i}$.

We then apply Gaussian elimination to A to find row vectors $v_1, ..., v_s \in F^{1\times(\ell+1)}$ that are coordinate vectors corresponding to a basis for the kernel of ρ. Now, the coordinate vector of the minimal polynomial of is a linear combination of $v_1, ..., v_s$. To find it, we form the $s \times (\ell + 1)$ matrix B whose rows consist of $v_1, ..., vs$, and apply Gaussian elimination to B, obtaining an s $(\ell + 1)$ matrix B' in reduced row echelon form whose row space is the same as that of B. Let ϕ be the polynomial whose coordinate vector is the last row of B'.

Because of the choice of basis for $F[X]_{\leq \ell}$, and because B' is in reduced row echelon form, it is clear that no non-zero polynomial in Ker ϕ has degree less than that of ρ. Moreover, as ϕ is already monic (again, by the fact that B' is in reduced row echelon form), it follows that is in fact the minimal polynomial of α over F.

The total amount of work performed by this algorithm is $O(\ell^3)$ operations in F to build the matrix A (this just amounts to computing ℓ successive powers of , that is, $O(\ell)$ multiplications in E, each of which takes $O(\ell^2)$ operations in F), and $O(\ell^3)$ operations in F to perform both Gaussian elimination steps.

EUCLID'S ALGORITHM

In this section, F denotes a field, and we consider the computation of greatest common divisors in $F[X]$.

The Euclidean algorithm for integers is easily adapted to compute gcd(g, h) for polynomials $g, h \in F[X]$. Analogous to the integer case, we assume that $\deg(g) \geq \deg(h)$; however, we shall also assume that $g \neq 0$. This is not a serious restriction, of course, as gcd(0, 0) = 0, and making this restriction will simplify the presentation a bit. Recall that we defined gcd(g, h) to be either zero or monic, and the assumption that $g \neq 0$ means that gcd(g, h) is non-zero, and hence monic.

The following is the analog of Theorem, and is based on the division with remainder property for polynomials.

Theorem: Let $g, h \in F[X]$, with $\deg(g) \geq \deg(h)$ and $g \neq 0$. Define the polynomials $r_0, r_1, ..., r_{\lambda+1} \in F[X]$ and $q_1, ..., q_\lambda \in F[X]$, where $\lambda \geq 0$, as follows:

$$g = r_0,$$

$$h = r_1,$$

$$r_0 = r_1 q_1 + r_2 \ (0 \leq \deg(r_2) < \deg(r_1)),$$

$$\vdots$$

$$r_{i-1} = r_i q_i + r_{i+1} \ (0 \leq \deg(r_{i+1}) < \deg(r_i)),$$

$$\vdots$$

$$r_{\lambda-2} = r_{\lambda-1} q_{\lambda-1} + r_\lambda \ (0 \leq \deg(r_\lambda) < \deg(r_{\lambda-1})),$$

$$r_{\lambda-1} = r_\lambda q_\lambda \ (r_{\lambda-1} = 0).$$

Note that by definition, $\lambda = 0$ if $h = 0$, and $\lambda > 0$ otherwise. Then we have

$$r_\lambda / \mathrm{lc}(r_\lambda) = \gcd(g, h), \text{ and if } h \neq 0, \text{ then } \lambda \leq \deg(h) + 1.$$

Proof. Arguing as in the proof of Theorem, one sees that

$$\gcd(g, h) = \gcd(r_0, r_1) = ... = \gcd(r_\lambda, r_{\lambda+1}) = \gcd(r_\lambda, 0) = r_\lambda / \mathrm{lc}(r_\lambda).$$

That proves the first statement.

For the second statement, if $h \neq 0$, then the degree sequence

$$\deg(r_1), \deg(r_2), ..., \deg(r_\lambda)$$

is strictly decreasing, with $\deg(r_\lambda) \geq 0$, from which it follows that $\deg(h) = \deg(r_1) \geq \lambda - 1$.

This gives us the following polynomial version of the Euclidean algorithm:

EUCLID'S ALGORITHM

On input g, h, where $g, h \in F[X]$ with $\deg(g) \geq \deg(h)$ and $h \neq 0$, compute $d = \gcd(g, h)$ as follows:

$r \leftarrow g, r' \leftarrow h$

while $r' \neq 0$ do

$r'' \leftarrow r \bmod r'$

$(r, r') \leftarrow (r', r'')$

$d \leftarrow r/\mathrm{lc}(r)$ // make monic

output d

Theorem: Let $g, h, r_0, ..., r_{\lambda+1}$ and $q_1, ..., q_\lambda$ be as in Theorem 17.2. Define polynomials $s_0, ..., s_{\lambda+1} \in F[X]$ and $t_0, ..., t + 1 \in F[X]$ as follows:

$$s_0 := 1, t_0 := 0,$$

$$s_1 := 0, t_1 := 1,$$

and for $i = 1, ..., \lambda$,

$$s_{i+1} := s_{i-1} - s_i q_i, \ t_{i+1} := t_{i-1} - t_i q_i.$$

Then:

(i) for $i = 0, ..., \lambda + 1$, we have $gs_i + ht_i = r_i$; in particular, $gs_\lambda + ht_\lambda = \mathrm{lc}(r_\lambda) \gcd(g, h)$;

(ii) for $i = 0, ..., \lambda$, we have $s_i t_{i+1} - t_i s_{i+1} = (-1)^i$;

(iii) for $i = 0, ..., \lambda + 1$, we have $\gcd(s_i, t_i) = 1$;

(iv) for $i = 1, ..., \lambda + 1$, we have

$\deg(t_i) = \deg(g) - \deg(r_{i-1})$,

and for $i = 2, \ldots, \lambda + 1$, we have

$\deg(s_i) = \deg(h) - \deg(r_{i-1})$;

(v) for $i = 1, \ldots, \lambda + 1$, we have $\deg(t_i) \leq \deg(g)$ and $\deg(s_i) \leq \deg(h)$; if $\deg(g) > 0$ and $h \neq 0$, then $\deg(t_\lambda) < \deg(g)$ and $\deg(s_\lambda) < \deg(h)$.

Proof: (i), (ii), and (iii) are proved just as in the corresponding parts of Theorem.

For (iv), the proof will hinge on the following facts:

- For $i = 1, \ldots, \lambda$, we have $\deg(r_{i-1}) \geq \deg(r_i)$, and since q_i is the quotient in dividing r_{i-1} by r_i, we have $\deg(q_i) = \deg(r_{i-1}) - \deg(r_i)$.
- For $i = 2, \ldots, \lambda$, we have $\deg(r_{i-1}) > \deg(r_i)$.

We prove the statement involving the t_i's by induction on i, and leave the proof of the statement involving the si's to the reader.

One can see by inspection that this statement holds for $i = 1$, since $\deg(t_1) = 0$ and $r_0 = g$. If $\lambda = 0$, there is nothing more to prove, so assume that $\lambda > 0$ and $h \neq 0$.

Now, for $i = 2$, we have $t_2 = 0 - 1 \,.\, q_1 = -q_1$. Thus, $\deg(t_2) = \deg(q_1) = \deg(r_0) - \deg(r_1) = \deg(g) - \deg(r_1)$.

Now for the induction step. Assume $i \geq 3$. Then we have

$\deg(t_{i-1}\, q_{i-1}) = \deg(t_{i-1}) + \deg(q_{i-1})$

$= \deg(g) - \deg(r_{i-2}) + \deg(q_{i-1})$ (by induction)

$= \deg(g) - \deg(r_{i-1})$

(since $\deg(q_{i-1}) = \deg(r_{i-2}) - \deg(r_{i-1})$)

$> \deg(g) - \deg(r_{i-3})$ (since $\deg(r_{i-3}) > \deg(r_{i-1})$)

$= \deg(t_{i-2})$ (by induction).

By definition, $t_i = t_{i-2} - t_{i-1}\, q_{i-1}$, and from the above reasoning, we see that

$$\deg(g) - \deg(r_{i-1}) = \deg(t_{i-1}\, q_{i-1}) > \deg(t_{i-2}).$$

from which it follows that $\deg(t_i) = \deg(g) - \deg(r_{i-1})$.

Theorem : Euclid's algorithm for polynomials uses $O(\text{len}(g)\ \text{len}(h))$ operations in F.

THE EXTENDED EUCLIDEAN ALGORITHM

On input g, h, where $g, h \in F[X]$ with $\deg(g) \geq \deg(h)$ and $g \neq 0$, compute d, s, and t, where $d, s, t \in F[X]$, $d = \gcd(g, h)$ and $gs + ht = d$, as follows:

$r \leftarrow g, r' \leftarrow h$

$s \leftarrow 1, s' \leftarrow 0$

$t \leftarrow 0, t' \leftarrow 1$

while $r' \neq 0$ do

compute q, r'' such that $r = r'q + r''$, with $\deg(r'') < \deg(r')$

$$(r, s, t, r', s', t') \leftarrow (r', s', t', r'', s - s'q, t - t'q)$$

$c \leftarrow \mathrm{lc}(r)$

$d \leftarrow r/c, s \leftarrow s/c, t \leftarrow t/c$ // make monic

output d, s, t

Theorem : The extended Euclidean algorithm for polynomials performs $O(\mathrm{len}(g)\,\mathrm{len}(h))$ operations in F.

EXERCISE

1. Assume that integers of length at most ℓ can be multiplied in time $\overline{M}(\ell)$, where $\overline{M}$ is a well-behaved complexity function. Let $g, h \in \mathbb{Z}[X]$ with $g = \sum_{i=0}^{m-1} a_i X^i$ and $h = \sum_{i=0}^{m-1} b_i X^i$, where each a_i and b_i is a non-negative integer, strictly less than 2^k. The product $f := gh \in \mathbb{Z}[X]$ may be written $f = \sum_{i=0}^{2m-2} c_i X^i$, where each c_i is a non-negative integer. Show how to compute f, given g and h, using $O(\overline{M}((k + \mathrm{len}(m))m))$ operations in R.

The following exercises develop an important algorithm for multiplying polynomials in almost-linear time. For an integer $n \geq 0$, let us call $\omega \in R$ a primitive 2^nth root of unity if $n \geq 1$ and $\omega^{2^{n-1}} = -1_R$, or $n = 0$ and $\omega = 1_R$; if $2_R \neq 0_R$, then in particular, ω has multiplicative order 2^n. For $n \geq 0$, and $\omega \in R$ a primitive 2^nth root of unity, let us define the R-linear map $\varepsilon_{n,\omega} : R^{\times 2^n} \to R^{\times 2^n}$ that sends the vector $(a_0, ..., a_{2^n-1})$ to the vector $(g(1_R), g(\omega), ..., g(\omega^{2^n-1}))$, where $\sum_{i=0}^{2^n-1} a_i X^i \in R[X]$.

2. Let $g, h \in R[X, Y]$ with $g = \sum_{i=0}^{m-1} g_i Y^i$ and $h = \sum_{i=0}^{m-1} h_i Y^i$ where each g_i and h_i is a polynomial in X of degree less than k. The product $f := gh \in R[X, Y]$ may be written $f = \sum_{i=0}^{2m-2} f_i Y^i$, where each f_i is a polynomial in X. Show how to compute f, given g and h, using $O(M(km))$ operations in R.

3. Suppose $2_R \in R^*$ and $\omega \in R$ is a primitive 2^nth root of unity.

   (a) Show that $\omega^k - 1_R \in R^*$ for all integers $k \not\equiv 0 \pmod{2^n}$.

   (b) Let k be any integer, and consider $\gcd(k, 2^n)$, which must be of the form 2^m for some $m = 0, \ldots, n$. Show that ω^k is a primitive 2^{n-m} th root of unity.

   (c) Show that if $n \geq 1$, then $\omega - 1_R \in R$.

   (d) Let M_2 be the 2-multiplication map on $R^{\times 2^n}$, which is a bijective, R-linear map. Show that

   $$\varepsilon_{n,\omega} \circ \varepsilon_{n,\omega^{-1}} = M_2^n = \varepsilon_{n,\omega^{-1}} \circ \varepsilon_{n,\omega},$$

   and conclude that $\varepsilon_{n,\omega}$ is bijective, with $M_2^{-n} \circ \varepsilon_{n,\omega^{-1}}$ being its inverse.

   (e) Show that for every integer k, we have

   $$\sum_{i=0}^{2^n-1} \omega^{ki} = \begin{cases} 2_R^r & \text{if } k \equiv 0 \pmod{2^n} \\ 0_R & \text{if } k \not\equiv 0 \pmod{2^n} \end{cases}.$$

4. Assume $2_R \in R$. Suppose that we are given two polynomials $g, h \in R[X]$ of length at most ℓ, along with a primitive 2^nth root of unity $\omega \in R$, where $2\ell \leq 2^n < 4\ell$. Let us "pad" g and h, writing $g = \sum_{i=0}^{2^n-1} a_i X^i$ and $h = \sum_{i=0}^{2^n-1} b_i X^i$, where a_i and b_i are zero for $i \geq \ell$. Show that the following algorithm correctly computes the product of g and h using $O(\ell \operatorname{len}(\ell))$ operations in R:

   $(\alpha_0, \ldots, \alpha_{2^n-1}) \leftarrow \mathrm{FFT}(n, \omega; a_0, \ldots, a_{2^n-1})$

   $(\beta_0, \ldots, \beta_{2^n-1}) \leftarrow \mathrm{FFT}(n, \omega; b_0, \ldots, b_{2^n-1})$

   $(\gamma_0, \ldots, \gamma_{2^n-1}) \leftarrow (\alpha_0\beta_0, \ldots, \alpha_{2^n-1}\beta_{2^n-1})$

   $(c_0, \ldots, c_{2^n-1}) \leftarrow 2_R^{-n}\ \mathrm{FFT}(n, \omega^{-1}; \gamma_0, \ldots, \gamma_{2^n-1})$

   output $\sum_{i=0}^{2\ell-2} c_i X^i$

   Also, argue more carefully that the algorithm performs $O(\ell \operatorname{len}(\ell))$ additions and subtractions in R, $O(\ell \operatorname{len}(\ell))$ multiplications in R by powers of ω, and $O(\ell)$ other multiplications in R.

5. This exercise develops a fast algorithm, called the fast Fourier transform or FFT, for computing the function $\varepsilon_{n,\omega}$. This is a recursive algorithm FFT(n, ω; a_0, ..., a_{2^n-1}) that takes as input an integer $n \geq 0$, a primitive 2^nth root of unity $\omega \in R$, and elements $a_0, ..., a_{2^n-1} \in R$, and runs as follows:

   if $n = 0$ then

   return a_0

   else

   $(\alpha_0, ..., \alpha_{2^{n-1}-1}) \leftarrow \text{FFT}(n-1, \omega^2; a_0, a_2, ..., a_{2^n-2})$

   $(\beta_0, \ldots, \beta_{2^{n-1}-1}) \leftarrow \text{FFT}(n-1, \omega^2; a_1, a_3, ..., a_{2^n-1})$

   for $i \leftarrow 0$ to $2^{n-1} - 1$ do

   $\gamma_i \leftarrow \alpha_i + \beta_i\omega^i$, $\gamma_{i+2^{n-1}} \leftarrow \alpha_i - \beta_i\omega^i$

   return $(\gamma_0, ..., \gamma_{2^n-1})$

   Show that this algorithm correctly computes $\varepsilon_{n,\omega}(a_0, ..., a_{2^n-1})$ using $O(2^n n)$ operations in R.

6. Assume $2_R \in R^*$. In this exercise, we use the FFT to develop an algorithm that multiplies polynomials over R of length at most ℓ using $O(\ell \text{ len}(\ell)^\beta)$ operations in R, where is a constant. Unlike the previous exercise, we do not assume that R contains any particular primitive roots of unity; rather, the algorithm will create them "out of thin air." Suppose that $g, h \in R[X]$ are of length at most ℓ. Set $k :=$ $\lfloor \ell/2 \rfloor$, $m := \lfloor \ell/k \rfloor$. We may write $g = \sum_{i=0}^{m-1} g_i X^{ki}$ and $h = \sum_{i=0}^{m-1} g_i X^{ki}$, where the g_i's and h_i's are polynomials of length at most k. Let n be the integer determined by $2m \leq 2^n < 4m$. Let $q := X^{2^{n-1}} + 1_R \in R[X]$, $E := R[X]/(q)$, and $\omega := [X]_q \in E$.

   (a) Show how to compute the coefficients of $f := gh \in R[X]$ from the value $\bar{f} \in E[Y]$ computed in part (b), using $O(\ell)$ operations in R.

   (b) Show that ω is a primitive 2^nth root of unity in E, and that given an element $\zeta \in E$ and an integer i between 0 and $2^n - 1$, we can compute $\omega^i \in E$ using $O(\ell^{1/2})$ operations in R.

   (c) Let $\bar{g} := \sum_{i=0}^{m-1} [g_i]_q Y^i$ and $\bar{h} := \sum_{i=0}^{m-1} [h_i]_q Y^i \in E[Y]$. Using the FFT (over E), show how to compute $\bar{f} := \bar{g}\,\bar{h} \in E[Y]$ by computing $O(\ell^{1/2})$ products in $R[X]$ of polynomials of length $O(\ell^{1/2})$, along with $O(\ell \text{ len}(\ell))$ additional operations in R.

7. This exercise uses the FFT to develop a linear-time algorithm for integer multiplication; however, a rigorous analysis depends on an unproven conjecture (which follows from a generalization of the Riemann hypothesis). Suppose we want to multiply two positive integers a and b, each of length at most ℓ. Throughout this exercise, assume that all computations are done on a RAM, and that arithmetic on integers of length $O(\text{len}(\ell))$ takes time $O(1)$. Let k be an integer parameter

with $k = \Theta(\text{len}(\ell))$, and let $m := \lfloor \ell/k \rfloor$. We may write $a = \sum_{i=0}^{m-1} a_i 2^{ki}$ and $b = \sum_{i=0}^{m-1} b_i 2^{ki}$, where $0 \leq a_i < 2^k$ and $0 \leq b_i < 2^k$. Let n be the integer determined by $2m \leq 2^n < 4m$.

   (a) Assuming you have computed $f \in \mathbb{Z}_p[X]$ as in part (b), show how to obtain $c := ab$ in time $O(\ell)$.

   (b) Assuming you have computed p and ω as in part (a), let $g := \sum_{i=0}^{m-1} [a_i]X^i \in \mathbb{Z}_p[X]$ and $h :=$ Pm−1i=0 [bi]pXi $\in \mathbb{Z}_p[X]$, and show how to compute $f := gh \in \mathbb{Z}_p[X]$ in time $O(\ell)$ using the FFT (over $\mathbb{Z}_p$). Here, you may store elements of $\mathbb{Z}_p$ in single memory cells, so that operations in $\mathbb{Z}_p$ take time $O(1)$.

   (c) Show how to efficiently generate a prime $p \equiv 1 \pmod{2^n}$ and an element $\omega \in \mathbb{Z}_p^*$ of multiplicative order $2n$, such that

   $2^{2k} m < p \leq \ell^{O(1)}$.

   Your algorithm should be probabilistic, and run in expected time polynomial in $\text{len}(\ell)$.

8. Continuing with the previous exercise:

   (a) Using Theorem, show how to make this algorithm fully deterministic and rigorous, provided that on inputs of length ℓ, it is provided with a certain bit string ℓ of length $O(\text{len}(\ell))$ (this is called a non-uniform algorithm).

   (b) Show how the algorithm presented there can be implemented on a RAM that has only built-in addition, subtraction, and branching instructions, but no multiplication or division instructions, and still run in time $O(\ell)$. Also, memory cells should store numbers of length at most $\text{len}(\ell) + O(1)$.

9. Let n be a large, positive integer. We can factor n using trial division in time $n^{1/}$

$^{2+o(1)}$; however, using fast polynomial arithmetic in $\mathbb{Z}_n[X]$, one can get a simple, deterministic, and rigorous algorithm that factors n in time $n^{1/4+o(1)}$. Note that all of the factoring algorithms discussed, while faster, are either probabilistic, or deterministic but heuristic. Assume that we can multiply polynomials in $\mathbb{Z}_n[X]$ of length at most ℓ using $M(\ell)$ operations in $\mathbb{Z}_n$, where M is a well-behaved complexity function, and $M(\ell) = \ell^{1+o(1)}$ (the algorithm from Exercise 17.24 would suffice).

(a) Let ℓ be a positive integer, and for $i = 1, ..., \ell$, let

$$\sum_{j=0}^{\ell-1}(i\ell - j) \bmod n.$$

Using fast polynomial arithmetic, show how to compute $(a_1, ..., a_\ell)$ in time $\ell^{1+o(1)} \operatorname{len}(n)^{O(1)}$.

(b) Using the result of part (a), show how to factor n in time $n^{1/4+o(1)}$ using a deterministic algorithm.

CHINESE REMAINDERING

We will explore various applications of Euclid's algorithm for polynomials. Most of these applications are analogous to their integer counterparts, although there are some differences to watch for.

Throughout this section, F denotes a field.

We begin with the obvious application of the extended Euclidean algorithm for polynomials to the problem of computing multiplicative inverses in $F[X] / (f)$.

Theorem: Suppose we are given polynomials $g, f \in F[X]$, where $\deg(g) < \deg(f)$. Then using $O(\operatorname{len}(f)^2)$ operations in F, we can determine if g is relatively prime to f, and if so, compute $g^{-1} \bmod f$.

Proof: We may assume $\deg(f) > 0$, since when $\deg(f) = 0$, we have $g = 0$ and $g^{-1} \bmod f = 0$. We run the extended Euclidean algorithm on input f, g, obtaining polynomials d, s, t such that $d = \gcd(f, g)$ and $fs + gt = d$. If $d \neq 1$, then g does not have a multiplicative inverse modulo f. Otherwise, if $d = 1$, then t is a multiplicative inverse of g modulo f. Moreover, by part (v) of Theorem 17.4, we have $\deg(t) < \deg(f)$, and so $t = g^{-1} \bmod f$. Based on Theorem, it is clear that all the computations can be performed using $O(\operatorname{len}(f)^2)$ operations in F.

Theorem: Suppose we are given polynomials $f_1, ..., f_k \in F[X]$ and $g_1, ..., gk \in F[X]$, where the family $\{f_i\}_{i=1}^k$ is pairwise relatively prime, and where $\deg(f_i) > 0$ and $\deg(g_i) < \deg(f_i)$ for $i = 1, ..., k$. Let $f := \prod_{i=1}^k f_i$. Then using $O(\operatorname{len}(f)^2)$ operations in F, we can compute the unique polynomial $g \in F[X]$ satisfying $\deg(g) < \deg(f)$ and $g \equiv g_i \pmod{fi}$ for $i = 1, ..., k$.

POLYNOMIAL INTERPOLATION

We remind the reader of the discussion following Theorem, where the point was made that when $f_i = X - x_i$ and $g_i = y_i$, for $i = 1, ..., k$, then the Chinese remainder theorem for polynomials reduces to Lagrange interpolation. Thus, Theorem says that given distinct elements $x_1, ..., xk \in F$, along with elements $y_1, ..., yk \in F$, we can compute the unique polynomial $g \in F[X]$ of degree less than k such that

$$g(x_i) = y_i \ (i = 1, ..., k),$$

using $O(k^2)$ operations in F.

It is perhaps worth noting that we could also solve the polynomial interpolation problem using Gaussian elimination, by inverting the corresponding Vandermonde matrix . However, this algorithm would use $O(k^3)$ operations in F. This is a specific instance of a more general phenomenon: there are many computational problems involving polynomials over fields that can be solved using Gaussian elimination, but which can be solved more efficiently using more specialized algorithmic techniques.

MODULAR COMPUTATION

We have discussed how the Chinese remainder theorem could be used to speed up certain types of computations involving integers. The example we gave was the multiplication of integer matrices. We can use the same idea to speed up certain types of computations involving polynomials. For example, if one wants to multiply two matrices whose entries are elements of $F[X]$, one can use the Chinese remainder theorem for polynomials to speed things up. This strategy is most easily implemented if F is sufficiently large, so that we can use polynomial evaluation and interpolation directly, and do not have to worry about constructing irreducible polynomials.

Chapter 6

Complex Numbers

Complex analysis is the study of complex numbers together with their derivatives, manipulation, and other properties. Complex analysis is an extremely powerful tool with an unexpectedly large number of practical applications to the solution of physical problems. They also have applications in a variety of areas, ranging from fluid flow, to electric circuits, to the mysterious quantum world. Generally, they are described as belonging to the area of mathematics known as complex analysis.

The key result in complex analysis is the Cauchy integral theorem, which is the reason that single-variable complex analysis has so many nice results. A fundamental result of complex analysis is the Cauchy-Riemann equations, which give the conditions a function must satisfy in order for a complex generalization of the derivative, the so-called complex derivative, to exist. When the complex derivative is defined "everywhere," the function is said to be analytic.

The complex numbers are the field $\mathbb{C}$ of numbers of the form $x + iy$, where x and y are real numbers and i is the imaginary unit equal to the square root of -1.

COMPLEX NUMBERS

Complex numbers are useful abstract quantities that can be used in calculations and result in physically meaningful solutions. However, recognition of this fact is one that took a long time for mathematicians to accept.

THE ORIGIN OF COMPLEX NUMBERS C

Complex analysis can roughly be thought of as the subject that applies the theory of calculus to imaginary numbers. But what exactly are imaginary numbers? Usually, students learn about them in high school with introductory remarks from their teachers along the following lines: "We can't take the square root of a negative number. But let's pretend we can and begin by using the symbol $i = \sqrt{-1}$."

If $i = \sqrt{-1}$, it stands to reason that $i^2 = -1$.

ALGEBRA OF COMPLEX NUMBERS

We have seen that complex numbers came to be viewed as ordered pairs of real numbers. That is, a complex number z is defined to be

$$z = (x, y),$$

where x and y are both real numbers.

The reason we say ordered pair is because we are thinking of a point in the plane. The point (2, 3), for example, is not the same as (3, 2). The order in which we write x and y in the equation makes a difference. Clearly, then, two complex numbers are equal if and only if their x coordinates are and their y coordinates are equal. In other words,

$(x, y) = (u, v)$ iff $x = u$ and $y = v$.

A meaningful number system requires a method for combining ordered pairs. The definition of algebraic operations must be consistent so that the sum, difference, product, and quotient of any two ordered pairs will again be an ordered pair. The key to defining how these numbers should be manipulated is to follow Gauss's lead and equate (x, y) with $x + iy$. Then, if $z_1 = (x_1, y_1)$ and $z_2 = (x_2, y_2)$ are arbitrary complex numbers, we have

$$\begin{aligned} z_1 + z_2 &= (x_1, y_1) + (x_2, y_2) \\ &= (x_1 + iy_1) + (x_2 + iy_2) \\ &= (x_1 + x_2) + i\,(y_1 + y_2) \\ &= (x_1 + x_2, y_1 + y_2) \end{aligned}$$

Thus, the following definitions should make sense.

ADDITION OF COMPLEX NUMBERS

Two complex numbers $z = x + iy$ and $z' = x' + iy'$ are added together component-wise,

$$z + z' = (x + x') + i\,(y + y').$$

In component form,

$$(x, y) + (x', y') = (x + x', y + y')$$

$$z_1 + z_2 = (x_1 + iy_1) + (x_2 + iy_2) = (x_1 + x_2, y_1 + y_2).$$

$$z_1 = x_1 + iy_1,\ z_2 = x_2 + iy_2$$

$$z_1 + z_2 = x_1 + x_2 + i\,y_1 + i\,y_2)$$

$$z_1 + z_2 = x_1 + x_2 + i\,(y_1 + y_2)$$

SUBTRACTION OF COMPLEX NUMBERS

Formula (1-9) $z_1 - z_2 = (x_1 + i\,y_1) - (x_2 + i\,y_2) = (x_1 - x_2, y_1 - y_2)$.

$$z_1 = x_1 + i\,y_1, \qquad z_2 = x_2 + i\,y_2$$

$$z_1 - z_2 = x_1 - x_2 + i\,y_1 - i\,y_2$$

$$z_1 - z_2 = x_1 - x_2 + i\,(y_1 - y_2)$$

Example. Given $z_1 = 3 + 7\,i$ and $z_2 = 5 - 6\,i$.

(a) Find $z_1 + z_2$ and (b) $z_1 - z_2$.

$z_1 + z_2 = (3, 7) + (5, -6) = (3 + 5, 7 - 6) = (8, 1)$ and

$z_1 + z_2 = (3, 7) - (5, -6) = (3 - 5, 7 + 6) = (-2, 13)$.

We can also use the notation $z_1 = 3 + 7\,i$ and $z_2 = 5 - 6\,i$:

$z_1 + z_2 = 3 + 7\,i + 5 - 6\,i = 8 + i$ and $z_1 + z_2 = 3 + 7\,i - (5 - 6\,i) = -2 + 13\,i$.

COMPLEX MULTIPLICATION

Two complex numbers $x = a + i\,b$ and $c + i\,d$ are multiplied as follows:

$$x\,y = (a + i\,b)\,(c + i\,d) = ac + i\,b\,c + i\,a\,d - bd$$

$$= (ac - bd) + i\,(ad + bc).$$

In component form,

$$(x, y)\,(x', y') = (x\,x' - y\,y', x\,y' + y\,x')$$

The special case of a complex number multiplied by a scalar is then given by

$$a\,(x, y) = (a, 0)\,(x, y) = (a\,x, a\,y).$$

Surprisingly, complex multiplication can be carried out using only three real multiplications, ac, bd, and $(a + b)\,(c + d)$ as

$$R\,[(a + i\,b)\,(c + i\,d)] = a\,c - b\,d$$

$$\mathrm{I}\,[(a + i\,b)\,(c + i\,d)] = (a + b)\,(c + d) - ac - bd.$$

Complex multiplication has a special meaning for elliptic curves.

$$z_1 z_2 = (x_1, y_2)\,(x_2, y_2) = (x_1 x_2 - y_1 y_2,\ x_1 y_2 + x_2 y_1).$$

Example. Given $z_1 = 3 + 7\,i$ and $z_2 = 5 - 6\,i$. Find $z_1\,z_2$.

Solution.

Ist Method

$$z_1\,z_2 = (3, 7)\,(5, -6)$$

$$= (3 * 5 - 7\,(-6), 3\,(-6) + 5 * 7)$$

$$= (15 + 42, -18 + 35)$$

$$= (57, 17)$$

IInd Method

$z_1 = 3 + 7\,i$ and $z_2 = 5 - 6\,i$:

$$z_1 z_2 = (3, 7)(5, -6) = (3 + 7i)(5 - 6i)$$
$$= 15 - 18i + 35i - 42i^2$$
$$= 15 - 42(-1) + (-18 + 35)\,i$$
$$= 57 + 17i$$
$$= (57, 17)$$

Of course, it makes sense that the answer came out as we expected because we used the notation $x + iy$ as motivation for our definition in the first place.

To motivate our definition for division, we will proceed along the same lines as we did for multiplication, assuming $z_2 \neq 0$:

Complex Division

$$\frac{z_1}{z_2} = \frac{(x_1, y_1)}{(x_2, y_2)} = \frac{x_1 + i\,y_1}{x_2 + i\,y_2}$$

$$\frac{z_1}{z_2} = \frac{(x_1, i\,y_1)(x_2 - i\,y_2)}{(x_2, i\,y_2)(x_2 - i\,y_2)}$$

$$= \frac{x_1, x_2 + y_1 y_2 + i(-x_1 y_2 + x_2 y_1)}{x_2^2 + y_2^2}$$

$$= \frac{x_1 x_2 + y_1 y_2}{x_2^2 + y_2^2} + i\frac{-x_1 y_2 + x_2 y_1}{x_2^2 + y_2^2}$$

$$= \left(\frac{x_1 x_2 + y_1 y_2}{x_2^2 + y_2^2}, \frac{-x_1 y_2 + x_2 y_1}{x_2^2 + y_2^2}\right)$$

COMPLEX DIVISION

The division of two complex numbers can be accomplished by multiplying the numerator and denominator by the complex conjugate of the denominator, for example, with $z_1 = a + b\,i$ and $z_2 = c + d\,i$, $z = z_1/z_2$ is given by

$$z = \frac{a + bi}{c + di} = \frac{(a + bi)\overline{c + di}}{(c + di)\,\overline{c + di}}$$

$$= \frac{(a + bi)(c - di)}{(c + di)(c - di)} = \frac{(ac + bd) + i(bc - ad)}{c^2 + d^2},$$

where $\bar{z}$ denotes the complex conjugate. In component notation with $(x, y) = x + i\,y$,

$$\frac{(a + b)}{(c + d)} = \left(\frac{ac + bd}{c^2 + d^2}, \frac{bc - ad}{c^2 + d^2}\right).$$

$$\frac{z_1}{z_2} = \frac{(x_1, y_1)}{(x_2, y_2)} = \left(\frac{x_1x_2 + y_1y_2}{x_2^2 + y_2^2}, \frac{-x_1d_2 + x_2y_1}{x_2^2 + y_2^2}\right), \text{ for } z_2 \neq 0.$$

Example. Given $z_1 = 3 + 7$ II and $z_2 = 5 - 6$ II. Find $\frac{z_1}{z_2}$.

$$\frac{z_1}{z_2} = \frac{(3, 7)}{(5, -6)} = \left(\frac{15-42}{25+36}, \frac{18+35}{25+36}\right) = \left(-\frac{27}{61}, \frac{53}{61}\right)$$

As with the example for multiplication, we also get this answer if we use the notation x + IIY :

$$\frac{z_1}{z_2} = \frac{(3, 7)}{(5, -6)} = \left(\frac{3+7\text{II}}{5-6\text{II}}\right)$$

$$= \frac{(3+7\pi)(5+6\pi)}{(5-6\pi)(5+6\pi)} = \frac{15+18\pi+35\pi+42\pi^2}{25+30\pi-30\pi-36\pi^2}$$

$$= \frac{15-42+(18+35)\pi}{(25+3)} = -\frac{27}{61} + \frac{53}{61}\pi$$

$$= \left(-\frac{27}{61}, \frac{53}{61}\right)$$

To perform operations on complex numbers, most mathematicians would use the notation $x + i\,y$ and engage in algebraic manipulations, as we did here, rather than apply the complicated-looking definitions we gave for those operations on ordered pairs.

It turns out that our algebraic definitions give complex numbers all the properties we normally ascribe to the real number system. Taken together, they describe what algebraists call a field. In formal terms, a field is a set (in this case, the complex numbers) together with two binary operations (in this case, addition and multiplication) having the following properties.

LAWS ASSOCIATED WITH ADDITION WITH COMPLEX NUMBER

1. Commutative Law

$$z_1 + z_2 = z_2 + z_1.$$

2. Associative Law

$$z_1 + (z_2 + z_3) = (z_1 + z_2) + z_3.$$

3. Additive Identity.

There is a complex number ω such that

$z + \omega = z$ for all complex numbers z.

The number ω is obviously the ordered pair (0, 0).

4. Additive Inverses

Given any complex number z, there is a complex number n (depending on z) with the property that

$$z + n = (0, 0).$$

Obviously, if $z = (x, y) = x + i\,y$, the number n will be $n = (-x, -y) = -x - i\,y = -z$.

LAWS ASSOCIATED WITH MULTIPLICATION 2WITH COMPLEX NUMBER

1. Commutative Law

$$z_1 z_2 = z_2 z_1.$$

2. Associative Law

$$z_1 (z_2 z_3) = (z_1 z_2)\, z_3.$$

3. Multiplicative Identity

There is a complex number ξ such that

$\xi z = z$ for all complex numbers z.

As one might expect, it turns out that (1, 0) is the unique complex number ξ with this property.

4. Multiplicative Inverses

Given any number z other than the number (0, 0), there is a complex number (depending on z) which we shall denote by z^{-1} with the property that

$z\,z^{-1} = (1, 0) = 1.$

Based on our definition for division, it seems reasonable that the number z^{-1} would be

$$z^{-1} = \frac{(1,0)}{z} = \frac{1}{x+iy}$$

$$= \frac{x-iy}{(x+iy)(x-iy)} = \frac{x-iy}{x^2+y^2}$$

$$= \frac{x}{x^2+y^2} + i\frac{-y}{x^2+y^2}$$

$$= \left(\frac{x}{x^2+y^2}, \frac{-y}{x^2+y^2}\right)$$

5. The Distributive Law

$$z_1(z_2 + z_3) = z_1(z_2 + z_3).$$

•**Real Part of z :** The real part of $z = x + \dot{\pi} y$, denoted by Re(z), is the real number x.

•**Imaginary Part of z:** The imaginary part of $z = x + \dot{\pi} y$, denoted by Im(z), is the real number y.

Conjugate of z

The complex conjugate of a complex number $z \equiv a + bi$ is defined to be

$$\bar{z} \equiv a - bi.$$

The conjugate matrix of a matrix $A = (a_{ij})$ is the matrix obtained by replacing each element a_{ij} with its complex conjugate, $\overline{A} = (\bar{a}_{ij})$.

By definition, the complex conjugate satisfies

$$\bar{\bar{z}} = z.$$

The complex conjugate is distributive under complex addition,

$$\overline{z_1 + z_2} = \overline{z_1} + \overline{z_2},$$

since

$$\overline{(a_1 + ib_1) + (a_2 + ib_2} = \overline{(a_1 + a_2) + i(b_1 + b_2)}$$
$$= (a_1 + a_2) - i(b_1 + b_2)$$
$$= (a_1 - ib_1) + (a_2 - ia_2)$$
$$= \overline{a_1 + ib_1} + \overline{a_1 + ib_2},$$

and distributive over complex multiplication,

$$\overline{z_1 z_2} = \overline{z_1}\ \overline{z_2},$$

since

$$\overline{(a_1 + b_1 i)(a_2 + b_2 i)} = \overline{(a_1 a_2 - b_1 b_2 + i(a_1 b_2 + a_2 b_1)}$$
$$= (a_1 a_2 - b_1 b_2) - i(a_1 b_2 + a_2 b_1)$$
$$= (a_1 - ib_1)(a_2 - ib_2)$$
$$= \overline{a_1 + ib_1}\ \overline{a_2 + ib_2}.$$

The conjugate of $z = x + i\,y$, denoted by $\overline{z}$, is the complex number $(x, -y) = x - \overline{z}\,y$.

The following theorem gives some important facts relating to these operations. You will be asked for a proof in the exercises

Theorem: Suppose z_1, z_2 and z_3 are arbitrary complex numbers. Then if $z_2 \neq 0$

$$\overline{\overline{z}} = z$$

$$\overline{z_1 + z_2} = \overline{z_1} + \overline{z_2}$$

$$\overline{z_1 z_2} = \overline{z_1}\;\overline{z_2}$$

$$\overline{\left(\frac{z_1}{z_2}\right)} = \frac{\overline{z_1}}{\overline{z_2}} \text{ if } z_2 \neq 0$$

$$\text{Im}[z] = \frac{z - \overline{z}}{2i}$$

$$\text{Re}[i\,z] = -\,\text{Im}[z]$$

$$\text{Im}\,[i\,z] = \text{Re}[z]$$

EXERECISE

1. Find each of the following:
   (a) $(1 + 2i)(3 + 4i)$
   (b) $(1 - 2i)^2/(1 - i)$
   (c) $\{1/(1 + i) + 1/(1 - 2i)\}$
2. Find all the roots of the equation $(z^8 - 1)(z^3 + 8) = 0$.
3. Suppose that $z = x + iy$, where $x, y \in R$. Express each of the following in terms of x and y:
   (a) $\left|\frac{z+1}{z-1}\right|$ (b) $\left|\frac{z+i}{1-iz}\right|$
4. For each of the following, compute all the values and plot them on the plane:
   (a) $(1+i)^{\frac{1}{2}}$ (b) $(-4)^{\frac{3}{4}}$ (c) $(1-i)^{\frac{3}{8}}$
5. Solve the equation $z^2 + z + 1 = 0$ for $z = (x, y)$ by writing $(x, y)(x, y) + (x, y) + (0, 0)$ and then solving a pair of simultaneous equations in x and y.

COMPLEX FUNCTIONS AND LINEAR MAPPINGS

A complex-valued function *f* of the complex variable *z* is a rule that assigns to each complex number *z* in a set *D* one and only one complex number *w*. We write $w = f(z)$ and call *w* the image of *z* under *f*. A simple example of a complex-valued function is given by the formula $w = f(z) = z^4$. The set *D* is called the domain of *f*, and the set of all images $\{w = f(z) : z \in D\}$ is called the range of *f*. When the context is obvious, we omit the phrase complex-valued, and simply refer to a function *f*, or to a complex function *f*.

We can define the domain to be any set that makes sense for a given rule, so for $w = f(z) = z^4$, we could have the entire complex plane for the domain D, or we might artificially restrict the domain to some set such as $D = D_1(0) = \{z : |z| < 1\}$. Determining the range for a function defined by a formula is not always easy, but we will see plenty of examples later on. In some contexts functions are referred to as mappings or transformations.

In Section, we used the term domain to indicate a connected open set. When speaking about the domain of a function, however, we mean only the set of points on which the function is defined. This distinction is worth noting, and context will make clear the use intended.

Just as z can be expressed by its real and imaginary parts, $z = x + i\,y$, we write $f(z) = w = v + i\,v$, where u and v are the real and imaginary parts of w, respectively. Doing so gives us the representation.

$$w = f(z) = f(x, y) = f(x + i\,y) = u + i\,v.$$

Because u and v depend on x and y, they can be considered to be real-valued functions of the real variables x and y; that is,

$$u = u(x, y) \text{ and } v = v(x, y).$$

Combining these ideas, we often write a complex function f in the form

$$f(z) = f(x + i\,y) = u(x, y) + i\,v(x, y).$$

Figure illustrates the notion of a function (mapping) using these symbols.

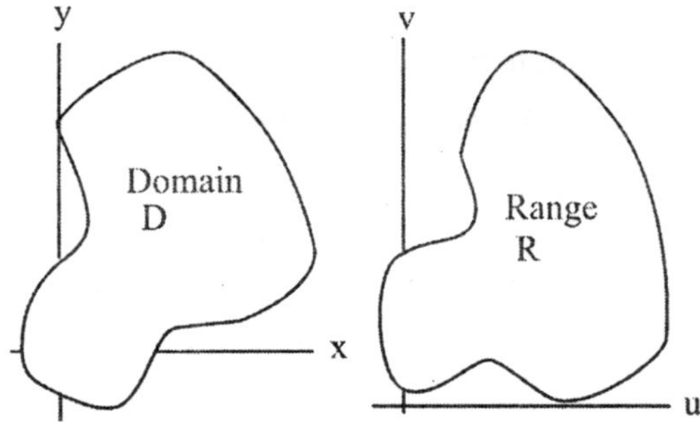

Fig. The mapping $w = f(z) = u(x, y) + i\,v(x, y)$.

Example . Write $f(z) = z^4$ in the for $f(z) = u(x, y) + i\,v(x, y)$.

Solution. Using the binomial formula, we obtain

$$f(z) = f(x + i\,y) = (x + i\,y)4$$
$$= x^4 + 4x^3(i\,y) + 6x^2(i\,y)^2 + 4x(i\,y)^3 + (i\,y)^4$$
$$= x^4 + 4\,i\,x^3\,y - 6\,x^2y^2 - 4\,i\,x\,y^3 + y^4$$
$$= x^4 - 6\,x^2\,y^2 + y^4 + i\,(4\,x^3\,y - 4\,x\,y^3)$$
$$= u(x, y) + i\,v(x, y)$$

so that $u(x, y) = x^4 - 6\,x^2\,y^2 + y^4$ and $v(x + y) = 4\,x^3\,y - 4\,x\,y^3$.

Example. Express the function $f(z) = \bar{z}\,\text{Re}[z] + z^2\,\text{Im}[z]$ in the form $f(z) = u(x, y) + v(x, y)$.

Solution. Using the elementary properties of complex numbers, it follows that

$$f[z] = (x - iy)x + (x + iy)^2 + y = (2x^2 - y^2 + y) + i(xy)$$

so that $u(x, y) = 2x^2 - y^2 + y$ and $(x, y) = xy$.

Examples 1 and 2 show how to find $u(x, y)$ and $v(x, y)$ when a rule for computing f is given. Conversely, if $u(x, y)$ and $v(x, y)$ are two real-valued functions of the real variables x and y, they determine a complex-valued function

$f(z) = u(x, y) + i\,v(x, y)$, and we can use the formulas

$$x = \frac{z + \bar{z}}{2} \text{ and } y = \frac{z - \bar{z}}{2i}$$

to find a formula for f involving the variables z and $\bar{z}$.

Example. Express $f(z) = 4x^2 + i\,4y^2$ by a formula involving the variables z and $\bar{z}$.

Solution. Calculation reveals that

$$f(z) = 4\left(\frac{z + \bar{z}}{2}\right)^2 + i4\left(\frac{z - \bar{z}}{2i}\right)^2$$

$$= z^2 + 2z\bar{z} + \bar{z}^2 - i(z^2 - 2z\bar{z} + \bar{z}^2)$$

$$= (1 - i)z^2 + (2 + 2i)z\bar{z} + (1 - i)\bar{z}^2$$

Using $z = re^{i\theta}$ in the expression of a complex function f may be convenient. It gives us the polar representation

$$f(z) = f(re^{i\theta}) = u(r, \theta) + iv(r, \theta),$$

where U and V are real functions of the real variables r and θ.

Remark. For a given function f, the functions u and v defined above are different from those used previously in $f(z) = f(x + iy) = u(x, y) + i\,v(x, y)$ which used Cartesian coordinates instead of polar coordinates.

Example. Express $f(z) = z^2$ in both Cartesian and polar form.

Solution. For the Cartesian form, a simple calculation gives

$$f(z) = f(x + iy) = (x + iy)^2$$

$$x^2 - y^2 + 2ixy$$

$u(x, y) + i\,v(x, y)$

so that $u(x, y) = x^2 - y^2$ and $v(x, y) = x\,y$.

For the polar form, we get v

$$f(z) = f(re^{i\theta}) = (re^{i\theta})^2 = r^2 e^{i2\theta}$$

$$= r^2(\cos 2\theta + i \sin 2\theta)$$

$$= r^2 \cos 2\theta + i r^2 \sin 2\theta$$

$$= U(r,\theta) + i\,V(r,\theta)$$

so that $= u(r,\theta) + r^2 \cos 2\theta$ and $v(r,\theta) = r^2 \sin 2\theta$.

Remark. Once we have defined u and v for a function f in Cartesian form, we must use different symbols if we want to express f in polar form. As is clear here, the functions u and U are quite different, as are v and V. Of course, if we are working only in one context, we can use any symbols we choose.

For a given function f, the functions u and v defined here are different from those defined by equation (1), because equation (2) involves Cartesian coordinates and equation (3) involves polar coordinates.

Example. Express $f(z) = z^5 + 4z^2 - 6$ in polar form.

Solution. We obtain

$$f(z) = f(re^{i\theta}) = (re^{i\theta})^5 + 4(re^{i\theta})^2 - 6$$

$$= r^5 e^{i5\theta} + 4r^2 e^{i2\theta} - 6$$

$$= r^5 \cos 5\theta + 4r^2 \cos 2\theta - 6 + i(r^5 \sin 5\theta + 4r^2 \sin 2\theta)$$

$$= U(r, \theta) + i\,V(r, \theta)$$

so that $U(r, \theta) = r^2 \cos 5\theta + 4r^2 \cos 2\theta - 6$ and $V(r, \theta) = r^2 \sin 5\theta + 4r^2 \sin 2\theta$.

We now look at the geometric interpretation of a complex function. If D is the domain of real-valued functions $u(x, y)$ and $v(x, y)$, the equations

$$u = u(x, y) \text{ and } v = v(x, y)$$

describe a transformation (or mapping) from D in the xy plane into the uv plane, also called the w plane. Therefore, we can also consider the function

$$w = f(z)\,u(x, y) + i\,v(x, y)$$

to be a transformation (or mapping) from the set D in the z plane onto the range R in the w plane. In the following paragraphs we present some additional key ideas. They are staples for any kind of function, and you should memorize all the terms in bold.

If A is a subset of the domain D of f, the set $B = \{w = f(z) : z \in A\}$ is called the image of the set A, and f is said to map A onto B. The image of a single point is a single point, and the image of the entire domain, D, is the range, R. The mapping $w = f(z)$ is said to be from A into S if the image of A is contained in S. Mathematicians use the notation $f: A \rightarrow S$ to indicate that a function maps A into S. Figure illustrates a function f whose domain is D and whose range is R. The shaded areas depict that the function maps A onto B. The function also maps A into R, and, of course, it maps D onto R.

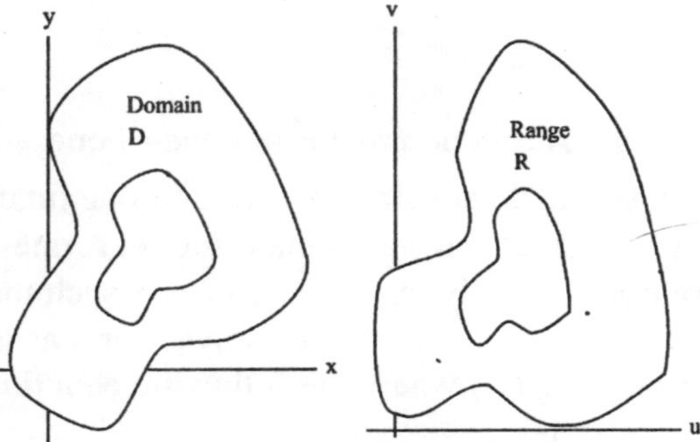

Fig. $w = f(z)$ maps A onto B; $w = f(z)$ maps A into R.

The inverse image of a point w is the set of all points z in D such that $w = f(z)$. The inverse image of a point may be one point, several points, or nothing at all. If the last case occurs then the point w is not in the range of f. For example, if $w = f(z) = i\,z$, the inverse image of the point -1 is the single point i, because $w = f(i) = i\,(i) = -1$, and i is the only point that maps to -1. In the case of $w = f(z) = z^2$, the inverse image of the point -1 is the set $\{i, -i\}$. You will learn that, if $w = f(z) = e^z$, the inverse image of the point 0 is the empty set---there is no complex number z such that $e^z = 0$.

The inverse image of a set of points, S, is the collection of all points in the domain that map into S. If f maps D onto R it is possible for the inverse image of R to be function as well, but the original function must have a special property: a function f is said to be one-to-one if it maps distinct points $z_1 \neq z_2$, onto distinct points $f(z_1) \neq f(z_2)$. Many times an easy way to prove that a function f is one-to-one is to suppose $f(z_1) = f(z_2)$, and from this assumption deduce that z_1 must equal z_2. Thus, $f(z) = i\,z$ is one-to-one because if $f(z_1) = f(z_2)$, then $i\,z_1 = i\,z_2$. Dividing both sides of the last equation by i gives $z_1 = z_2$. Figure illustrates the idea of a one-to-one function: distinct points get mapped to distinct points.

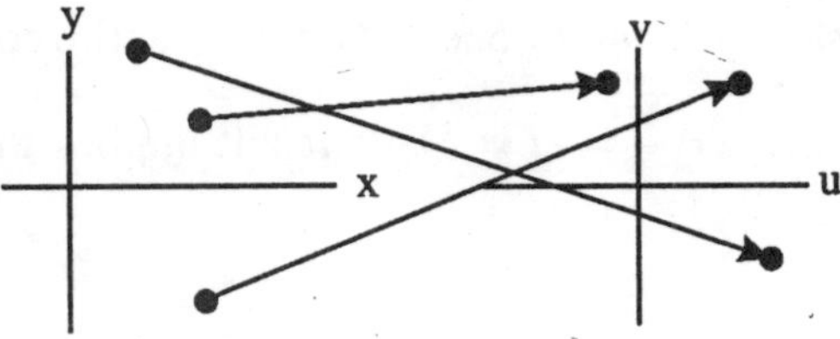

Fig. A function $w = f(z)$ that is one-to-one.

The function $f(z) = z^2$ is not one-to-one because $-i \neq i$, but $f(i) = f(-i) = -1$. Figure depicts this situation: at least two different points get mapped to the same point.

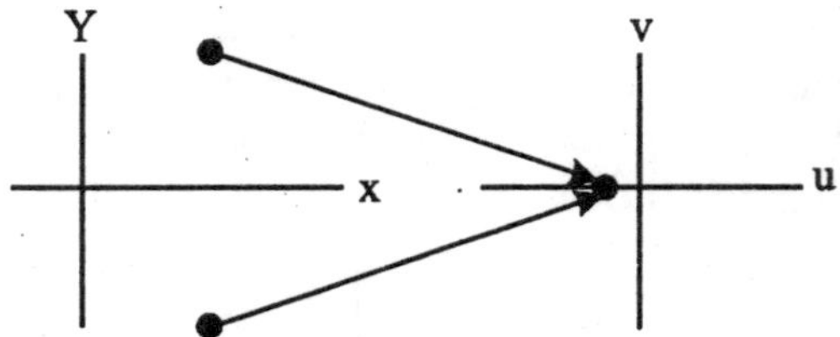

Fig. A function that is not one-to-one.

In the exercises we ask you to demonstrate that one-to-one functions give rise to inverses that are functions. Loosely speaking, if $w = f(z)$ maps the set A one-to-one and onto the set B, then for each w in B there exists exactly one point z in A A such that $w = f(z)$. For any such value of z we can take the equation $w = f(z)$ and "solve" for z as a function of w. Doing so produces an inverse function $z = g(w)$ where the following equations hold:

$$g(f(z)) = z \text{ for all } z \in A, \text{ and}$$

$$f(g(w)) = w \text{ for all } z \in B$$

Conversely, if $w = f(z)$ and $z = g(w)$ are functions that map A into B and B into A, respectively, and the above hold, then f maps the set A one-to-one and onto the set B.

Further, if f is a one-to-one mapping from D onto T and if A is a subset of D, then f is a one-to-one mapping from A onto its image B. We can also show that, if $\zeta = f(z)$ is a one-to-one mapping from A onto B and $w = g(\zeta)$ is a one-to-one mapping from B onto S, then the composite mapping $w = g(f(z))$ is a one-to-one mapping from A onto S.

We usually indicate the inverse of f by the symbol f^{-1}. If the domains of f and f^{-1} are A and B respectively, then we write

$$f^{-1}(f(z)) = z \text{ for all } z \in A, \text{ and } f(f^{-1}(w)) = w \text{ for all } w \in B.$$

Also, for $z_0 \in A$ and $w_0 \in B$.

$$w_0 = f(z_0) \text{ iff } f^{-1}(w_0) = z_0, \text{ and}$$

$$z_0 = f^{-1}(w_0) \text{ iff } f(w_0) = z_0.$$

Example. If $w = f(z) = i\,z$ for any complex number z, find $f^{-1}(w)$. Also find $f(f^{-1}(w))$

Solution. We can easily show f is one-to-one and onto the entire complex plane. We solve for z, given $w = f(z) = i$, to get $z = \frac{w}{i} = -i\,w$. This result implies that $f^{-1}(w) = -i\,w$ for all complex numbers w.

Since,

$$w = f[z] = i\,z$$

$\{\{z \to i\,(-4\,x^2 - 4\,i\,y^2)\}\}$

$z = f^{-1}\,[w] = i\,(-4\,x^2 - 4\,i\,y^2)$

$f^{-1}\,[f\,[z]] = -\,i\,(i\,z)$

$f^{-1}\,[f\,[z]] = i\,(-4\,x^2 - 4\,i\,y^2)$

$f\,[f^{-1}\,[w]] = i\,(i\,(-4\,x^2 - 4\,i\,y^2))$

$f\,[f^{-1}\,[w]] = 4\,x^2 + 4\,i\,y^2$

Remark. Once we have specified $f^{-1}\,(w) = -\,i\,w$ for all complex numbers w, we note that there is nothing magical about the symbol w. We could just as easily write $f^{-1}\,(z) = -\,i\,z$ for all complex numbers z.

We now show how to find the image B of a specified set A under a given mapping $u + iv = w = f(z)$.

The set A is usually described with an equation or inequality involving x and y. Using inverse functions, we can construct a chain of equivalent statements leading to a description of the set B in terms of an equation or an inequality involving u and v.

Example. Show that the function $f\,[z] = i\,z$ maps the line $y = x + 1$ in the xy plane onto the line $v = -\,u - 1$ in the w plane.

Solution.

First Method: With $A = \{(x, y)\} : y = x + 1$, we want to describe $B = f\,(A)$. We let $z = x + i\,y \in A$ and get

$$u + i\,v = w = f\,(z) \in B$$
$$\Leftrightarrow f^{-1}\,(w) = z = x + i\,y \in A$$
$$\Leftrightarrow i\,w \in A$$
$$\Leftrightarrow v - i\,u \in A$$
$$\Leftrightarrow (v, -\,u) \in A$$
$$\Leftrightarrow u = v + 1$$
$$\Leftrightarrow v = -\,u - 1$$

where $\Leftrightarrow$ is the notation for "if and only if." Note what this result says: $u + i\,v = w + B \Leftrightarrow v = -\,u - 1$. The image of A under f, therefore, is the set $B = \{(u, v) : v = -\,u - 1\}$.

Second Method: We write $u + i\,v = w = f\,(z) = i\,(x + i\,y) = -\,y + i\,x$ and note that the transformation can be given by the equations $u = -\,y$ and $v = x$. Because A is described by $A = \{x + i\,y : y = x + 1\}$, we can substitute $u = -\,y$ and $v = x$ into the equation $y = x + 1$ to obtain

$-u = v + 1$, which we can rewrite as $v = -u - 1$. If you use this method, be sure to pay careful attention to domains and ranges.

We now look at some elementary mappings. If we let $B = a + ib$ denote a fixed complex constant, the transformation $w = T(z) = z + B = x + a + i(y + b)$ is a one-to-one mapping of the z-plane onto the w-plane and is called a translation.

TRANSFORMATION

A transformation consisting of a constant offset with no rotation or distortion. In n-dimensional Euclidean space, a translation may be specified simply as a vector giving the offset in each of the $n$coordinates. This transformation can be visualized as a rigid translation whereby the point z is displaced through the vector $B = a + ib$ to its new position $w = T(z)$. The inverse mapping is given by

$$z = T^{-1}(w) = w - B = u - a + i(v - b)$$

and shows that T is a one-to-one mapping from the z-plane onto the w-plane. The effect of a translation is depicted in Figure.

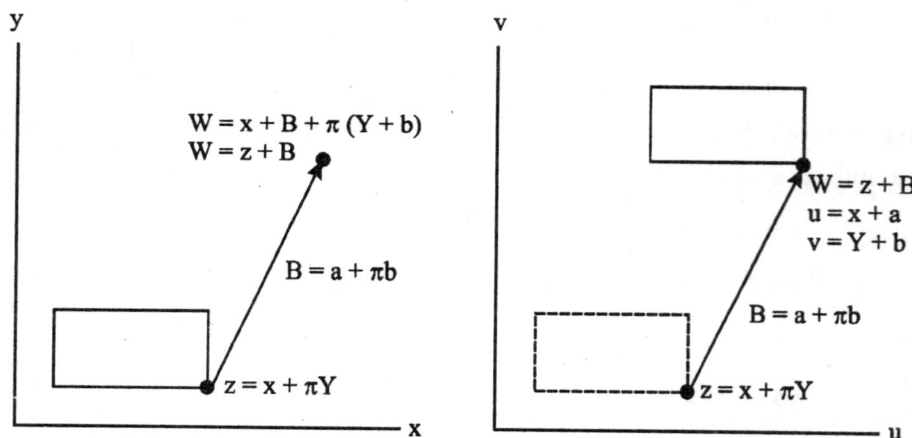

Fig. The translation $w = T(z) = z + B = x + a + i(y + b)$.

If we let α be a fixed real number, then for $z = re^{i\theta}$, the transformation

$$w = R(z) = ze^{i\alpha} = re^{i\theta}e^{i\alpha} = re^{i(\theta+\alpha)}$$

is a one-to-one mapping of the z-plane onto the w-plane and is called a rotation.

ROTATION TRANSFORMATION

The symmetry operation corresponding to a rotation followed by an inversion operation, also called a roto-inversion. This operation is denoted $\bar{n}$ for an improper rotation by 360°/*n*, so the crystallography restriction gives only $\bar{1}, \bar{2}, \bar{3}, \bar{4}, \bar{6}$ for crystals. The mirror plane symmetry operation is $(x, y, z) \to (x, y, -z)$, etc., which is equivalent to $\bar{2}$.

The turning of an object or coordinate system by an angle about a fixed point. A rotation is an orientation-preserving orthogonal transformation. Euler's rotation theorem states that an arbitrary rotation can be parameterized using three parameters. These parameters are commonly taken as the Euler angles. Rotations can be implemented using rotation matrices.

Rotation in the plane can be concisely described in the complex plane using multiplication of complex numbers with unit modulus such that the resulting angle is given by arg (*z*). For example, multiplication by *i* represents a rotation to the right by 90° and by – *i* represents rotation to the left by 90°. So starting with $z = 1$ and rotating left twice gives $z = 1\,(-i)\,(-i) = i^2 = -1$, which is the same as rotating right twice, $z = 1\,(i)\,(i) = i^2 = -1$, and $\arg(-1) = \pi = 180°$. For multiplication by multiples of 90°, the possible positions are then concisely represented by $z = 1, i, -1$, and $-i$.

The rotation symmetry operation for rotation by 360°/*n* is denoted "*n*." For periodic arrangements of points ("crystals"), the crystallography restriction gives the only allowable rotations as 1, 2, 3, 4, and 6.

It can be visualized as a rigid rotation whereby the point *z* is rotated about the origin through an angle α to its new position $w = R(z)$. If we use polar coordinates and designate $w = \rho e^{i\phi}$ in the *w*-plane, then the inverse mapping is

$$z = R^{-1}(w) = we^{-i\alpha} = \rho e^{i\phi} e^{-i\alpha} \rho e^{i(\phi-\alpha)}.$$

This analysis shows that *R* is a one-to-one mapping from the *z*-plane onto the *w*-plane. The effect of rotation is depicted in Figure.

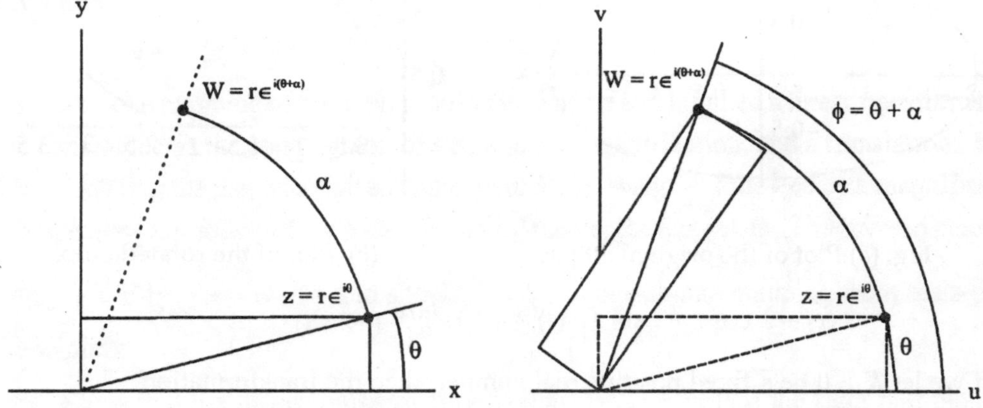

Fig. The rotation $w = R(z) = re^{i(\theta+\alpha)}$.

Example. The ellipse centered at the origin with a horizontal major axis of 4 units and vertical minor axis of 2 units can be represented by the parametric equation

$$s(t) = 2\cos t + i\sin t = (2\cos t, \sin t), \text{ for } 0 \le t \le 2\pi.$$

Solution:

Suppose we wanted to rotate the ellipse by an angle of $\frac{\pi}{6}$ radians and shift the center of the ellipse 2 units to the right and 1 unit up. Using complex arithmetic, we can easily generate a parametric equation $r(t)$ that does so:

$$r[t] = s[t]e^{-i\pi/6} + (2+i)$$

$$= (2\cos t + i\sin t)\left(\cos\frac{\pi}{6} + i\sin\frac{\pi}{6}\right) + (2+i)$$

$$= \left(2\cos t\cos\frac{\pi}{6} + -\sin t\sin\frac{\pi}{6}\right) + i\left(2\cos t\sin\frac{\pi}{6} + \sin t\cos\frac{\pi}{6}\right) + (2+i)$$

$$= \left(\sqrt{3}\cos t - \frac{1}{2}\sin t + 2\right) + i\left(\cos t + \frac{\sqrt{3}}{2}\sin t + 1\right)$$

$$\sqrt{3}\cos t - \frac{1}{2}\sin t + 2,\ \cos t + \frac{\sqrt{3}}{2}\sin t + 1$$

for $0 \le t \le 2\pi$. Figure shows parametric plots of these ellipses.

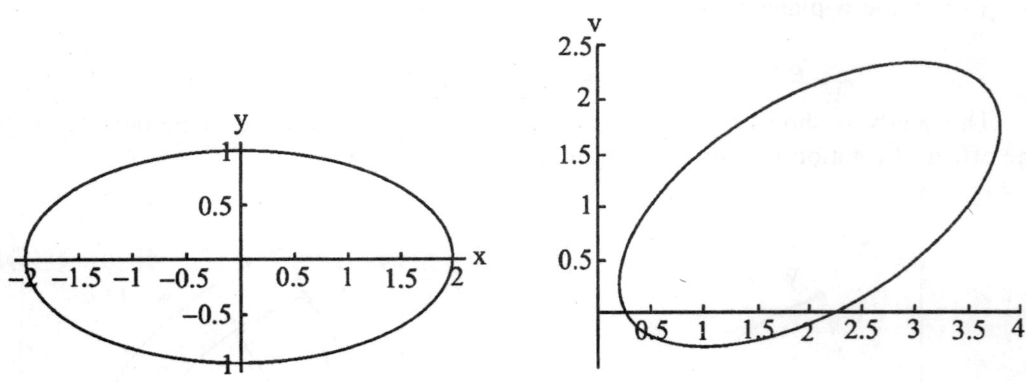

Fig. (*a*) Plot of the original ellipse (*b*) Plot of the rotated ellipse

$$s(t) = 2\cos t + i\sin t \quad r(t) = s(t)e^{i\pi/6} + (2+i)$$

If we let $K > 0$ be a fixed positive real number, then the transformation

$$w = S(z) = Kz = Kx + iKy$$

is a one-to-one mapping of the z-plane onto the w-plane and is called a magnification. If $K > 1$, it has the effect of stretching the distance between points by the factor K. If $K > 1$, then it reduces the distance between points by the factor K. The inverse transformation is given by

$$z = S^{-1}(w) = \frac{1}{K}w = \frac{1}{K}u + i\frac{1}{k}v$$

and shows that S is a one-to-one mapping from the z-plane onto the w-plane. The effect of magnification is shown in Figure.

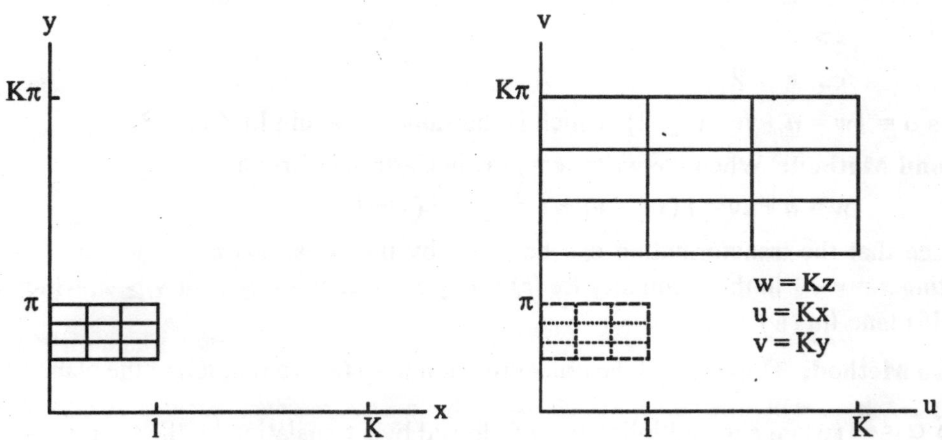

Fig. The magnification $w = S(z) = Kz = Kx + i\,Ky$.

Finally, if we let $A = Ke^{i\alpha}$ and $B = a + i\,b$, where $K > 0$ is a positive real number, then the transformation

$$w = L(z) = A\,z + B$$

is a one-to-one mapping of the z-plane onto the w-plane and is called a linear transformation. It can be considered as the composition of a rotation, a magnification, and a translation. It has the effect of rotating the plane though an angle given by α = Agr A, followed by a magnification by the factor $K = |A|$, followed by a translation by the vector $B = a + ib$. The inverse mapping is given by $z = L^{-1}(w) = \frac{1}{A}w - \frac{B}{A}$ and shows that L is a one-to-one mapping from the z-plane onto the w-plane.

Example. Show that the linear transformation $w = f(z) = i\,z + i$ maps the right half plane Re $(z) > 1$ onto the upper half plane Im $(w) > 2$.

Solution.

First Method: Let A = {(x, y) : x ≥ 1}. To describe $B = F(A)$, we solve $w = iz + i$ for z to get $z = \dfrac{w-i}{i} = -iw - 1 = f^{-1}(w)$. We have the following

$$\begin{aligned} & u + i\,v = w = f(z) \in B \\ & \Leftrightarrow f^{-1}(w) = z \in A \\ & \Leftrightarrow -i\,w - 1 \in A \\ & \Leftrightarrow v - 1 - i\,u \in A \\ & \Leftrightarrow (v - 1, -u) \in A \\ & \Leftrightarrow v - 1 \geq 1 \\ & \Leftrightarrow v \geq 2 \end{aligned}$$

Thus $B = \{w = u + iv : v \geq 2\}$, which is the same as saying Im $(w) > 2$.

Second Method: When we write $w = f(z)$ in Cartesian form as

$$w = u + iv = i\,(x + i\,y) + i = -y + i\,(x + 1),$$

we see that the transformation can be given by the equations $u = -y$ and $v = x + 1$. Substituting $x = v - 1$ in the inequality Re $(z) = x \geq 1$ gives $v - 1 \geq 1$, or $v \geq 2$, which is the upper half-plane Im $(w) > 2$.

Third Method: The effect of the transformation $w = f(z)$ is a rotation of the plane through the angle $\alpha = \dfrac{\pi}{2}$ (when z is multiplied by i) followed by a translation by the vector $B = i$. The first operation yields the set Im $(w) \geq 1$. The second shifts this set up 1 unit, resulting in the set Im $(w) > 2$. We illustrate this result in Figure.

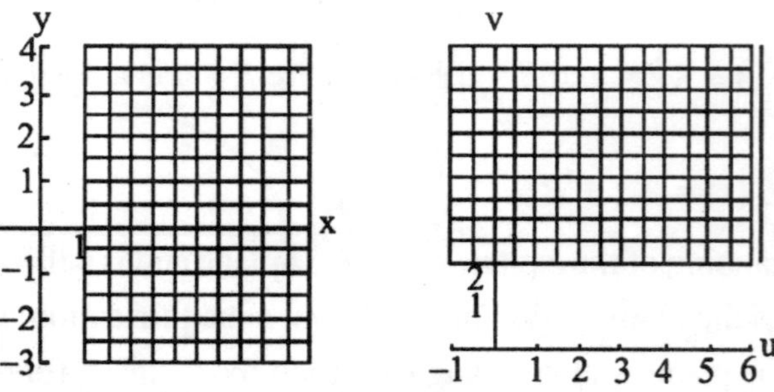

Fig. The linear transformation $w = f(z) = i\,z + i$.

Translations and rotations preserve angles. First, magnifications rescale distance by a factor K, so it follows that triangles are mapped onto similar triangles, preserving angles. Then, because a linear transformation can be considered to be a composition of a rotation, a magnification, and a translation, it follows that linear transformations preserve angles. Consequently, any geometric object is mapped onto an object that is similar to the original object; hence linear transformations can be called similarity mappings.

Example. Show that the image of the open disk $D_1(-1-i) = \{z : |z+1+i| < 1\}$ under the linear transformation $w = f(z) = (3-4i)z + 6 + 2i$ is the open disk $D_5(-1+3i) = \{w : |w + 1 - 3i| < 5\}$.

Solution. The inverse transformation is $z = \dfrac{w-6-2i}{3-4i}$, so if we designate the range of f as B, then

$$w = f(z) \in B \Leftrightarrow f^{-1}(w) = z \in D_1(-1-i)$$

$$\Leftrightarrow \frac{w-6-2i}{3-4i} \in D_1(-1-i)$$

$$\Leftrightarrow \left|\frac{w-6-2i}{3-4i} + 1 + i\right| < 1$$

$$\Leftrightarrow \left|\frac{w-6-2i}{3-4i} + 1 + i\right| |3-4i| < 1|3-4i|$$

$$\Leftrightarrow |w - 6 - 2i + (1+i)(3-4i)| < 5$$

$$\Leftrightarrow |w + 1 - 3i| < 5$$

Hence the disk with center $-1-i$ and radius 1 is mapped one-to-one and onto the disk with center $-1+3i$ and radius 5 as shown in Figure.

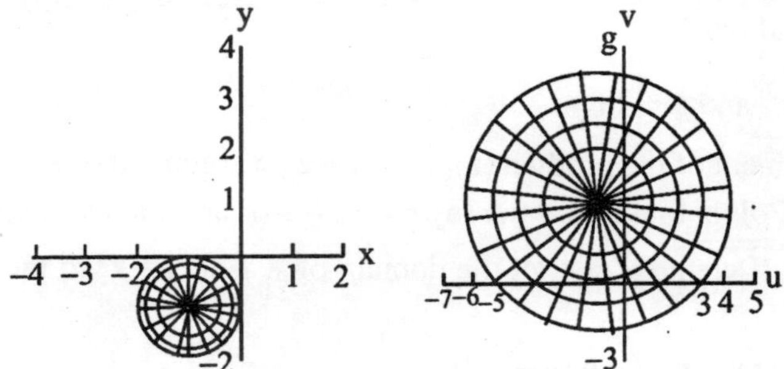

Fig. The mapping $w = f(z) = (3 - 4i)z + 6 + 2i$.

Example. Show that the image of the right half plane Re $(z) = x > 1$ under the linear transformation $w = (-1+i)z - 2 + 3i$ is the half plane $v > u + 7$.

Solution. The inverse transformation is given by

$$z = \frac{w+2-3i}{-1+i} = \frac{u+2+i(v-3)}{-1+i},$$

which we write as

$$x + iy = \frac{-u+v-5}{2} + i\frac{-u-v+1}{2}$$

Substituting $x = \frac{-u+v-5}{2}$ into e $Re(z) = x > 1$ gives $\frac{-u+v-5}{2} > 1$, which simplifies v > u + 7. Figure illustrates the mapping.

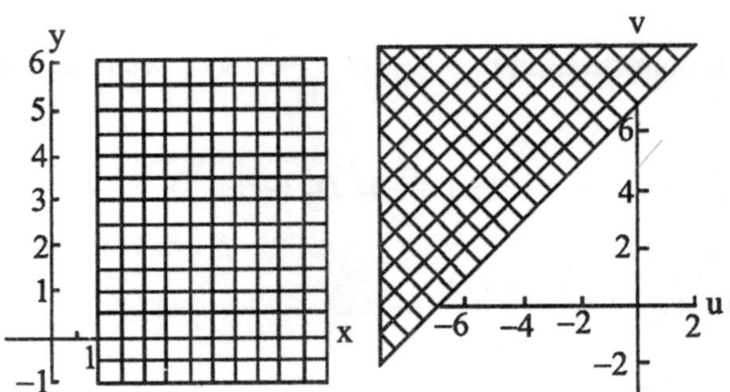

Fig. The the linear transformation

THE MAPPINGS $w = z^n$ AND $w = z^{1/n}$

In this section we turn our attention to power functions. For $z = re^{i\theta}$, we can express the function $w = f(z) = z^2$ in polar coordinates by

$$w = f(z) = z^2 r^2 e^{i2\theta}$$

If we also use polar coordinates for $w = \rho e^{i\theta}$ in the w–plane, we can express this mapping by the system of equations

$$\rho = r^2 \text{ and } \phi = 2\theta.$$

Because an argument of the product (z)(z) is twice an argument of z, we say that f doubles angles at the origin. Points that lie on the ray $r > 0,\ \theta = \alpha$ are mapped onto points that lie on the ray $\rho = 0,\ \phi = 2\alpha$. If we now restrict the domain of $w = f(z) = z^2$ to the region

$$A = \left\{ re^{i\theta} : r > 0 \text{ and } \frac{-\pi}{2} < \theta \leq \frac{\pi}{2} \right\},$$

then the image of A under the mapping $w = z^2$ can be described by the set

$$B = \left\{ \rho e^{i\theta} : \rho > 0 \text{ and } -\pi < \phi \leq \pi \right\},$$

which consists of all points in the w-plane except the point $w = 0$.

The inverse mapping of f, which we shall denote by g, is then

$$z = g(w) = w^{1/2} \rho^{1/2} e^{i\phi/2}$$

where $w \in B$. That is

$$z = g(w) = w^{1/2} = |w|^{1/2} e^{iArg(w)/2},$$

where $w \neq 0$. The function g is so important that we call special attention to it with a formal definition.

SQUARE ROOT FUNCTION

A square root, also called a radical or surd, of x is a number r such that $r^2 = x$. The function $\sqrt{x}$ is therefore the inverse function of $f(x) = x^2$ for $x \geq 0$.

There are a number of square root algorithms that can be used to approximate the square root of a given number. These include the Bhaskara-Brouckner algorithm and Wolfram's iteration. The simplest algorithm for $\sqrt{n}$ is Newton's iteration

$$x_{k+1} = \frac{1}{2}\left(x_k + \frac{n}{x_k}\right)$$

with $x_0 = 1$.

Note that any positive real number has two square roots, one positive and one negative. For example, the square roots of 9 are – 3 and + 3, since $(-3)^2 = (+3)^2 = 9$. Any nonnegative real number x has a unique nonnegative square root r; this is called the principal square root and is written $r = x^{1/2}$ or $r = \sqrt{x}$. For example, the principal square root of 9 is $\sqrt{9} = +3$, while the other square root of 9 is $-\sqrt{9} = -3$.

The square roots of a complex number $z = x + i\,y$ are given by

$$\sqrt{x+iy} = \pm(x^2+y^2)^{1/4}\left\{\cos\left[\frac{1}{2}\tan^{-1}(x,y)\right] + i\sin\left[\frac{1}{2}\tan^{-1}(x,y)\right]\right\}$$

In addition,

$$\sqrt{x+iy} = \frac{1}{2}\sqrt{2}\left[\sqrt{\sqrt{x^2+y^2}+x} + x + i\sin(y)\sqrt{\sqrt{x^2+y^2}-x}\right]$$

As can be seen in the above figure, the imaginary part of the complex square root function has a branch cut along the negative real axis.

A nested radical of the form $\sqrt{a \pm b\sqrt{c}}$ can sometimes be simplified into a simple square root by equating

$$\sqrt{a \pm b\sqrt{c}} = \sqrt{d} \pm \sqrt{e}$$

Squaring gives

$$a \pm b\sqrt{c} = d + e \pm 2\sqrt{de}$$

so

$$a = d + e$$

$$b^2 c = 4\, d\, e.$$

Solving for d and e gives

$$d, e = \frac{a \pm \sqrt{a^2 - b^2 c}}{2}.$$

For example,

$$\sqrt{5 + 2\sqrt{6}} = \sqrt{2} + \sqrt{3}$$

$$\sqrt{3 - 2\sqrt{2}} = \sqrt{2} - 1$$

A counterintuitive property of inverse functions is that

$$\sqrt{z}\sqrt{\frac{1}{z}} = \begin{cases} -1 & \text{for } J[z] = 0 \text{ and } R[z] < 0 \\ \text{undefined} & \text{for } z = 0 \\ 1 & \text{otherwise,} \end{cases}$$

so the expected identity (i.e., canceling of the $\sqrt{z}$ s) does not hold along the negative real axis.

PRINCIPAL SQUARE ROOT FUNCTION

The function

$$z = g\,(w) = w^{1/2} = |\, w\,|^{1/2}\, e^{iArg\,(w)/2}, \text{ for } w \neq 0.$$

is called the principal square root function.

It is left as an exercise to show that f and g satisfy equations $g\,(f\,(z)) = z$ and $f\,(g(w)) = w$, and thus are inverses of each other that map the set A one-to-one and onto the set B and the set B one-to-one and onto the set A, respectively. Figure illustrates this relationship.

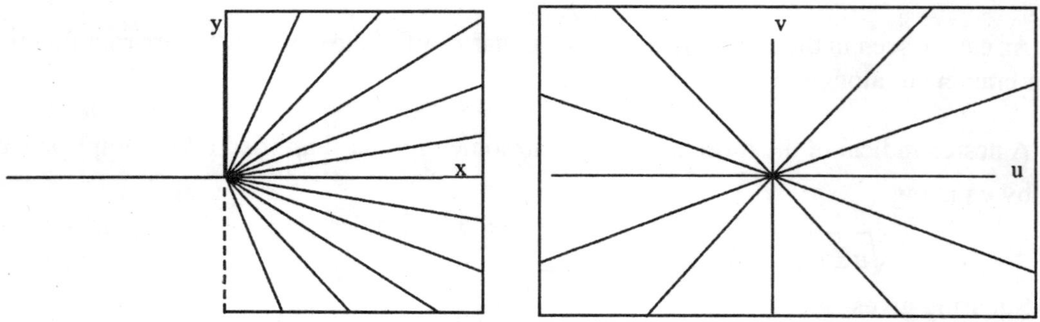

Fig. The mapping $w = f\,(z) = z^2$ (and $z = g\,(w) = w^{1/2}$).

What are the images of rectangles under the mapping $w = z^2$? To find out, we make use of the Cartesian form

$$w = u + i\,v = f(z) = z^2 = x^2 - y^2 + i\,2\,x\,y = (x^2 - y^2, 2\,x\,y) = (u, v)$$

and the resulting system of equations

$$u = x^2 - y^2, \text{ and}$$

$$v = 2\,x\,y.$$

Example. Show that the transformation $w = f(z) = z^2$, usually maps vertical and horizontal lines onto parabolas and use this fact to find the image of the rectangle $R = \{(x, y)\} : 0 < x < a, 0 < x < a, 0 < y < b\}$. (a) Find the image of the vertical line $x = a$. (b) Find the image of the horizontal line y = b.

Solution. Using the above equations, we determine that the vertical line $x = a$ is mapped onto the set of points given by the equations $u = a^2 - y^2$ and $v = 2\,a\,y$. If a $\neq$ 0, then $y = \dfrac{v}{2a}$ and $u = a^2 - \dfrac{v^2}{2a^2}$,

this equation represents a parabola with vertex at $(a^2, 0)$, oriented horizontally, and opening to the left. If $a > 0$, the set $\{(u, v) : u = a^2 - y^2, v = 2\,a\,y\}$ has $v > 0$ precisely when $y > 0$, so the part of the line $x = a$ lying above the x axis is mapped to the top half of the parabola.

The horizontal line $y = b$ is mapped onto the parabola given by the equations $u = x^2 - b^2$ and $v = 2\,x\,b$. If $b \neq 0$, then as before we get

$$v = -b^2 + \frac{v^2}{4b^2},$$

this equation represents a parabola with vertex at $-b^2$, oriented horizontally and opening to the right. If $b > 0$, the part of the line $y = b$ to the right of the y axis is mapped to the top half of the parabola because the set $\{(u, v) : u = x^2 - b^2, v = 2\,b\,x\}$ has $v > 0$ precisely when $x > 0$.

Quadrant I is mapped onto quadrants I and II by $w = z^2$, so the rectangle $R = \{(x, y) : 0 < x < a, 0 < y < b\}$ is mapped onto the region bounded by the top halves of the parabolas given by $u = a^2 - \dfrac{v^2}{4a^2}$ and $v = -b^2 + \dfrac{v^2}{4b^2}$ and the u axis. The vertices 0, a^2, $a^2 + 2\,i\,a\,b - b^2$, and $-b^2$ of the rectangle are mapped onto the four points 0, a^2, $a^2 - b^2 + 2\,i\,a\,b$, and $-b^2$ respectively, as indicated in Figure.

Finally, we can easily verify that the vertical line $\{(0, y) : -\infty < y < \infty$ is mapped to the set $\{(-y^2, 0) : -\infty < y < \infty\}$. This is simply the set of non-positive real numbers. Similarly, the horizontal line $\{(x, 0) : -\infty < x < \infty\}$ is mapped to the set $\{(x^2, 0) : -\infty < x < \infty\}$, which is the set of non-negative real numbers.

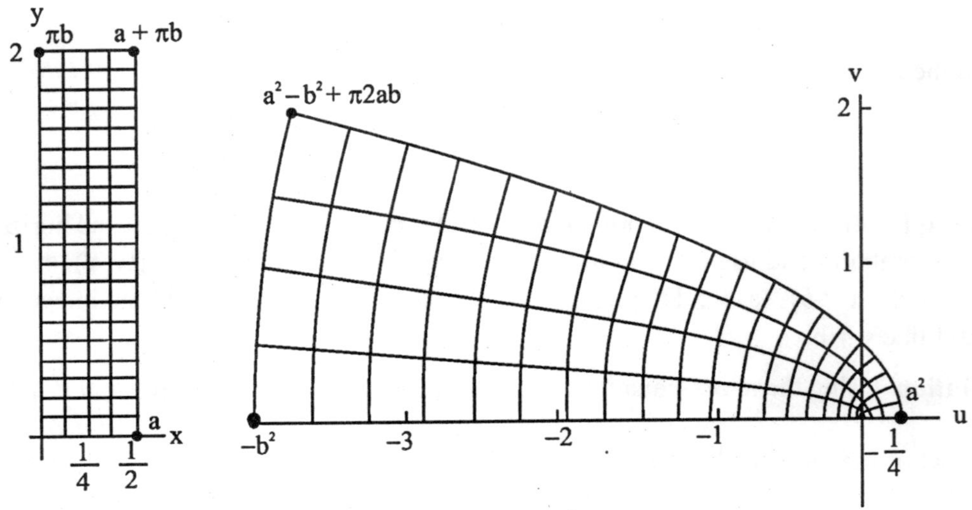

Fig. The transformation $w = f(z) = z^2$.

Example. Consider the mapping $w = z^2$. Find the image of the ray $r > 0$, $\theta = \alpha$ and circle $r = c$.

Solution

Here we find the image of the ray $r > 0$, $\theta = \alpha$ and circle $r = c$.

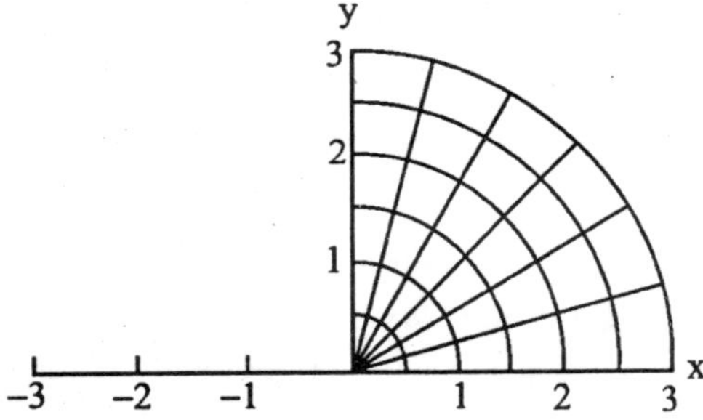

The rays $r > 0$, $\theta = \alpha$ and circles $r = c$.

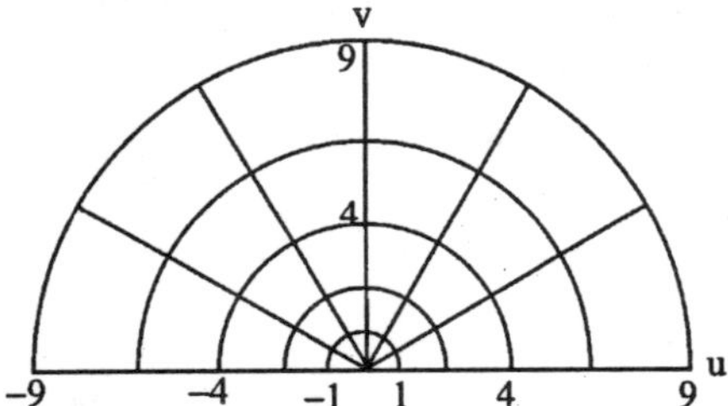

The image rays $r > 0$, $\theta = 2\alpha$, and circles $r = c^2$, under the mapping $w = z^2$

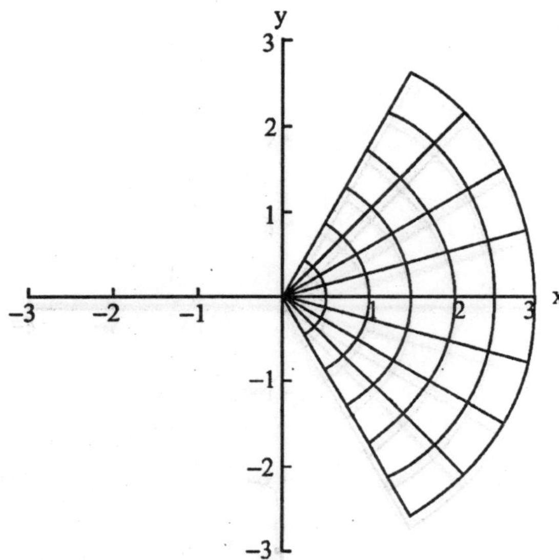

The rays $r > 0$, $\theta = \alpha$ and circles $r = c$.

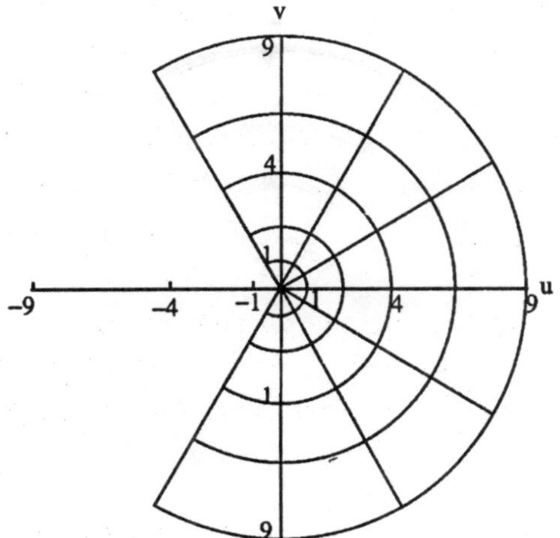

The image $r > 0,\ \theta = 2\alpha$, and circles $r = c^2$, under the mapping $w = z^2$

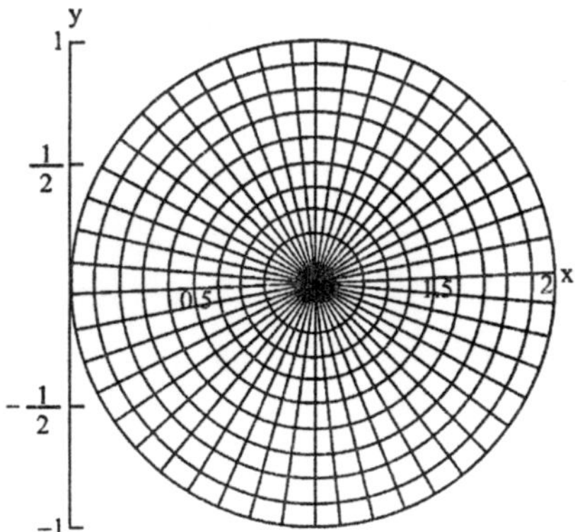

The radii, and circles.

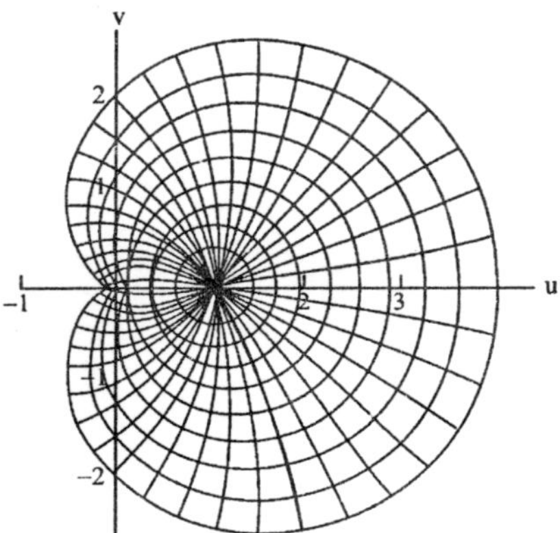

The image curves under the mapping $w = z^2$

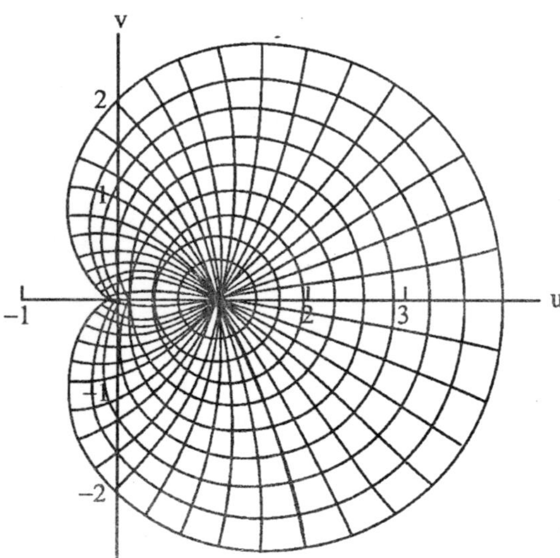

The outer boundary curve is a cardioid.

$(u^2 + v^2 - 2u)^2 = 4(u^2 + v^2)$

(u, v) = (2 cos [t] (1 + cos [t]), 2 sin [t] (1 + cos [t]))

We can use knowledge of the inverse mapping $z = w^2$ to get further insight into how the mapping $w = z^{1/2}$ acts on rectangles. If we let $z = x + iy \neq 0$, then

$$z = x + iy = w^2 = u^2 - v^2 + i\,2uv,$$

and we note that the point $z = x + iy$ in the z plane is related to the point $w = u + iv = z^{1/2}$ in the w-plane by the system of equations

$$x = u^2 - v^2, \text{ and } y = 2uv.$$

Example. The transformation $w = f(z) = z^{1/2}$ usually maps vertical and horizontal lines onto portions of hyperbolas.

(a) Find the image of the vertical line $x = a$.

(b) Find the image of the horizontal line $y = b$.

Solution. Let $a > 0$. The above equations map the right half-plane given by Re $(z) > a$ (i.e., $x > a$) onto the region in the right half-plane satisfying $u^2 - v^2 > a$ and lying to the right of the hyperbola $u^2 - v^2 = a$. If $b > 0$, the equations $x = u^2 - v^2$ and $y = 2uv$ map the upper half-plane Im $(z) > b$ (i.e., $y > b$) onto the region in quadrant I satisfying $2uv > b$ and lying above the hyperbola $2uv = b$. This situation is illustrated in Figure.

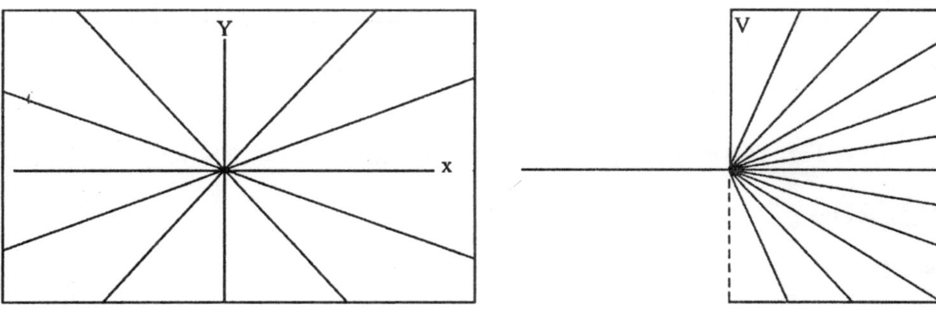

Fig. The mapping $w = f(z) = z^{1/2}$ and $(z = f^{-1}(w) = w^2)$.

The images of regions under the mapping

$$w = f(z) = z^{1/2} = |z|^{1/2} e^{i\,Arg(z)/2} = r^{1/2} e^{i\theta/2} \text{ for } re^{i\theta} \neq 0,$$

where $-\pi < \theta \le \pi$. By the use of polar coordinates for w = $\rho e^{i\phi}$ in the *w* plane, we can represent this mapping by the system

$$\rho = r^{1/2} \text{ and } \phi = \frac{\theta}{2}.$$

This Equation indicate that the argument of $f(z)$ is half the argument of *z* and that the modulus of $f(z)$ is the square root of the modulus of *z*. Points that lie on the ray $r > 0,\ \theta = \alpha$ are mapped onto the ray $\rho > 0, \phi = \frac{\alpha}{2}$. The image of the *z* plane (with the point $z = 0$ deleted) consists of the right half-plane Re $(w) > 0$ together with the positive *v* axis, which is shown in Figure.

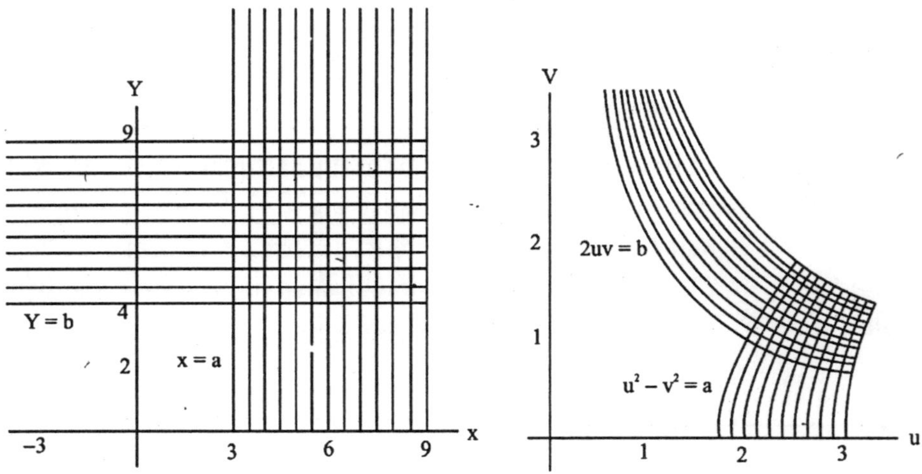

Fig. The mapping $w = f(z) = z^{1/2}$ (and $z = f^{-1}(w) = w^2$).

PRINCIPAL ROOT FUNCTION

The function

$$z = g(w) = w^{1/n} = |w|^{1/n} e^{i\,Arg(\omega)/n}, \text{ for } w \ne 0$$

is called the principal n^{th} root function.

We leave as an exercise to show that *f* and *g* are inverses of each other that map the set *E* one-to-one and onto the set *F* and the set *F* one-to-one and onto the set *E*, respectively. Figure illustrates this relationship.

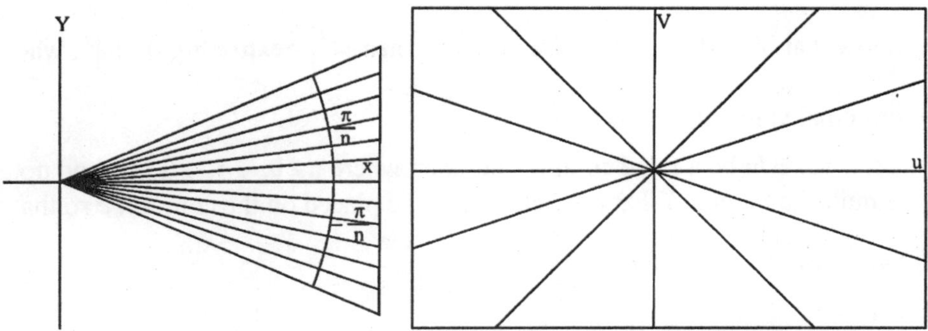

Fig. The mapping $w = f(z) = z^n$ (and $z = g(w) = w^{1/n}$).

EXERCISE

1. Suppose z and w are complex numbers. What is the curve described by $\gamma(t) = (1-t)w + t\,z \quad 0 \le t \le 1$?
2. Find a function γ that describes that part of the curve $y = 4x^3 + 1$ between $x = 0$ and $x = 10$.
3. Find the image of the strip {z: |Rez| < 1} and of the disc {z : |z| < 1} under each of the following mappings:

   (a) $w = (1+i)z + 1$ (b) $w = 2z^2$

   (c) $w = z^{-1}$ (d) $w = \left|\dfrac{z+1}{z-1}\right|$
4. A function f(z) is said to be an isometry if $|f(z_1) - f(z_2)| = |z_1 - z_2|$ for every $z_1, z_2 \in C$; in other words, if it preserves distance.

   (a) Suppose that f(z) is an isometry. Show that for every a, b$\in$C with |a| = 1, the function g(z) = af(z) + b is also an isometry.

   (b) Show that the function $h(z) = \dfrac{f(z) - f(0)}{f(1) - f(0)}$ is an isometry with h(0) = 0 and h(1) = 1.
5. Show that when n = 0, ± 1, ± 2, ...

   (a) $(1 + i)i = \exp\left(-\dfrac{\pi}{4} + 2n\pi\right)\exp\left(\dfrac{i}{2} ln 2\right)$;

   (b) $(-1)^{1/\pi} = e^{(2n+1)i}$
6. Find the principal value of

   (a) 1^i

   (b) $\left[\dfrac{e}{2}\left(-1-\sqrt{3i}\right)\right]^{3\pi i}$

   (c) $(1 - i)^{4i}$

7. Show that if $z \neq 0$ and a is a rea. number, then $\left|z^a\right| = \exp\left(a \ln|z|\right) = |z|^a$, where the principal value of $|z|^a$ is to be taken.
8. Let $c = a + bi$ be a fixed complex number, where $c \neq 0, \neq 1, \neq 2, \ldots$, and note that i^c is multiple-valued. What restriction must be placed on the constant c so that the values of $\left|i^c\right|$ areall teh same?

Ans. C is real.

9. Find all the values of
   (a) $\tan^{-1}(2i)$; (b) $\tan^{-1}(1 + i)$; (c) $\cosh^{-1}(-1)$; (d) $\tanh^{-1}0$.

Ans. (a) $\left(n+\frac{1}{2}\right)\pi + \frac{i}{2}\ln 3 \left(n = 0, \pm1, \pm2 \ldots\right)$;

(b) $n\pi i \left(n = 0, \pm1, \pm2 \ldots\right)$.

10. Solve the equation $\sin z = 2$ for z by equating real parts and imaginary parts in that equation;
11. Solve the equation $\cos z = \sqrt{2}$ for z.
12. $f(x) = (z + 2)/z$ and C is
    (a) the semicircle $z = 2e^{i\theta}$ $(0 \leq \theta \leq \pi)$;
    (b) the semicircle $z = 2e^{i\theta}$ $(\pi \leq \theta \leq 2\pi)$;
    (c) the semicircle $z = 2e^{i\theta}$ $(0 \leq \theta \leq 2\pi)$;
13. $f(z) = z - 1$ and C is the arc from $z = 0$ to $z = 2$ consisting of
    (a) the semicircle $z = 1 + e^{i\theta}$ $(\pi \leq \theta \leq 2\pi)$;
    (b) the segment $0 \leq x \leq 2$ of the real axis.
14. $f(z)$ is defined by the equations

$$f(x) = \begin{cases} 1 & \text{wheny } < 0, \\ 4 & \text{ywheny} > 0, \end{cases}$$

ANALYTIC FUNCTIONS

FUNCTIONS OF A COMPLEX VARIABLE

Functions of (x, y) i.e., $f(x, y)$ that depend only on the combination $(x + i\,y)$ are called functions of a complex variable. The combination $(x + i\,y)$ is generally called z, and we can define such functions as z^n, $\exp(z)$, $\sin z$ and all the standard functions of z as well as of x.A symbol, such as z, which can stand for any one of a set of complex numbers is called a *complex variable*.

They are defined in exactly the same way the only difference being that they are actually complex valued functions, that is, they are vectors in this two dimensional complex number space, each with a real and an imaginary part (or component).

Since we can multiply z by itself and by any other complex number, we can form any polynomial in z and any power series as well. We define the exponential and sine functions of z by their power series expansions which converge everywhere in the complex plane.

We can produce all the standard functions of a complex variable by the same steps as go to producing standard functions of real variables.If to each value which a complex variable z can assume there corresponds one or more values of a complex variable w, say that w is a function of z and write $w = f(z) = f(x+iy) = u(x,y)+iv(x,y) = u+iv$.

Single-valued Function

A single-valued complex function of a complex variable is a complex function $f: C \to C$ that has the same value (unique value) at every point z_0 independent of the path along which it is reached by analytic continuation

Example. If $w = f(z) = z$, then to each value of z there is only one value of w. Hence $w = f(z)$ single-valued function.

Multiple-Valued Function

A multiple-valued function, also known as a multiple-valued function, is a "function" that assumes two or more distinct values in its range for at least one point in its domain. While these "functions" are, not functions in the normal sense of being one-to-one or many-to-one.

The trigonometric, hyperbolic, exponential, and integer power functions are all single-valued functions, their inverses are multiple-valued.

Example. If $w = z^{\frac{1}{2}}$, then to each value of z there are two values of w. Hence $w = f(z) = z^{\frac{1}{2}}$ is a multiple-valued function.

The discontinuities of multiple-valued functions in the complex plane are commonly handled through the adoption of branch cuts, but use of Riemann surfaces is another possibility.

Inverse Functions

If $w = f(z)$, then there exist the function f^{-1} is often called the inverse function corresponding to f. Thus $w = f(z) =$ and $w = f^{-1}(z)$ are inverse function of each other.

MAPPING

If $w = u + i\,v$ (where u and v are real) is a single-valued function of $z = x + i\,y$, we can write $w = f(z) = u + i\,v = f(x + i\,y)$. By equating real and imaginary parts we get,

$$u = u\,(x, y),\ v = v\,(x, y).$$

Thus, for given a point P(x, y) in the z plane such as P in Figure there corresponds a point P'(u, v) in the z plane. We say that z is mapped or transformed into (u, v) by means of the mapping (transformation) and call P' the image of P.

Fig. Mapping

LIMIT AND CONTINUITY OF COMPLEX NUMBER

COMPLEX NUMBER SYSTEM

The complex plane with metric $d(x,y)\colon = |\, x - y \,|$ is also a metric space. There are two different types of limits when we consider complex-valued functions.

LIMIT OF A FUNCTION AT A POINT

Suppose f is a complex-valued function, then we write

$$\lim_{x \to p} f(x) = L$$

if and only if

for every $\varepsilon > 0$ there exists a $\delta > 0$ such that for all real numbers x with $0 < |x-p| < \delta$, we have $|f(x)-L| < \varepsilon$

It is just a particular case of functions over metric spaces with both M and N are the complex plane.

LIMIT OF A FUNCTION OF MORE THAN ONE VARIABLE

By noting that $|x-p|$ represents a distance, the definition of a limit can be extended to functions of more than one variable.

$$\lim_{(x,y) \to (p,q)} f(x, y) = L$$

if and only if

for every $\varepsilon > 0$ there exists a $\delta > 0$ such that for all real numbers x with $0 < \|(x, y)-(p,q)\| < \delta$, we have $|f(x, y)-L| < \varepsilon$

where $\|(x, y)-(p, q)\|$ represents the Euclidean distance. This can be extended to any number of variables.

FUNCTIONS OF A COMPLEX VARIABLE

The real excitement begins when we consider function f: $D \to C$ in which the domain D is a subset of the complex numbers. In some sense, these too are familiar to us from elementary calculus—they are simply functions from a subset of the plane into the plane:

$$f(z) = f(x,y) = u(x,y) + iv\,(x,y) = (u(x,y), y(x,y))$$

Thus $f(z) = z^2$ looks like $f(z) = z^2 = (x + iy)^2 = x^2 - y^2 + 2xyi$. In other words, $u(x,y) = x^2 - y^2$ and $v(x,y) = 2xy$. The complex perspective, as we shall see, generally provides richer and more profitable insights into these function.

The definition of the limit of a function f at a point $z = z_0$ is essentially the same as that which we learned in elementary calculus:

$$\lim_{z \to z0} f(z) = L$$

means that given and $\varepsilon > 0$, there is a δ so that $|f(z) - L| < e$ whenever $0 < |z - z_0| < \delta$. As you could guess, we say that is continuous at z_0 if it is true $\lim_{z \to z0} f(z) = f(z_0)$. If f is continuous at each point of its domain, we say simply that f is continuous.

PROPERTIES OF LIMITS

Suppose both $\lim_{z \to z0} f(z)$ and $\lim_{z \to z_0} g(z)$ exist. Then the following properties are easy to establish:

$$\lim_{z \to z_0} [f(z) \pm g(z)] = \lim_{z \to z_0} f(z) \pm \lim_{z \to z_0} g(z)$$

$$\lim_{z \to z_0} f(z)g(z) = \lim_{z \to z_0} f(z) \lim_{z \to z_0} g(z)$$

and,
$$\lim_{z \to z_0} \frac{f(z)}{g(z)} = \frac{\lim_{z \to z_0} f(z)}{\lim_{z \to z_0} g(z)}$$

provided, of course, that $\lim_{z \to z_0} g(z) \neq 0$.

It now follows at once from these properties that the sum, difference, product, and quotient of two functions are continuous at z_0 are z_0. (We must, as usual, except the dereaded 0 in the denominator.)

It should not be too difficult to convince yourself that if $z = (x,y)$, $z_0 = (x_0,y_0)$, and $f(z) = u(x,y) + iv(x,y)$, then

$$\lim_{z \to z_0} f(z) = \lim_{(x,y) \to (x_0 \cdot y_0)} u(x,y) + i \lim_{(x,y) \to (x_0 \cdot y_0)} v(x,y)$$

Thus f is continuous at $z_0 = (x_0,y_0)$ precisely when u and v are.

Let u = u(x,y) be a real-valued function of the two real variables x and y. Recall that u has the limit u_0 as (x,y) approaches (x_0, y_0) provided that the value of u (x,y) can be made to get as close as we please to the value u_0 by taking (x, y) to be sufficiently close to (x_0, y_0). When this happens we write

$$\lim_{(x,y)\to(x_0,y_0)} u(x, y) = u_0.$$

In more technical language, u has the limit u_0 as (x,y) approaches (x_0, y_0) iff $|u(x, y) - u_0|$ can be made arbitrarily small by making both $|x-x_0|$ and $|y-y_0|$ small. This condition is like the definition of a limit for functions of one variable. The point (x, y) is in the x-y plane, and the distance between (x, y) and (x_0, y_0) is $\sqrt{(x-x_0)^2+(y-y_0)^2}$. With this perspective we can now give a precise definition of a limit.

LIMIT OF U(*X*, *Y*)

The expression $\lim_{(x,y)\to(x_0,y_0)} u(x, y)$ means that for each number $\in > 0$, there corresponds a number $\delta > 0$ such that

$|u(x, y) - u_0| < \in$ whenever

$0 < \sqrt{(x-x_0)^2+(y-y_0)^2} < \delta$.

Example. Show, if $u(x, y) = \dfrac{2x^2}{x^2+y^2}$,

then $\lim_{(x,y)\to(0,0)} u(x, y) = 0$.

Solution: If $x = r\cos\theta$ and $y = r\sin\theta$, and $z = x + iy \neq 0$ then

$$u(x, y) = \frac{2r^3\cos\theta^3\theta}{r^2\cos^2\theta + r^2\sin^2\theta} = 2r\cos^3\theta.$$

Because $\sqrt{(x-0)^2+(y-0)^2} = r$ and

because $|\cos^3\theta| < 1$, we have

$|u(r, \theta)| = 2r|\cos^3\theta < \in$ whenever

$\sqrt{(x-0)^2+(y-0)^2} = r < \dfrac{\in}{2}$.

Hence, for any $\in > 0$, Inequality is satisfied for $\delta = \dfrac{\in}{2}$; that is, u (x, y) has the limit $u_0 = 0$ as (x, y) approaches (0, 0).

Example. Show that the function u (x, y) $\frac{xy}{x^2+y^2}$ does not have a limit as (x, y) approaches (0, 0).

Solution: If we let (x,y) approach (0,0) along the x axis, then

$$\lim_{(x,0)\to(0,0)} u(x, y) = \lim_{(x,0)\to(0,0)} \frac{(x)(o)}{x^2+o^2} = 0$$

But if we let (x,y) approach (0, 0) along the line y = x, then

$$\lim_{(x,y)\to(0,0)} u(x, y) = \lim_{(x,x)\to(0,0)} \frac{(x)(x)}{x^2+x^2} = \frac{1}{2}$$

Because the value of the limit differs depending on how (x,y) approaches (0, 0), we conclude that u (x, y) does not have a limit as (x, y)approaches (0, 0).

Let f(z) be a complex function of the complex variable z that is defined for all values of z in some neighborhood of z_0, except perhaps at the point z_0. We say that f has the limit w_0 as z approaches z_0 provided that the value f(z) can be made as close as we please to the value w_0 by taking z to be sufficiently close to z_0. When this happens we write $\lim_{z\to z_0}$ f(z) $=w_0$

The distance between the points z and z_0 can be expressed by $|z - z_0|$, so we can give a precise definition similar to the one for a function of two variables.

LIMIT OF *F(Z)*

The expression $\lim_{z\to z_0} f(z)=w_0$ means that for each number $\in > 0$, there exists a real number $\delta > 0$ such that

$|f(z)-w_0| < \in$ whenever $0 < |z - z_0| < \delta$.

We can also express the above relationship as

$f(z) \in D_\in(w_0)$ whenever $z \in D_\delta^*(z_0)$.

The formulation of limits in terms of open disks provides a good context for looking at this definition. It says that for each disk of radius $\in > 0$ about the point w_0 (represented by $D\in(w_0)$ there is a punctured disk of radius $\delta > 0$ about the point z_0 (represented by $D_\delta^*(z_0)$ such that the image of each point in the punctured δ-disk lies in the $\in$-disk. The image of the δ-disk does not have to fill up the entire $\in$-disk; but if z approaches z_0 along a curve that ends at z_0, then w = f(z) approaches w_0.

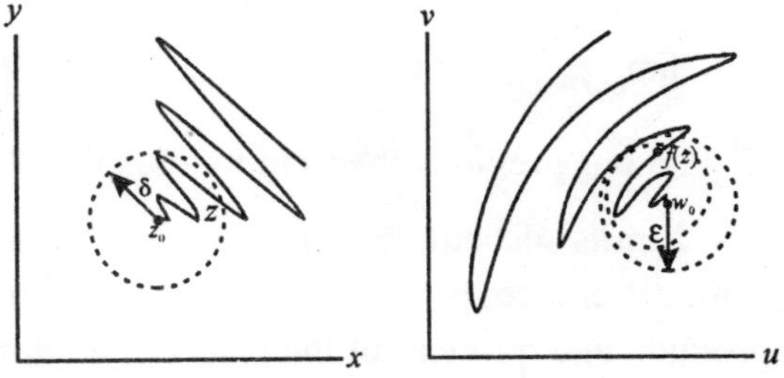

Fig. The limit f (z) $\to w_0$ as $z \to z_0$

Example. Show that if $f(z) = \bar{z}$, then

$\lim_{z \to z_0} f(z) = \bar{z}_0$, where z_0 is any complex number.

Solution: As f merely reflects points about the y axis, we suspect that any $\in$-disk about the point $\bar{z}_0$ would contain the image of the punctured δ-disk about z_0 if $\delta = \in$. To confirm thisconjecture, we let $\in$ be any positive number and set $\delta = \in$. Then we suppose that $z \in D_\delta^*(z_0) = D_\in^*(z_0)$, which means that $0 < |\overline{z - z_0}| < \in$. The modulus of a conjugate is the same as the modulus of the number itself, so the last inequality implies that $0 < |\overline{z - z_0}| < \in$. This is the same as $0 < |\bar{z} - \bar{z}_0| < \in$. Since $f(z) = \bar{z}$ and $w_0 = \bar{z}_0$, this is the same as $0 < |f(z) - w_0| < \in$, which in turn is the same as $f(z) \in D_\in(w_0)$, which is what we needed to show.

Theorem. Let $f(z) = u(x, y) + \dot{n}\, v(x, y)$ be a complex function that is defined in some neighborhood of z_0, except perhaps at $z_0 = x_0 + \dot{n} y_0$. Then $\lim_{z \to z_0} f(z) = w_0 = u_0 + \dot{n} v_0$

iff both

$\lim_{(x,y) \to (x_0, y_0)} u(x, y) = u_0$ and $\lim_{(x,y) \to (x_0, y_0)} v(x, y) = v_0$

Example. Show that $\lim_{z \to 1+i} (z^2 - 2z + 1) = -1$

Solution: We have

$$f(z) = 1 - 2z + z^2$$

$$f(x + \dot{n}y) = (x^2 - y^2 - 2x + 1) + \dot{n}(2xy - 2y)$$

$$u(x, y) = x^2 - y^2 - 2x + 1$$

$$v(x, y) = 2xy - 2y$$

Computing the limits for u and v, we obtain

$$\lim_{(x,y) \to (1,1)} u(x, y) = \lim_{(x,y) \to (1,1)} (x^2 - y^2 - 2x + 1) = -1$$

and

$$\lim_{(x,y) \circledR (1,1)} v(x, y) = \lim_{(x,y) \to (1,1)} (2xy) - 2y = 0,$$

so our previous theorem implies that $\lim_{z \to 1+i} f(z) = -1$.

Limits of complex functions are formally the same as those of real functions, and the sum, difference, product, and quotient of functions have limits given by the sum, difference, product, and quotient of the respective limits. We state this result as a theorem and leave the proof as an exercise.

Theorem. Suppose that $\lim_{z\to z_0} f(z) = A$ and $\lim_{z\to z_0} g(z) = B$ Then

(i) $\lim_{z\to z_0} f(z)+g(z) = A + B$, *and*

(ii) $\lim_{z\to z_0} f(z)-g(z) = A - B$,

(iii) $\lim_{z\to z_0} f(z)\, g(z) = A\,B$,

(iv) $\lim_{z\to z_0} \frac{f(z)}{g(z)} = \frac{A}{B}$ *where* $B \neq 0$.

CONTINUITY

The graph of f is unbroken at a i.e., it's continuous there. It has a hole at b i.e., it's broken, or discontinuous (ie, not contnuous) there; we notice that $f(b)$ doesn't exit. The graph jumps abruptly at c; it's discontinuous there; we notice that $f(c)$ exists but $\lim_{x\to c} f(x)$ doesn't exist (both $\lim_{x\to c+} f(x)$ and $\lim_{x\to c-} f(x)$) exist but aren't equal). The graph has a hole at d; it's discontinuous there; we notice that $f(d)$ exists and $\lim_{x\to d} f(x)$ exists, but they aren't equal.

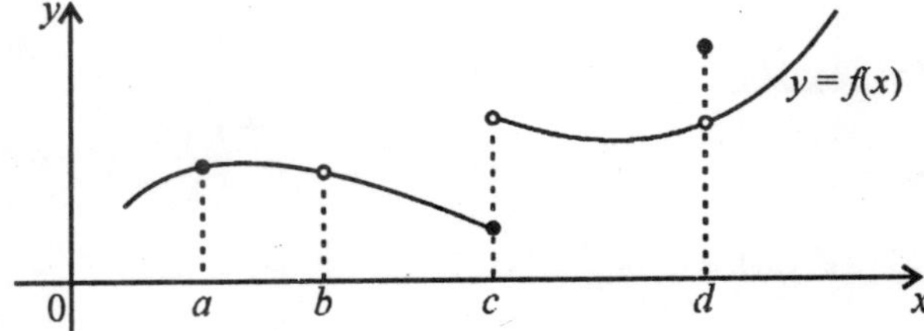

Fig. f is Continuous at a, discontinuous at b, c, and d.

This means that:

$\lim_{x\to a} f(x)$, exists and equals $f(a)$.

All that's just been examined forms the basis for the definition of continuity.

***Definition of Continuity*:** Three conditions for continuity of $f(x)$ at a point $x = c$:

a. the function must be defined at c, or $f(c)$ must be defined.

b. $\lim_{x\to c} f(x)$ must exist.

c. $\lim_{x\to c} f(x) = f(c)$

CONTINUITY OF *U(X, Y)*

Let u(x,y) be a real-valued function of the two real variables x and y. We say that u is continuous at the point (x_0, y_0) if the three conditions are satisfied:

(i) $\lim_{(x,y)\to(x_0,y_0)} u(x,y)$ exists.

(ii) $u(x_0, y_0)$ exists,

(iii) $\lim_{(x,y)\to(x_0,y_0)} u(x,y) = u(x_0 y_0)$

Condition (iii) actually implies Conditions (i) and (ii) because the existence of the quantity on each side of Equation (iii) is implicitly understood to exist. For example, if $u(x, y) = \frac{x^2}{x^2 + y^2}$ when $(x, y) \neq (0, 0)$ and if $u(0, 0) = 0$, then $\lim_{(x,y)\to(0,0)} u(x, y) = 0$ so that Conditions (i), (ii), and (iii) are satisfied. Hence u (x, y) is continuous at (0, 0).

There is a similar definition for complex valued functions.

CONTINUITY OF COMPLEX VALUED FUNCTION

Definition 6. **(Continuity of *f*(*z*)):** Let f(z) be a complex function of the complex variable z that is defined for all values of z in some neighborhood of z_0. We say that f is continuous at z_0 if three conditions are satisfied:

(i) $\lim_{z\to z_0} f(z)$ exists,

(ii) $f(z_0)$ exists,

(iii) $\lim_{z\to z_0} f(z) = f(z_0)$ exists.

THEOREMS ON CONTINUITY

Theorem. Let $f(z) = u(x, y) + iv(x, y)$ be a defined in some neighborhood of z_0. Then $f(z)$ is continuous at $z_0 = x_0 + i y_0$ iff $u(x, y)$ and $v(x, y)$ are continuous at (x_0, y_0).

Theorem. Suppose that f (z) and g (z) are continuous at the point z_0. Then the following functions are continuous at z_0.

(i) The sum f (z) + g (z),

(ii) The difference f (z) – g (z),

(iv) The product f (z) g (z),

(v) The quotient $\frac{f(z)}{g(z)}$, provided that $g(z_0) \neq 0$.

The composition f (g(z)), provided that f (z) is continuous in a neighborhood of the point $g(z_0)$.

Example. Show that the polynomial p (z) = $1 - z - z^2 + z^3 - z^4 + z^5$ is continuous at the point $z_0 = 1$ in the complex plane.

Solution: We have p(z) = $1 - z - z^2 + z^3 - z^4 + z^5$ and find the limit.

$$p(z) = 1 - z - z^2 + z^3 - z^4 + z^5$$

$$p(1) = 0$$

$$\lim_{z \to 1} p(z) = \lim_{x \to 1} (1 - z - z^2 + z^3 - z^4 + z^5)$$

$$\lim_{z \to 1} p(z) = 0$$

Therefore, f(z) is continous at z0 = 1.

Example. Show that the polynomial p (z) = $1 - z - z^2 + z^3 - z^4 + z^5$ is continuous at the point $z_0 = \frac{1}{2}$ in the complex plane.

Solution. We have, p (z) = $1 - z - z^2 + z^3 - z^4 + z^5$ and find the limit.

$$p(z) = 1 - z - z^2 + z^3 - z^4 + z^5$$

$$p\left[\frac{1}{2}\right] = \frac{11}{32}$$

$$\lim_{z \to \frac{1}{2}} p(z) = \lim_{z \to \frac{1}{2}} (1 - z - z^2 + z^3 - z^4 + z^5)$$

$$\lim_{z \to \frac{1}{2}} p(z) = \frac{11}{32}$$

Therefore, f(z) is continous $z_0 = \frac{1}{2}$

If we let P and Q be polynomials and if $Q(z_0) \neq 0$, then

$$\lim_{z \to z_0} \frac{P(z)}{Q(z)} = \frac{p(z_0)}{Q(z_0)}.$$

Another technique, involves factoring polynomials. If both p $(z_0) = 0$ and Q $(z_0) = 0$, then P and Q can be factored as

$$p(z) = (z - z_0)\, p_1(z)$$

and Q (z) = $(z - z_0)\, Q_1(z)$. If Q $(z_0) \neq 0$

Q (z) = $(z - z_0)\, Q_1(z)$. If Q $(z_0) \neq 0$, then the limit is

$$\lim_{z \to z_0} \frac{P(z)}{Q(z)} = \frac{(z - z_0)P_1(z)}{(z - z_0)Q_1(z)} = \frac{P_1(z_0)}{Q_1(z_0)}.$$

Example. Show that $\lim\limits_{z\to 1+i} \dfrac{z^2-2i}{z^2-2z+2} = 1-i$

***Solution*:** Here P and Q can be factored in the form

$$P(z) = (z-1-i)(z+1+i),$$

and

$$Q(z) = (z-1-i)(z-1+i),$$

so that the limit is obtained by the calculation

$$\lim_{x\to 1+i} \frac{z^2-2i}{z^2-2z+2} = \lim_{z\to 1+i} \frac{(z-1-i)(z+1+i)}{(z-1-i)(z-1+i)}$$

$$= \lim_{z\to 1+i} \frac{z+1+i}{z-1+ia}$$

$$= \frac{(1+i)+1+i}{(1+i)-1+i}$$

$$= \frac{2+2i}{2i} = 1-i$$

DIFFERENTIATION OF COMPLEX NUMBER

Earlier we have learnt about the derivative of real valued functions, here we study the derivative of complex-valued functions.A complex function is a function in which the independent variable and the dependent variable are both complex numbers. More precisely, a complex function is a function whose domain Ù is a subset of the complex plane and whose range is also a subset of the complex plane.

For any complex function, both the independent variable and the dependent variable may be separated into real and imaginary parts:

$$z = x+iy \text{ and } w = f(z) = f(x+iy) = u(x,y)+iv(x,y) = u+iv$$

where $x,\ y \in R$ and $u(z), v(z)$ are real-valued functions.

In other words, the components of the function $f(z)$, $u(x,y)$ and $v(x,y)$ can be interpreted as real valued functions of the two real variables, x and y.

THE DERIVATIVE OF A FUNCTION OF A COMPLEX VARIABLE

Let f be a function whose domain of definition contains a neigbourhood of a point z_0.

The derivative of f at z_0 written $f'(z_0)$ is defined by the equation

$$f'(z_0) = \lim_{z \to z_0} \frac{f(z) - f(z_0)}{z - z_0}$$

provided this limits exists. The function *f* is said to be differentiable at z_0 when its derivable at z_0 exists.

To begin with, the usual calculus definition of a derivative can be applied to complex functions just as well as to real functions. It is just that the norm $|\Delta z|$ of an increment must be sent to zero, rather than the absolute value of a real difference, $|\Delta x|$.

$$\frac{d}{dz} f(z) = c \lim_{|\Delta z| \to 0} \frac{f(z + \Delta z) - f(z)}{\Delta z},$$

supposing that the limit exists.

From this, the usual rules governing the derivatives of sums, products, differences and quotients can be deduced. Derivatives of powers can be obtained from the binomial theorem

$$(z + \Delta z)^n = \sum_{i=0}^{n} \frac{n!}{i!(n-i)!} z^{n-i} \Delta z^i$$

in the usual way with the result

$$\frac{d}{dz} z^n = n\, z^{n-1}$$

The chain rule for composite functions is applicable, so that the derivatives of inverse functions can be obtained. For the exponenial, the derivative of products and the definition holding that

$$e^z = \lim_{x \to \infty} \left(1 + \frac{z}{n}\right)^n$$

SUFFICIENT CONDITIONS FOR DIFFRENTIABILITY

Theorem: Let the function

$$f(z) = u(x, y) + iv(x, y)$$

be defined throughout some ε neighborhood of a point $z_0 = x_0 + iy_0$. Suppose that the first-order partial derivatives of the functions *u* and *v* with respect to *x* and *y* exist everywhere in that neighborhood and that they are continuous at (x_0, y_0). Then if those partial derivatives satisfy the Cauchy-Riemann equations $u_x = v_y$, $u_y = -v_x$

at (x_0, y_0), the derivative $f'(z_0)$ exists.

To check that the derivative is well-defined, separate the complex function into the sum of two real functions, just as *z* can be written as the sum of a real and an imaginary part:

$z = x + i\,y$

$f = u(x,y) + i\;v(x,y)$.

By first taking Δz real,

$$\frac{df}{dz} = \lim_{\Delta z \to 0} \frac{f(z + \Delta x)}{\Delta x}$$

$$= \lim_{\Delta x \to 0} \frac{u(x+\Delta x,y)+iv(x+\Delta x,y)-(u(x,y)+iv(x,y))}{\Delta x}$$

$$= \underset{\Delta x \to 0}{lt} \; \underset{\Delta x \to 0}{lt} \; \frac{u(x + \Delta xy) - u(x, y)}{\Delta x + iu}$$

$$\underset{\Delta x \to 0}{lt} \; \frac{(x + \Delta x, y) - u(x, y)}{\Delta x}$$

$$f'(t) = \frac{\partial u}{\partial x} + i\frac{\partial v}{dx},$$

and then imaginary,

$$\frac{df}{dz} = \lim_{\Delta z \to 0} \frac{f(z + i\Delta y)}{i\Delta y}$$

$$= \lim_{\Delta y \to 0} \frac{u(x,y+\Delta y)+iv(x,y+\Delta y)-(u(x,y)+iv(x,y))}{i\Delta y}$$

$$\underset{\Delta y \to 0}{lt} \; \frac{u(x, y + \Delta y) - u(x, y)}{i\Delta y} + \underset{\Delta y \to 0}{lt} \; \frac{u(x, y + \Delta y) - u(x, y)}{\Delta y}$$

$$f'(t) = -i\frac{\partial u}{\partial y} + \frac{dv}{dy}.$$

CAUCHY RIEMAN EQUATIONS

Supposing that the derivative is independent of phase, at least in the context taken here, comparison shows

$$\frac{\partial u}{\partial x} = \frac{\partial v}{\partial y}$$

$$-\frac{\partial v}{\partial x} = \frac{\partial u}{\partial y}$$

The members of this pair of equations are known as the Cauchy - Riemann equations, relating the real and imaginary parts of a complex function which is supposed to have a derivative.

Consider the Jacobian matrix for the real functions $u(x,y)$ and $v(x,y)$ of the real variables x and y,

$$\frac{d(u,v)}{d(x,y)} = \begin{bmatrix} \frac{\partial u}{\partial x} & \frac{\partial v}{\partial x} \\ \frac{\partial u}{\partial y} & \frac{\partial v}{\partial y} \end{bmatrix} = \begin{bmatrix} \frac{\partial u}{\partial x} & \frac{\partial v}{\partial x} \\ -\frac{\partial v}{\partial x} & \frac{\partial v}{\partial y} \end{bmatrix}$$

$$= \frac{\partial u}{\partial x}1 + \frac{\partial v}{\partial x}i$$

The impact of the Cauchy-Riemann equations is to give the Jacobian matrix the form of a complex number in quaternion.

By knowing one of the pair, one effectively knows the other. That is,

$$v(x,y) - v(a,y) = \int_a^x \frac{\partial v(\sigma,y)}{\partial \sigma} d\sigma$$
$$= \int_a^x \frac{\partial u(\sigma,y)}{\partial y} d\sigma$$

Knowing u, take its derivative and integrate to get v. For example, according to Euler's formula,

$e^z = e^x(\cos y + i \sin y)$

Then

$u = e^x \cos y$

$$\frac{\partial u}{dy} = -e^x \sin y$$

$$-\int_a^b e^\sigma \sin y d\sigma = e^\sigma \sin y\Big|_a^x$$

$v(x,y) - v(a,y) = e^x \sin y - e^a \sin a$

which is consistent with

$v(x,y) = e^x \sin y$

So trying to think of something like $e^x (\cos y + i \tan y)$, as an analytic function just wouldn't work.

HARMONIC FUNCTIONS OF REAL VARIABLES

Statement: If D is region in the plane, a real-valued function $u(x,y)$ having continuous second partial derivatives is said to be harmonic on D if it satisfies Laplace's equation on D:

$$\frac{\partial^2 u}{\partial x^2}+\frac{\partial^2 u}{\partial y^2}=0.$$

There is an intimate relationship between harmonic functions and analytic functions. Suppose f is analytic on *D*, and let $f(z) = u(x,y) + iv(x,y)$. Since, we know

$$\frac{\partial u}{\partial x}=\frac{\partial v}{\partial y}$$

and, $\frac{\partial u}{\partial y}=\frac{\partial v}{\partial x}.$

If we differentiate the first and the second with respect to x and respectively y, and then add the two results, we have

$$\frac{\partial^2 u}{\partial x^2}+\frac{\partial^2 u}{\partial y^2}=\frac{\partial^2 y}{\partial x dy}-\frac{\partial^2 y}{\partial y \partial x}=0.$$

Thus the real part of any analytic function is harmonic! Next, if we differentiate the first of the Cauchy-Riemann equations with respect to *y* and the second with respect to *x*, and then substract the second from the first, we have

$$\frac{\partial^2 v}{\partial x^2}+\frac{\partial^2 v}{\partial y^2}=0,$$

and we see that the imaginary part of an analytic function is also harmonic.

There is even more excitement. Suppose we are given a function φ harmonic in a simply connected region *D*. Then there is a function *f* analytic on *D* which is such that Re*f* $= \varphi$. First, define *g* by

$$g(z)=\frac{\partial \varphi}{\partial x}-i\frac{\partial \varphi}{\partial y}.$$

We'll show that g is analytic by verifying that the real and imaginary parts satisfy the Cauchy-Riemann equations: $\frac{\partial}{\partial x}\left(\frac{\partial \varphi}{\partial x}\right)=\frac{\partial^2 \varphi}{\partial x^2}=-\frac{\partial^2 \varphi}{\partial y^2}=\frac{\partial}{\partial y}\left(-\frac{\partial \varphi}{\partial y}\right),$

since φ is harmonic. Next, $\frac{\partial}{\partial y}\left(\frac{\partial \varphi}{\partial x}\right)=\frac{\partial^2 \varphi}{\partial y \partial x}=\frac{\partial^2 \varphi}{\partial x \partial y}=-\frac{\partial}{\partial x}\left(-\frac{\partial \varphi}{\partial y}\right).$

since *g* is analytic on the simply connected region *D*, we know that the intgral of *g* around any closed curve is zero, and so it has an antiderivative $G(z) = u + iv$. This antiderivative is, of course, analytic on *D*, and we know that $G'(z)=\frac{\partial u}{\partial x}-i\frac{\partial u}{\partial y}=\frac{\partial \varphi}{\partial x}-i\frac{\partial \varphi}{\partial y}.$

Thus, $u(x, y) = \varphi(x, y)\ h(y)$. From this, $\frac{\partial u}{\partial y} = \frac{\partial \varphi}{\partial y} + h'(y)$,

and so $h'(y) = 0$, or h = constant, from which it follows that $u(x, y) = \varphi(x, y) + c$. In other words, Re $G = u$, as we promised to show.

An important practical consequence is that such functions never have maxima nor minima. By writing the second derivatives in finite difference form

$$\frac{d^2 f}{dx^2} = \frac{f(x-\Delta x) - 2f(x) + f(x+\Delta x)}{\Delta x^2}$$

so that

$$u(x,y) = \frac{1}{4}\{u(x+\Delta x, y) + u(x-\Delta x, y) + u(x, y+\Delta y) + u(x, y+\Delta y)\}$$

or, in other words, each central value is the average of its neighbors. No average can exceed all the values it is averaging, nor fall below all of them, either.

Example: The Function $\varphi(x, y) = x^3 - 3xy^2$ is harmonic everywhere.

Solution. We shall find function G that Re $G = \varphi$. We know that $G(z) = x^3 - 3xy^2 + iv$, and so from the Cauchy-Riemann equations:

$$\frac{\partial v}{\partial x} = \frac{\partial u}{\partial y} = 6xy$$

Hence,

$$v(x, y) = 3x^2 y + k(y).$$

To find $k(y)$ differentiate with respect to y :

$$\frac{\partial v}{\partial y} = 3x^2 + k'(y) = \frac{\partial u}{\partial x} = 3x^2 - 3y^2,$$

and so,

$k'(y) = -3y^2$, or

$k(y) = -y^3$ + any constant.

If we choose the constant to be zero, this gives us

$$v = 3x^2 y + k(y) = 3x^2 y - y^3,$$

and finally,

$$G(z) = u + iv = (x^3 - 3xy^2) + i(3x^2 y - y^3).$$

THE SCHWARTZ DERIVATIVE

Consider the basic linear fractional transformation

$$w = \frac{az+b}{cz+d}$$

which is equivalent to the vanishing of the implicit expression

$cwz + dw - az - b = 0.$

To get a relation between w and z which does not depend on a, b, c and d, take three derivatives with respect to a still unspecified parameter:

$c(wz)' + dw' - az' = 0$

$c(wz)'' + dw'' - az'' = 0$

$c(wz)''' + dw''' - az''' = 0.$

In equivalent matrix form,

$$\begin{bmatrix} (wz)' & w' & -z' \\ (wz)'' & w'' & -z'' \\ (wz)''' & w''' & -z''' \end{bmatrix} \begin{bmatrix} c \\ d \\ a \end{bmatrix} = \begin{bmatrix} 0 \\ 0 \\ 0 \end{bmatrix}.$$

The determinant of this matrix, which is a Wronskian, must vanish to get a nontrivial solution. But

$$\begin{vmatrix} (wz)' & w' & -z' \\ (wz)'' & w'' & -z' \\ (wz)''' & w''' & -z''' \end{vmatrix}$$

$$= \begin{vmatrix} w'z + wz' & w' & -z' \\ w''z + 2w'z' + wz'' & w'' & -z'' \\ w'''z + 3w''z' + 3w'z'' + wz''' & w''' & -z''' \end{vmatrix}$$

Operate $C_1, (-z)z + w3$, we get

$$= \begin{vmatrix} 0 & w' & -z' \\ 2w'z' & w'' & -z'' \\ 3w''z' + 3w'z'' & w''' & -z''' \end{vmatrix}$$

Expowding along first raw, we get

$$= -w'z' \begin{vmatrix} 0 & w' & -z' \\ 2 & w'' & -z'' \\ 3\dfrac{w''}{w'} + 3\dfrac{z''}{z'z} & w''' & -z''' \end{vmatrix}$$

$$= -w'z' \left\{ -w' \begin{vmatrix} 2 & z'' \\ 3\left(\dfrac{w''}{w'} + \dfrac{z''}{z'}\right) & z''' \end{vmatrix} - z' \begin{vmatrix} 2 & w'' \\ 3\left(\dfrac{w''}{w'} + \dfrac{z''}{z'}\right) & w''' \end{vmatrix} \right\}$$

$$= -w'z'\left\{-w'\left(2z'''-3z''\left(\frac{w''}{w'}+\frac{z''}{z'}\right)\right)+z'\left(2w'''-3w''\left(\frac{w''}{w'}+\frac{z''}{z'}\right)\right)\right\}$$

$$= -w'z'\left\{-2w'z'''+3w'z''\left(\frac{w''}{w'}+\frac{z''}{z'}\right)+2z'w'''-3w''\left(\frac{w''}{w'}+\frac{z''}{z'}\right)\right\}$$

$$= -(w'z')^2\left\{-2\frac{z'''}{z'}+3\frac{z''}{z'}\left(\frac{w''}{w'}+\frac{z''}{z'}\right)+2\frac{w'''}{w'}-3w''\left(\frac{w''}{w'}+\frac{z''}{z'}\right)\right\}$$

Since this expression vanishes, the factor $(w'z')^2$ can be discarded (unless it happens to vanish everywhere) to leave

$$2\frac{w''}{w'}-3\left(\frac{w''}{w'}\right)^2=2\frac{z'''}{z'}-3\left(\frac{z''}{z'}\right)^2,$$

which is a relationship which is expected to hold after applying a linear fractional transformation to a variable, no matter what the linear fractional transformation. Besides, if z is the variable, the right hand is zero, which leaves the left hand side as a constraint on any function that supports invariance. Some rearrangement can make the invariant read

$$\frac{2w'w'''-3(w'')^2}{(w')^2}$$

half of which is called a *Schwartz derivative*. The symbol for the Schwartz derivative of w with respect to z is $\{w,z\}$, some variants of whose definition are:

$$\{w,z\}=\frac{w''}{w'}-\frac{3}{2}\left(\frac{w''}{w'}\right)^2,$$

$$=\left\{\frac{w''}{w'}\right\}'-\frac{1}{2}\left(\frac{w''}{w'}\right)^2,$$

$$=\frac{d^2}{dz^2}\left[\ln\frac{dw}{dz}\right]-\frac{1}{2}\left\{\frac{d}{dz}\left[\ln\frac{dw}{dz}\right]\right\}^2.$$

The Schwartz derivative is not really a derivative (it is a differential invariant) because it doesn't follow the product rule when applied to products, nor does it even work linearly on sums, and it is furthermore immune to non-zero multiplicative constants. As the derivative was derived, w and z depended on a parameter with respect to which two differential expressions were identical. When z itself was the parameter, one result was obviously zero, forcing the other to assume the same value; which of the two was chosen creates an asymmetry by which $\{w,z\}\neq\{z,w\}$.

Nevertheless, the invariant can be calculated for various w's all depending on the same z, for which the following rules and particular results can be verified through calculation:

$$\left\{\frac{aw+b}{cw+d}, z\right\} = \{w, z\},$$

$$\left\{w, \frac{az+b}{cz+d}\right\}\frac{(ad-bc)}{(cz+d)^4} = (w, z),$$

$$\{w, z\} = \{w, u\}\left(\frac{du}{dz}\right)^2 + \{u, z\},$$

$$\{w, z\} = -\{z, w\}\left(\frac{dw}{dz}\right)^2$$

$$\{w, z\} = 0 \Rightarrow w = \frac{az+b}{cz+d'}$$

$$\{z^n, z\} = \frac{1-n^2}{2z^2},$$

$$\{e^{\lambda z}, z\} = -\frac{1}{2}\lambda^2.$$

The last three lines suggest the task of inverting the Schwartz derivative; that is, of finding $w(z)$ for which $\{w, z\} = Q(z)$, for some prescribed function Q. The answer is that one should solve the differential equation

$$y''(z) + \frac{1}{2}Q(z)y(z) = 0,$$

for which one can expect to obtain two linearly independent solutions $y_1(z)$ and $y_2(z)$. Their quotient

$$w(z) \doteq \frac{y_2(z)}{y_1(z)}$$

will then solve the Schwartzian equation.

HIGHER ORDER DERIVATIVES

Note that a partial derivative is itself a function of two variables, and so further partial derivatives can be calculated. We write

$$\frac{\partial}{\partial x}\frac{\partial f}{\partial x} = \frac{\partial^2 f}{\partial x^2}, \quad \frac{\partial}{\partial x}\frac{\partial f}{\partial y} = \frac{\partial^2 f}{\partial x \partial y}, \quad \frac{\partial}{\partial y}\frac{\partial f}{\partial x} = \frac{\partial^2 f}{\partial y \partial x} \text{ and } \frac{\partial}{\partial y}\frac{\partial f}{\partial y} = \frac{\partial^2 f}{\partial y^2}$$

This notation generalises to more than two variables, and to more than two derivatives in the way you would expect. There is a complication that does not occur when dealing with functions of a single variable; there are *four* derivatives of second order, as follows:

$$\frac{\partial^2 f}{\partial x^2}, \frac{\partial^2 f}{\partial x \partial y}, \frac{\partial^2 f}{\partial y \partial x} \text{ and } \frac{\partial^2 f}{\partial y^2}.$$

Fortunately, when f has mild restrictions, the order in which the differentiation is done doesn't matter.

Proposition. Assume that all second order derivatives of f exist and are continuous. Then the mixed second order partial derivatives of f are equal. i.e.

$$\frac{\partial^2 f}{\partial x \partial y} = \frac{\partial^2 f}{\partial y \partial x}.$$

Example. Suppose that $f(x, y)$ is written in terms of u and v where $x = u + \log v$ and $y = u - \log v$. Show that, with the usual convention,

$$\frac{\partial^2 f}{\partial u^2} = \frac{\partial^2 f}{\partial x^2} + 2\frac{\partial^2 f}{\partial x \partial y} + \frac{\partial^2 f}{\partial y^2}$$

and

$$v^2\frac{\partial^2 f}{\partial u^2} = \frac{\partial f}{\partial y} - \frac{\partial f}{\partial x} + \frac{\partial^2 f}{\partial x^2} + 2\frac{\partial^2 f}{\partial x \partial y} + \frac{\partial^2 f}{\partial y^2}$$

You may assume that all second order derivatives of f exist and are continuous.

Solution: Using the chain rule, we have

$$\frac{\partial f}{\partial u} = \frac{\partial f}{\partial x}\frac{\partial x}{\partial u} + \frac{\partial f}{\partial y}\frac{\partial y}{\partial u} = \frac{\partial f}{\partial x} + \frac{\partial f}{\partial y}$$

and $$\frac{\partial f}{\partial v} = \frac{\partial f}{\partial x}\frac{\partial x}{\partial v} + \frac{\partial f}{\partial y}\frac{\partial y}{\partial v} = \frac{1}{v}\frac{\partial f}{\partial x} - \frac{1}{v}\frac{\partial f}{\partial y}$$

Thus using both these and their operator form, we have

$$\frac{\partial^2 f}{\partial v^2} = \frac{\partial}{\partial u}\left(\frac{\partial f}{\partial u}\right) = \frac{\partial}{\partial u}\left(\frac{\partial f}{\partial x} + \frac{\partial f}{\partial y}\right)$$

$$= \frac{\partial}{\partial x}\left(\frac{2f}{\partial x} + \frac{\partial f}{\partial y}\right)\frac{\partial x}{\partial u} + \frac{\partial}{\partial y}\left(\frac{\partial f}{\partial x} + \frac{\partial f}{\partial y}\right)\frac{\partial y}{\partial u}$$

$$= \left(\frac{\partial^2 f}{\partial x^2} + \frac{\partial^2 f}{\partial y}\right)^1 + \left(\frac{\partial^2 f}{\partial y \partial x} + \frac{\partial^2 f}{\partial y^2}\right)^1$$

$$\frac{\partial^2 f}{\partial x^2} + \frac{\partial^2 f}{\partial x \partial y} + \frac{\partial^2 f}{\partial y \partial x} + \frac{\partial^2 f}{\partial y^2}$$

While differentiating with respect to u, we have

$$\frac{\partial^2 f}{\partial v^2} = \frac{\partial}{\partial u}\left(\frac{\partial f}{\partial u}\right) = \frac{\partial}{\partial v}\left(\frac{1}{v}\frac{\partial f}{\partial x} - \frac{1}{v}\frac{\partial f}{\partial y}\right)$$

$$= \frac{-1}{v^2}\frac{\partial f}{\partial x} + \frac{1}{v}\frac{\partial}{\partial v}\left(\frac{\partial f}{\partial x}\right) + \frac{1}{v^2}\frac{\partial f}{\partial y} - \frac{1}{v}\frac{\partial}{\partial v}\left(\frac{\partial f}{\partial y}\right)$$

$$= \frac{-1}{v^2}\frac{2f}{2x} + \frac{1}{v}\left(\frac{\partial}{2x}\left(\frac{\partial f}{\partial x}\right)\frac{\partial x}{\partial v} + \frac{\partial}{\partial y}\left(\frac{\partial f}{\partial y}\right)\frac{\partial y}{\partial v}\right)$$

$$+ \frac{1}{v^2}\frac{\partial f}{\partial y}\frac{-1}{v}\left(\frac{\partial}{\partial x}\left(\frac{\partial f}{\partial y}\right)\frac{\partial x}{\partial v}\right)$$

$$= \frac{-1}{v^2}\frac{2f}{2x} + \frac{1}{v}\left(\frac{\partial \cdot}{2x}\left(\frac{\partial f}{\partial x}\right)\frac{\partial x}{\partial v} + \frac{\partial}{\partial y}\left(\frac{\partial f}{\partial y}\right)\frac{\partial y}{\partial v}\right)$$

$$= \frac{1}{v^2}\left(\frac{\partial f}{\partial y} - \frac{\partial f}{\partial x} + \frac{\partial^2 f}{\partial x^2} - 2\frac{\partial^2 f}{\partial x \partial y} + \frac{\partial^2 f}{\partial y^2}\right)$$

Theorem : If $f(z)$ is analytic in a region $\Re$, so also are $f'(z), f''(z), \ldots$ analytic in $\Re$, i.e., all higher derivatives exists in $\Re$.

L' HOSPITAL'S RULE

Let $f(z)$ and $g(z)$ be an analytic in a region containing the point z_0 and suppose that

$f(z_0) = g(z_0) = 0$, but $g'(z_0) \neq 0$. Then L' Hospital's rule state that $\lim_{z \to z_0} \frac{f(z)}{g(z)} = \frac{f'(z_0)}{g'(z_0)}$

In case $f'(z_0) = g'(z_0) = 0$, the rule may be extended.

Limits represented by so-called indeterminate forms $\frac{\infty}{\infty}, 0 \circ \infty, \infty^0, 0^0, 1^\infty$ and $\infty - \infty$ can be evaluated by appropriate modifications of L' Hospital's rule.

EXERCISE

1. Evaluate each of the following limits:

(a) $\lim_{z \to i}\left(4z^3 - 6z + 3\right)$ (b) $\lim_{w \to 3i}\left(\frac{w^2 + 9}{w - 3i}\right)$ (c) $\lim_{z \to i}\left(\frac{z^4 - 1}{z^2 + 1}\right)$

2. For each of the following, find $\lim_{n \to \infty} z_n$. Also, plot $z_1, z_2, z_3, \ldots, z_{15}$ in the complex plane:

   (a) $z_n = \left(\frac{3-n}{n}\right) + i\left(\frac{n+1}{2n+3}\right)$ (b) $z_n = \left(\frac{2-n}{n^2}\right) - i\left(4 + \frac{6}{n}\right)$

   (c) $z_n = 3e^{i\frac{\pi}{n}}$ (d) $z_n = e^{i\frac{\pi(n-1)}{n}}$

3. Suppose $f(z) = 3xy + i(x - y^2)$. Find $\lim_{z \to 3+2i} f(z)$ or explain carefully why it does not exist.
4. Prove that if *f* has a derivative at *z*, then *f* is continuous at *z*.
5. Suppose that $f(z) = \dfrac{a_0 + a_1 z + a_2 z^2}{b_0 + b_1 z + b_2 z^2}$

   where $a_0, a_1, a_2, b_0, b_1, b_2 \in C$. Examine the behaviour of f(z) at z = 0 and at z = ∞.
6. Each of the following functions is not defined at z = z0. What value must f(z0) take to ensure continuity at z = z0?

   (a) $f(z) = \dfrac{z - z_0}{z - z_0}$

   (b) $f(z) = \dfrac{z^3 - z_0^{\ 3}}{z - z_0}$

   (c) $f(z) = \dfrac{1}{z - z_0}\left(\dfrac{1}{z} - \dfrac{1}{z_0}\right)$
7. Prove that polynomial is continuous everywhere in the finite plane.
8. Find all points at which the valued function *f* defined by $f(z) = \bar{z}$ has a derivative.
9. Find all points at which the valued function f defined by $f(z) = (2+i)z^3 - iz^2 + 4z - (1+7i)$ has a derivative.
10. At what points is the function *f* given by $f(z) = x^3 + i(1-y)^3$ analytic? Explain.
11. Do the real and imaginary parts of the function f satisfy the Cauchy-Riemann equations at $z = 0$? What do you make of your answer?

12. Find all points at which $f(z)=\frac{x}{x^2+y^2}-i\frac{y}{x^2+y^2}$ is differentiable. At what points is *f* analytic? Explain.

13. (a) Show that if f = u + iv is a complex function, where u and v are the real part and the imaginary part of f respectively, then the Cauchy-Riemann equation

$\frac{\partial f}{\partial z}=0$ is equivalent to $u_x = v_y$, $u_y =$ " v_x.

(b) Use to check that the function $f(x, y) = e^x \cos y + ie^x \sin y$ is analytic.

(c) Use the previous exercise or other methods to show that in polar coordinates

is $u_r = \frac{1}{r} v_\theta v_r = -\frac{1}{r} u_\theta$

14. For each of the following functions f(z), determine whether the Cauchy-Riemann equations are satisfied:

(a) $f(z) = x^3 - 3y^2 + 2x + i(3x^2y - y^3 + 2y)$

(b) $f(z) = \log(x2 - y2) + 2i \tan^{-1}(y/x)$

15. Show that a real valued analytic function is constant.

ELEMENTARY FUNCTIONS OF COMPLEX NUMBERS

Once we have mastered using *i* as an exponent, we can advance to functions of complex numbers. Exponential functions, of course, are simply functions which use *i* as an exponent.

POLYNOMIAL FUNCTIONS

It is defined by $w = a_0 z^n + a_1 z^{n-1} + \dots + a_{n-2} z^2 + a_{n-1} z + a_n = P(z)$

where the coefficients $a_0 \neq 0, a_1, \dots, a_{n-2}, a_{n-1}$ and a_n are all constants and *n* is a positive integer called degree of the polynomial P(*z*).

The transformation $w = az + b$ is called a linear transformation.

RATIONAL ALGEBRAIC FUNCTION

It is defined by $w=\frac{P(z)}{Q(z)}$

Where P(*z*) and Q(*z*) are polynomials. We sometimes call as a rational transformations

The special case $w=\frac{az+b}{cz+d}$, where $ad - bc \neq 0$ is often called a bilinear transformation.

EXPONENTIAL FUNCTIONS

It is defined by

$$z = x + iy = r(\cos\theta + i\sin\theta) = re^{i\theta} = re^{i(\theta+2n\pi)}$$

where n is an integer.

Common Operations of Complex Numbers

$$z_1 z_1 = r_1 e^{i\theta_1} r_2 e^{i\theta_2} = r_1 r_2 e^{i(\theta_1+\theta_2)}$$

$$z_1 / z_2 = r_1 e^{i\theta_1} / r_2 e^{i\theta_2} = \frac{r_1}{r_2} e^{i(\theta_1-\theta_2)}$$

$$z^p = \left(re^{i\theta}\right)^p = r^p e^{ip\theta}$$

$$\sqrt[p]{z} = \sqrt[p]{re^{i\theta}} = \left(re^{i\theta}\right)^{\frac{1}{p}} = \left[re^{i(\theta+2n\pi)}\right]^{\frac{1}{p}} = \sqrt[p]{r}e^{\frac{i(\theta+2n\pi)}{p}}$$

where n is an integer.

LOGARITHM FUNCTIONS

Since a complex number can be written in exponential form as re^{ip}, a logarithm is almost immediate. We get $\ln z = \ln\left(re^{i\theta}\right) = \ln\left(re^{i(\theta+2n\pi)}\right) = \ln r + i\theta + 2n\pi i$

which gives the principal value of the natural logarithm. But the angle P is not unique, as it can vary by multiples of 2π. Therefore, the formula for the natural logarithm of a complex number is $\ln z = \ln\left(re^{i\theta}\right) = \ln\left(re^{i(\theta+2n\pi)}\right) = \ln r + i\theta + 2n\pi i$

For example, we can find the natural logarithm of i. Its principal value is

$$\ln i = \ln\left(e^{i\pi/2}\right) = \frac{\pi}{2}i$$

Another interesting example is the natural logarithm of negative one. Its principal value is $\ln(-1) = \ln(1e^{\pi}) = \pi i$. (And you thought you couldn't take logarithms of negative numbers! You can, but the answers are not real numbers.)

We can use the change-of-base formula to find other base logarithms (including complex bases). For example,

$$\log_i e = \frac{\ln e}{\ln i} = \frac{1}{\frac{\pi}{2}i} = -\frac{2i}{\pi}.$$

TRIGONOMETRIC FUNCTIONS

From formula $e^{ix} = \cos x + i \sin x$, we can immediately obtain some formulas for the sine and cosine functions, in terms of complex exponentials. These formulas are often used as definitions of the trigonometric functions for complex numbers.

$$\cos x = \frac{e^{ix} + e^{-ix}}{2}, \sin x = \frac{e^{ix} - e^{-ix}}{2i}$$

Substituting the quantity ix in place of the variable x, we can produce formulas for pure imaginary numbers. Notice that the results are related to the hyperbolic functions.

$$\cos ix = \frac{e^x + e^{-x}}{2} = \cosh x, \ \sin ix = \frac{e^{-x} - e^x}{2i} i \sinh x$$

Then, using the formulas for the sine and cosine of the sum of two angles, we can obtain formulas for the sine and cosine of complex numbers.

$\cos (a + bi) = \cos a \cosh b - i \sin a \sinh b$

$\sin (a + bi) = \sin a \cosh b + i \cos a$ singh b

As an example, consider the cosine of i.

We get

$$\cos i = \cos 0 \cosh 1 - i \sin 0 \sinh 1 = \cosh 1 = \frac{e^1 + e^{-1}}{2} \approx 1.5431$$

HYPERBOLIC FUNCTIONS

Using the connection between hyperbolic functions and trigonometric functions, the results for hyperbolic functions are almost immediate.

$\cosh(a + bi) = \cosh a \cos b + i \sinh a \sin b$

$\sinh(a + bi) = \sinh a \cos b + i \cosh a \sin b$

As an example, consider the hyperbolic cosine of i. We get

$\cosh i = \cosh 0 \cos 1 + i \sinh 0 \sin 1 = \cos 1 \approx 0.5403$

INVERSE TRIGONOMETRIC FUNCTIONS

Since $\cos x = \dfrac{e^{ix} + e^{-ix}}{2}$, we can find the inverse by replacing the function with the variable y, swapping x and y, then solving for y. After the variable swap, we have $x = \dfrac{e^{iy} + e^{-iy}}{2}$. Multiplying both sides by $2e^{iy}$ gives the equation $0 = e^{2iy} - 2xe^{iy} + 1$. This can be solved using the quadratic formula, to get $e^{iy} = \dfrac{2x \pm \sqrt{4x^2 - 4}}{2}$. After simplifying, we can take the natural logarithm of both sides and divide by i. Then, the result is $\cos^{-1}x = -i \ln \left[x \pm i\sqrt{1 - x^2}\right]$. (A factor of i typically appears as the coefficient of the square root so that the similarities between the inverse cosine and inverse sine formulas are more apparent.)

By similar arguments, the inverse sine formula is found to be $\sin^{-1} x = -i \ln \left[ix \pm \sqrt{1 - x^2}\right]$, and the inverse tangent formula is $\tan^{-1} x = \dfrac{i}{2} \ln \dfrac{i + x}{i - x}$. As an example, consider the inverse cosine of i. We shall find only one of its many values, by using the plus sign in the formula. We get: $\cos^{-1} i = -i \ln\left[i + i\sqrt{1-i^2}\right] = -i\ln\left[\left(1+\sqrt{2}\right)i\right]$ Then we use the logarithm formula to continue:

$\cos^{-1} i = -i \ln\left[\left(1+\sqrt{2}\right)e^{i\pi/2}\right] = -i\left[\ln\left(1+\sqrt{2}\right)\right] + \frac{\pi}{2}i$ This can be simplified, and we get $\cos^{-1} i = \frac{\pi}{2} - i\ln\left(1+\sqrt{2}\right) \approx 1.57 - 0.88$

INVERSE HYPERBOLIC FUNCTIONS

The formulas are identical to those previously found for inverse hyperbolic functions:

$$\cosh^{-1} x = \ln\left[x \pm \sqrt{x^2-1}\right]$$

$$\sinh^{-1} x = \ln\left[x \pm \sqrt{x^2+1}\right]$$

For example, one of the values of the inverse hyperbolic cosine of i (using the plus sign in the formula) is:

$\cosh^{-1} i = \ln\left[i + \sqrt{i^2-1}\right] = \ln\left[\left(1+\sqrt{2}\right)i\right]$. Then using the logarithm formula, we get:

$$\cosh^{-1} i = \ln\left(1+\sqrt{2}\right) + \frac{\pi}{2}i \approx 0.88 + 1.57i$$

THE FUNCTION

z^{α}, α may be complex, is defined as $e^{\alpha \ln z}$. Similarly if $f(z)$ and $g(z)$ are two given functions of z, we can define $f(z)^{g(z)} = e^{g(z)\ln f(z)}$. In general such functions are multiple-valued.

Algebraic and Transcendental Functions

If w is a solution of the polynomial equation

$$P_0(z)w^n + P_1(z)w^{n-1} + P_2(z)w^{n-2} + P_3(z)w^{n-3} + \ldots + P_{n-1}(z)w^n + P_n(z) = 0 \ldots(2)$$

Where $P_0 \neq 0, P_1(z), P_2(z), \ldots, P_n(z)$ are polynomials in z and n is a positive integer then $w = f(z)$ is called an algebraic function of z.

Example. $w = z^{\frac{1}{2}}$ is a solution of the equation $w^2 - z = 0$ and so is an algebraic function of z.

Any function which cannot be expressed as a solution of (6) is called a transcendental function. The logarithmic, trigonometric and hyperbolic function and their corresponding inverses are example of transcendental functions.

BRANCH POINT AND BRANCH CUT

In complex analysis, a branch is a portion of the range of a multiple-valued function over which the function is single-valued. Combining all the branches gives the full structure of the function. It is often convenient to choose a particular branch of a function to work with, and this choice is often designated the principal branch.

In complex analysis, a branch point may be thought of informally as a point z_0 at which a "multiple-valued function" changes values when one winds once around z_0.

A branch point of an analytic function is a point in the complex plane whose complex argument can be mapped from a single point in the domain to multiple points in the range. For example, consider the behavior of the point z = 0 under the power function

$$f(z) = z^a$$

Examples:

- 0 is a branch point of the square root function. Suppose $w = \sqrt{2}$ and z starts at 4 and moves along a circle of radius 4 centered at 0. The dependent variable w changes while

  depending on z in a continuous manner. When z has made one full circle, going from 4 back to 4 again, w will have made one half-circle, going from the positive square root of 4, i.e., from 2, to the negative square root of 4, i.e., –2.
- 0 is also a branch point of the natural logarithm. Since e^0 is the same as $e^{\pi i}$, both 0 and $2\pi i$ are among the multiple values of Log (1). As z moves along a circle of radius 1 centered at 0, w = Log (z) goes from 0 to $2\pi i$.
- In trigonometry, since tan (π/4) and tan (5π/4) is both equal to 1, the two numbers π/4 and 5π/4 are among the multiple values of arc-tan (1). The imaginary units i and $-i$ are branch points of the arctangent function. That this is so may be seen by observing that the derivative $\frac{d}{dz}$ {arc-tan (z)} = $1/(1 + z^2)$ has simple poles at these two points i.e., at z = i, $-i$ $\frac{d}{dz}\left(\tan^{-1} z\right) = 0$ since the denominator is zero at those points.
- If the derivative f' of a function f has a simple pole at a point a, then f has a branch point at a.

The branch cut device may appear arbitrary (it is); but it is very useful, for example in the theory of special functions. An invariant explanation of the branch phenomenon is developed in Riemann surface theory (of which it is historically the origin), and more generally in the ramification and monodromy theory of algebraic functions and differential equations.

BRANCH CUT

A branch cut is a curve (with ends possibly open, closed, or half-open) in the complex plane across which an analytic multiple-valued function is discontinuous. For convenience, branch cuts are often taken as lines or line segments. Branch cuts (even those consisting of curves) are also known as cut lines, slits, or branch lines.

For example, consider the function z^2 which maps each complex number z to a well-defined number z^2. Its inverse function $\sqrt{z}$, on the other hand, maps, for example, the value $z = 1$ to $\sqrt{1} = \pm 1$. While a unique principal value can be chosen for such functions, the choices cannot be made continuous over the whole complex plane. Instead, lines of discontinuity must occur. The most common approach for dealing with these discontinuities is the adoption of so-called branch cuts. In general, branch cuts are not unique, but are instead chosen by convention to give simple analytic properties. Some functions have a relatively simple branch cut structure, while branch cuts for other functions are extremely complicated. An alternative to branch cuts for representing multiple-valued functions is the use of Riemann surfaces.

In addition to branch cuts, singularities known as branch points also exist. It should be noted, however, that the endpoints of branch cuts are not necessarily branch points.

Branch cuts do not arise for the single-valued trigonometric, hyperbolic, integer power, and exponential functions. However, their multiple-valued inverses do require branch cuts.

Function Name	Function	Branch cut(s)
inverse cosecant	$\csc^{-1} z$	$(-1, 1)$
inverse cosine	$\cos^{-1} z$	$(-\infty, -1)$ and $(1, \infty)$
inverse cotangent	$\cot^{-1}$ z	$(-i, i)$
inverse hyperbolic cosecant	csch^{-1}	$(-i, i)$
inverse hyperbolic cosine	$\cosh^{-1}$	$(-\infty, 1)$
inverse hyperbolic cotangent	$\coth^{-1}$	$(-1, 1)$
inverse hyperbolic secant	sech^{-1}	$(-\infty\ 0)$ and $(1, \infty)$
inverse hyperbolic sine	$\sinh^{-1}$	$(-i\ \infty, -i)$ and $(i, i\ \infty)$
inverse hyperbolic tangent	$\tanh^{-1}$	$(-\infty, -1)$ and $(1, \infty)$
inverse secant	$\sec^{-1} z$	$(-1, 1)$
inverse sine	$\sin^{-1} z$	$(-\infty, -1)$ and $(1, \infty)$
inverse tangent	$\tan^{-1} z$	$(-i\ \infty, -i)$ and $(i, i\ \infty)$
natural logarithm	$\ln z$	$(-\infty, 0)$
power	$z^n, n \notin \mathbb{Z}$ $(-\infty, 0)$ for R $[n] > 0$	$(-\infty, 0)$ for R $[n] \leq 0$;
square root	$\sqrt{z}$	$(-\infty, 0)$

Next we discuss about 'Riemann surface'.

RIEMANN SURFACE

A Riemann surface is a surface-like configuration that covers the complex plane with several, and in general infinitely many, "sheets." These sheets can have very complicated structures and interconnections. Riemann surfaces are one way of representing multiple-valued functions; another is branch cuts.

For example, they can look like a sphere or a torus or a couple of sheets glued together.

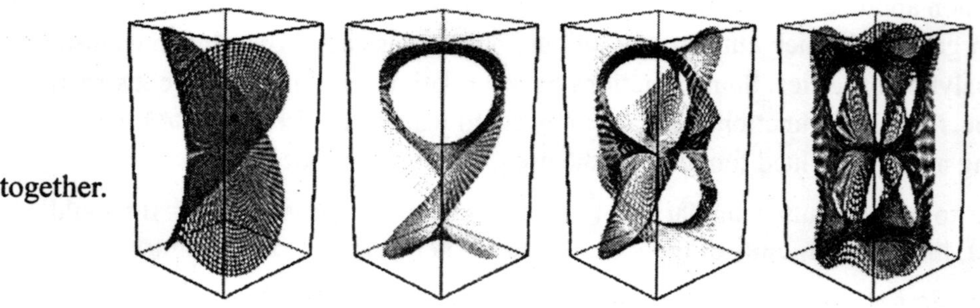

Fig. Riemann Surface.

The above plot shows Riemann surfaces for solutions of the equation

$[w(z)]^d + w(z) + z^{d-1} = 0$

with $d = 2, 3, 4$, and 5, where $w(z)$ is the Lambert W-function.

The main point of Riemann surfaces is that holomorphic functions may be defined between them. Riemann surfaces are nowadays considered the natural setting for studying the global behavior of these functions, especially multi-valued functions such as the square root or the logarithm.

Examples.

- The complex plane C is perhaps the most trivial Riemann surface. The map $f(z) = z$ (the identity map) defines a chart for C, and $\{f\}$ is an atlas for C. The map $g(z) = \bar{z}$ (the conjugate map) also defines a chart on C and $\{g\}$ is an atlas for C. The charts f and g are not compatible, so this endows C with two distinct Riemann surface structures. In fact, given a Riemann surface X and its atlas A, the conjugate atlas $B = \{f^* : f \to \text{A}\}$ is never compatible with A, and endows X with a distinct, incompatible Riemann structure.
- Every open subset of the complex plane can be viewed as a Riemann surface in a natural way. More generally, every open subset of a Riemann surface is a Riemann surface.

Every non-compact Riemann surface admits non-constant holomorphic functions (with values in). In fact, every non-compact Riemann surface is a Stein manifold.

In constrast, on a compact Riemann surface every holomorphic function with value in $\mathbb{C}$ is constant due to the maximum principle. However, there always exists non-constant meromorphic functions .

RIEMANN SPHERE

The Riemann sphere, is the unique way of viewing the extended complex plane (the complex plane plus a point at infinity) so that it looks exactly the same at the point infinity as at any complex number. The main application is to deal with extended complex functions (which may be defined at the point infinity and/or take the value infinity, in addition to complex numbers) in the same way at the point infinity as at any complex number, specifically with respect to continuity and differentiability.

From the geometrical view of the plane that deals with points, lines, circles and angles but not distances, the Riemann sphere is created by adding a point at infinity through which all lines cross, with parallel lines being tangent there and all other lines crossing at the same angle as they do at an existing point.

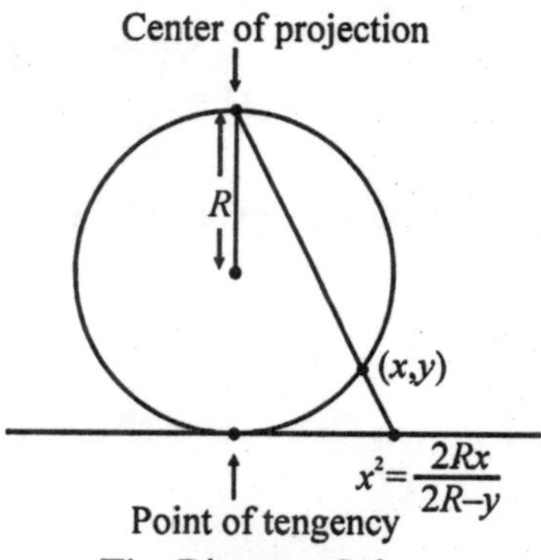

Fig. Riemann Sphere

Topologically, the Riemann sphere is the one-point compactification of the complex plane.

In all of these viewpoints, the point at infinity acquires an identical role to any point in the complex plane.

EXERCISE

1. Evaluate each of the following limits:

(a) $\lim\limits_{z \to i}\left(4z^3 - 6z + 3\right)$ (b) $\lim\limits_{z \to 1-i}\left(z^2 - 3z\right)$

(c) $\lim\limits_{w \to 3i}\left(\dfrac{w^2 + 9}{w - 3i}\right)$ (d) $\lim\limits_{z \to i}\left(\dfrac{z^4 - 1}{z^2 + 1}\right)$

2. Suppose $f(z) = 3xy + i(x - y^2)$. Find $\lim_{z \to 3+2i} f(z)$ or explain carefully why it does not exist.
3. Prove that if f has a derivative at z, then f is continuous at z.
4. Suppose that $f(z) = \dfrac{a_0 + a_1 z + a_2 z^2}{b_0 + b_1 z + b_2 z^2}$

   where $a_0, a_1, a_2, b_0, b_1, b_2 \in C$. Examine the behaviour of f(z) at z = 0 and at z = ∞.
5. Each of the following functions is not defined at z = z0. What value must f(z0) take to ensure continuity at z = z0?

   (a) $f(z) = \dfrac{z - z_0}{z - z_0}$ (b) $f(z) = \dfrac{z^3 - z_0^3}{z - z_0}$

   (c) $f(z) = \dfrac{1}{z - z_0}\left(\dfrac{1}{z} - \dfrac{1}{z_0}\right)$ (d) $f(z) = \dfrac{1}{z - z_0}\left(\dfrac{1}{z^3} - \dfrac{1}{z_0^3}\right)$
6. Prove that polynomial is continuous everywhere in the finite plane.

Chapter 7

Complex Topology

In this section we investigate some basic ideas concerning sets of points in the plane. The first concept is that of a curve. Intuitively, we think of a curve as a piece of string placed on a flat surface in some type of meandering pattern. More formally, we define a curve to be the range of a continuous complex-valued function $z(t)$ defined on the interval $[a, b]$. That is, a curve C is the range of a function given by $z(t) - (x(t), y(t)) = x(t) + i\,y(t)$, for $a \le t \le b$, where both $x(t)$ and $y(t)$ are continuous real-valued functions. If both $x(t)$ and $y(t)$ are differentiable, we say that the curve is smooth. A curve for which $x(t)$ and $y(t)$ are differentiable except for a finite number of points is called piecewise smooth.

CURVE

In analytic geometry, a curve is continuous map from a one-dimensional space to an n-dimensional space. Loosely speaking, the word "curve" is often used to mean the function graph of a two- or three-dimensional curve. The sim/plest curves can be represented parametrically in n-dimensional space as

$$\begin{aligned} x_1 &= f_1(t) \\ x_2 &= f_2(t) \\ \vdots &= \\ x_n &= f_n(t) \end{aligned}$$

Other simple curves can be simply defined only implicitly, i.e., in the form

$$f(x_1, x_2, \ldots) = 0$$

DEFINITION OF CURVE

We specify a curve C as

$$C : z(t) = x(t) + i\,y(t) \text{ for } a \le t \le b,$$

and say that $z(t)$ is a parameterization for the curve C. Notice that with this parametrization, we are specifying a direction the curve C, and we say that C is a curve that goes from the initial point $z(a) = (x(a), y(a)) = x(a) + i\,y(a)$ to the terminal point $z(b) = (x)(b), y(b) = x(b) + i\,y(b)$. If we had another function whose range was the same set of points as $z(t)$ but whose initial and final points were reversed, we would indicate the curve this function defines by $-C$.

Example. Find parameterizations for C and $-C$, where C is the straight line segment beginning at $z_0 = x_0 + i\,y_0$ and ending at $z_1 = x_1 + i\,y_1$.

Solution. Refer to Figure . The vector form of a line shows that the direction of C is $z_1 - z_0$. As z_0 is a point on C, its vector equation is

$$C : z(t) = z_0 + (z_1 - z_0)\,t \text{ for } 0 \le t \le 1, \text{ or}$$

$$C : z(t) = (x_0 + x_1 - x_0)\,t) + i\,(y_0 + (y_1 - y_0)\,t) \text{ for } 0 \le t \le 1.$$

Clearly one parameterization for $-C$ is

$$-C : \gamma(t) = z_1 + (z_0 - z_1)\,t \text{ for } 0 \le t \le 1, \text{ or}$$

Remark. Note that $\gamma(t) = z(1-t)$, which illustrates a general principle: If C is a curve parameterized by $z(t)$ for $0 \le t \le 1$, then one parameterization for $-C$ will be $\gamma(t) = z(1-t)$, for $0 \le t \le 1$.

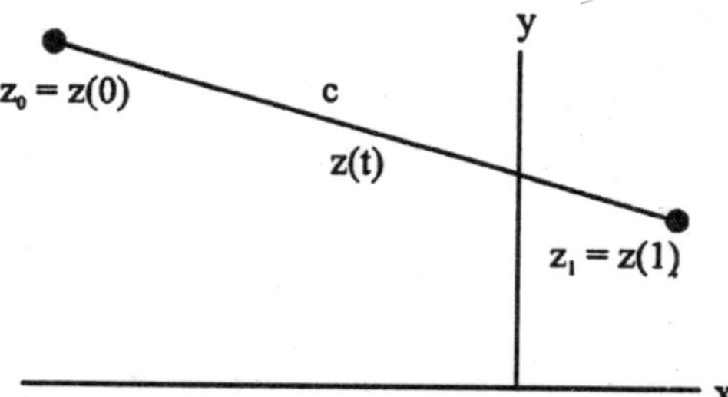

Fig. The straight-line segment C joining z_0 to z_1.

Example. Find the equation of the line segment with the initial point $z_0 = -3 + 2\,i$ and the terminal point $z_1 = 1 + i$.

Solution.

Enter the initial point $z_0 = -3 + 2\,i$ and the terminal point $z_1 = 1 + i$ and construct the line segment.

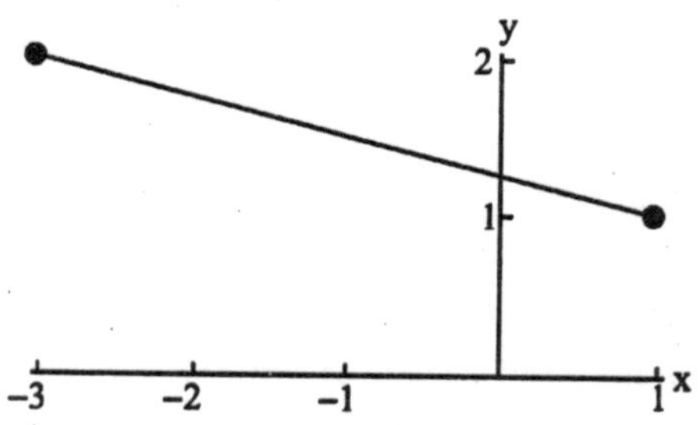

Equation of aline segment :

$$z[t] = -3 + i(2 - t) + 4t \text{ for } 0 \le t \le 1$$

Initial point (green) $z[0] = -3 + 2i$

Terminal point (red) $z[1] = 1 + i$

A curve C having the property that $z(a) = z(b)$ is said to be a closed curve. The line segment is not a closed curve. The range of $z(t) = x(t) + iy(t)$, where $x(t) = \sin 2t \cos t$, and $y(t) = \sin 2t \sin t$ for $0 \le t \le 2\pi$ is a closed curve because $z(0) = (0, 0) = z(2\pi)$. The range of is the four-leaved rose shown in Figure Note that, as t goes from 0 to $\frac{\pi}{2}$, the point is on leaf 1; from $\frac{\pi}{2}$ to π to, it is on leaf 2 ; between π and $\frac{3\pi}{2}$, it is on leaf 3; and finally, for t between $\frac{3\pi}{2}$ and 2π, it is on leaf 4.

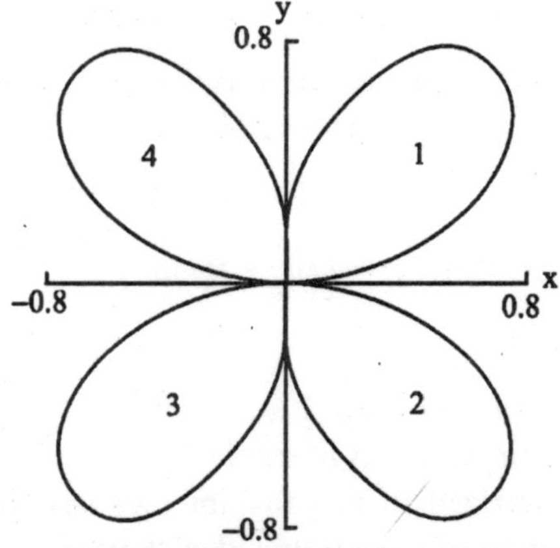

Fig. The curve $x(t) = \sin 2t \cos t$, and $y(t) = \sin 2t \sin t$ for $0 \le t \le 2\pi$, which forms a four-leaved rose.

Note further, in Figure, that at the point (0, 0), the curve has crossed over itself (at points other than those corresponding with $t = 0$ and $t = 2\pi$; we want to be able to distinguish when a curve does not cross over itself in this way. The curve C is called simple if it does not cross over itself, except possibly at its initial and terminal points. In other words, the curve $C : z(t)$, for $a \le t \le b$, is simple provided that $z(t_1) \ne z(t_2)$ whenever $t_1 \ne t_2$, except possibly when $t_1 = a$ and $t_2 = b$.

Example. The curve $x(t) = \sin 2t \cos t$, and $y(t) = \sin 2t \sin t$ for $0 \le t \le 2\pi$.

Solution.

The equation of the curve is $x(t) = \sin 2t \cos t$, and $y(t) = \sin 2t \sin t$.

Curve corresponding to the curve is.

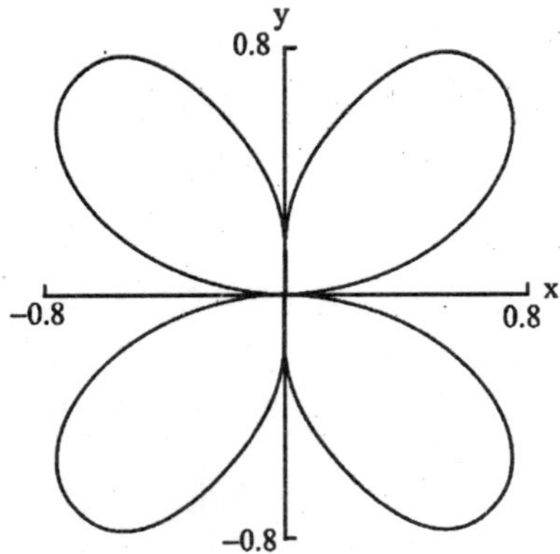

$x = \cos[t] \sin[2t]$

$y = \sin[t] \sin[2t]$ for $0 \le t \le 2\pi$

Remark. The curve looks like a "four leafed rose".

Example. Show that the circle C with center and radius can be parameterized to form a simple closed curve.

Solution. Note that the required parametrization is

$$C : z(t) = z_0 + Re^{it}$$

$$C : z(t) = x_0 + i\, y_0 + Re^{it}$$

$$C : z(t) = (x_0 + R \cos t) + i\,(y_0 + R \sin t),$$

for $0 \le t \le 2\pi$

Figure shows that, as t varies from 0 to 2π, the circle is traversed counter clockwise. If you were traveling around the circle in this manner, its interior would be on your left. When a simple closed curve is parametrized in this fashion, we say that the curve has a positive orientation. We will have more to say about this idea shortly.

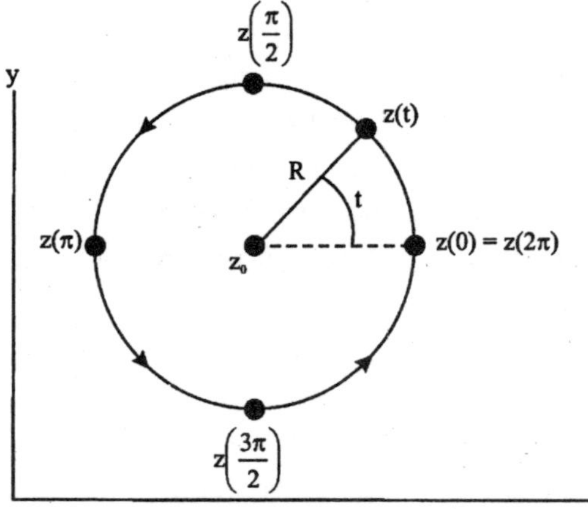

Fig. The simple closed curve $z(t) = z_0 + Re^{it}$, for $0 \le t \le 2\pi$.

We need to develop some vocabulary that will help describe sets of points in the plane. One fundamental idea is that of an $\in$ neighborhood of the point z_0. It is the open disk of radius $\in > 0$ about z_0 shown in Figure. Formally, it is the set of all points satisfying the inequality $D_\in(z_0) = \{z : |z - z_0| < \in\}$ and is denoted $D_\in(z_0)$.

That is, (1-49) $\quad D_\in(z_0) = \{z : |z - z_0| < \in\}$.

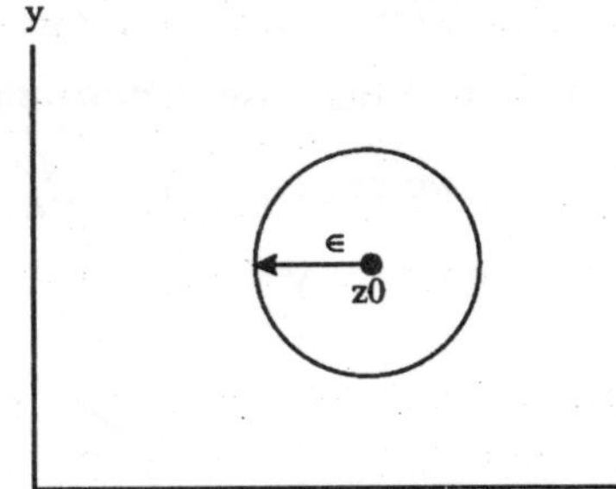

Fig. An $\in$ neighborhood of the point z_0.

Example. The solution sets of the inequalities $|z| < 1$, $|z - i| < 2$, and $|z + 1 + 2i| < 3$ are neighborhoods of the points 0, i, and $-1 - 2i$, with radii 1, 2, and 3 respectively. They can also be expressed as $D_1(0)$, $D_2(i)$, and $D_2(-1 - 2i)$.

We also define $\overline{D_\in}(z_0)$, the closed disk of radius $\in$ centered at z_0, and $D_\in^*(z_0)$, the punctured disk of radius $\in$ centered at z_0 as

$$\overline{D_\in}(z_0) = \{z : |z - z_0| \le \in\}, \text{ and}$$

$$D_\in^*(z_0) = \{z : 0 < |z - z_0| < \in\}.$$

DEFINITIONS OF INTERIOR POINT, EXTERIOR POINT, BOUNDARY POINT

The point z_0 is said to be an interior point of the set S provided that there exists an $\in$ neighborhood of z_0 that contains only points of S; z_0 is called an exterior point of the set S if there exists an $\in$ neighborhood of z_0 that contains no points of S. If z_0 is neither an interior point nor an exterior point of S, then it is called a boundary point of S and has the property that each $\in$ neighborhood of z_0 contains both points in S and points not in S. Figure illustrates this situation.

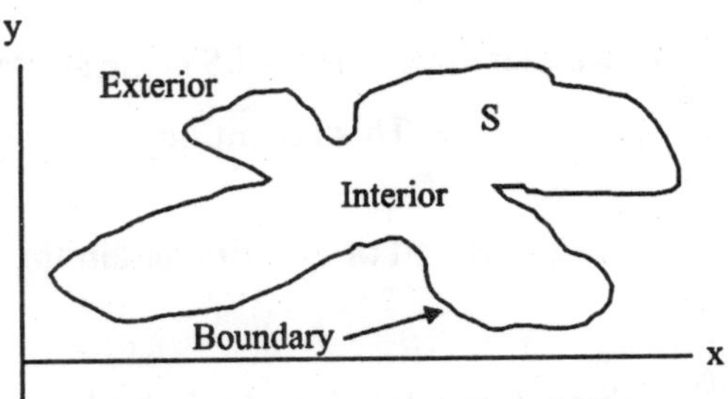

Fig. The interior, exterior, and boundary of a set S.

The boundary of $D_R(z_0)$ is the circle depicted in Figure. We denote this circle $C_R(z_0)$ and refer to it as the circle of radius R centered at z_0. Thus

$$C_R(z_0) = \{z : |z - z_0| R\}.$$

We use the notation $C_R^+(z_0)$ to indicate that the parametrization we chose for this simple closed curve resulted in a positive orientation; $C_R^-(z_0)$ denotes the same circle, but with a negative orientation. (In both cases, counterclockwise denotes the positive direction.) Using notation that we have already introduced, we get $C_R^-(z_0) = -C_R^+(z_0)$.

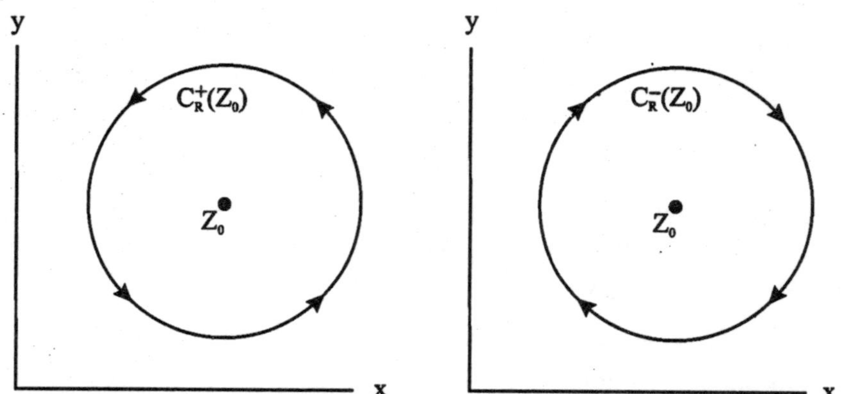

Fig. The positively oriented curve $C_R^+(z_0)$ and its opposite $C_R^-(z_0) = -C_R^+(z_0)$.

Example. Let S = {z : | z | < 1}.

(a) Find the interior of *S*. **(b)** Find the exterior of *S*. **(c)** Find boundary of S.

Solution: We show that every point of *S* is an interior point of *S*. Let be a point of *S*. Then $|z_0| < 1$, and we can choose $\in = 1 - |z_0| > 0$. We claim that $D_\in(z_0) c\, S$. If $z \in D_\in(z_0)$, then

$$|z| = |z - z_0 + z_0| \le |z - z_0| + |z_0| < \in + |z_0| = 1 - |z_0| + |z_0| = 1$$

Hence the $\in$ neighborhood of z_0 is contained in *S*, which shows that z_0 is an interior point of *S*.

It follows that the interior of *S* is the set *S* itself.

Similarly, it can be shown that the exterior of *S* is the set $\{z : |z| > 1\}$. The boundary of *S* is the unit circle $C_1(0) = \{z : |z| = 1\}$. This condition is true because if $z_0 = e^{i\theta_0}$ is any point on the circle, then any $\in$-neighborhood of z_0 will contain the point $\left(1 - \frac{\in}{2}\right)e^{i\theta_0}$, which belongs to *S*, and $\left(1 + \frac{\in}{2}\right)e^{i\theta_0}$, which does not belong to *S*. We leave the details as an exercise.

Solution :

(a). Find the interior of *S*.

Let z_0 be a point of *S*. Then $|z_0| < 1$ so that we can choose $\in = 1 - |z_0| > 0$. If *z* lies in the disk $|z - z_0| < \in$, then

$|z| = |z_0 + z - z_0| \leq |z_0| + |z - z_0| < |z_0| + \in < 1.$

Hence the $\in$-neighborhood of z_0 is contained in *S*, and z_0 is an interior point of *S*. It follows that the interior of *S* is the open unit disk.

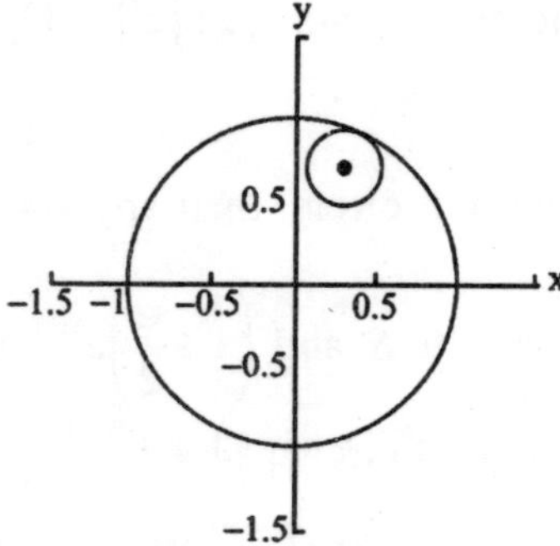

The $\in$-neighbourhood of z_0 is contained in $S = \{z : |z| < 1\}$.

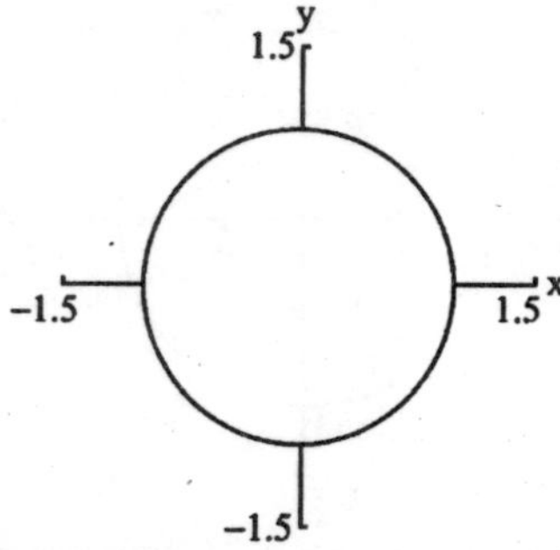

The interior of the set *S* is the unit disk $\{z : |z| < 1\}$.

(b). Find the exterior off *S*.

If $|z_0| > 1$ then a similar argument will produce an $\in$-neighborhood of z_0 that lies entirely in the region $|z| > 1$, hence z_0 is an exterior point of *S*. It follows that the exterior of *S* is the open region $|z| > 1$.

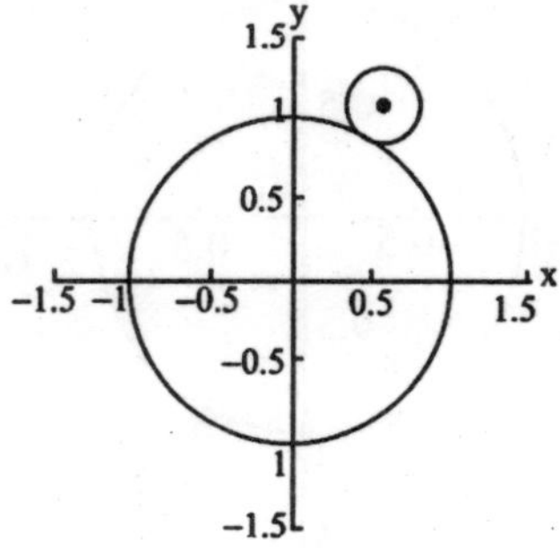

The $\in$-neighborhood of z_0 is contained in the region $|z| > 1$.

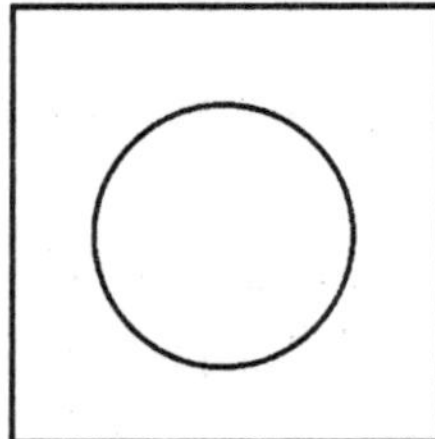

The exterior of the set S is the open region $\{z : |z| > 1\}$.

(c). Find the boundary of S.

If $z_0 = e^{i\theta_0}$ is any point on the unit circle, then any $\in$-neighborhood of z_0 will contain the point $\left(1-\frac{\in}{2}\right)e^{i\theta_0}$, which belongs to S, and $\left(1+\frac{\in}{2}\right)e^{i\theta_0}$ which does not belong to S. It follows that the boundary of S is the unit circle $|z| = 1$.

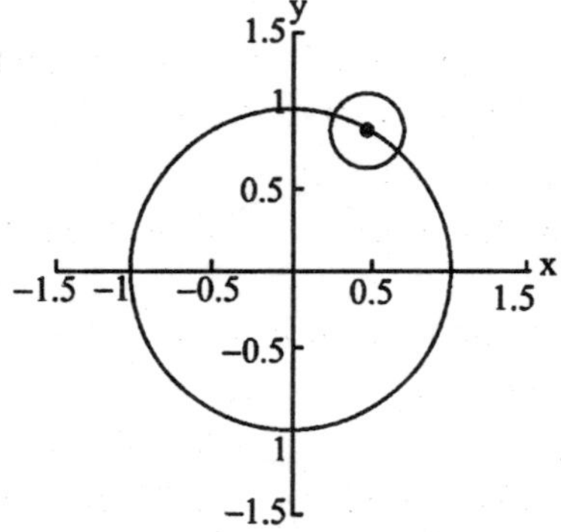

Every $\in$-neighborhood of z_0 contains points of S and points that do not belong to S.

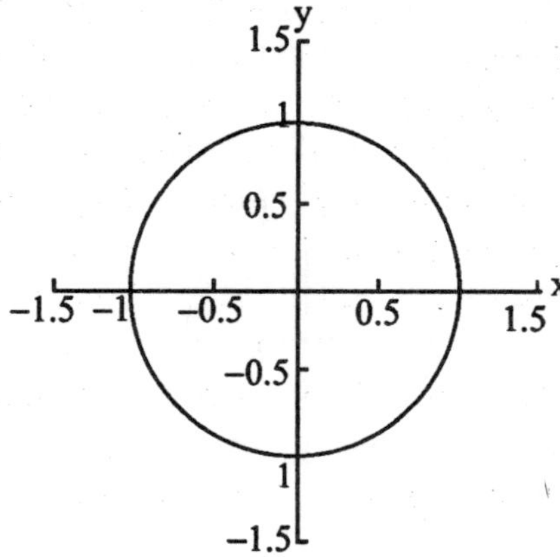

The boundary of the set $S = \{z : |z| < 1\}$ is the unit circle $\{z : |z| = 1\}$.

The point z_0 is called an accumulation point of the set S if, for each $\in$, the punctured disk contains at least one point of S. We ask you to show in the exercises that the set of accumulation points of $D_1(0)$ is $\overline{D_1}(0)$, and that there is only one accumulation point of $S = \left\{\frac{i}{n} : n = 1, 2, ...\right\}$, namely, the point 0.

We also ask you to prove that a set is closed if and only if it contains all of its accumulation points.

A set S is called an open set if every point of S is an interior point of S. Thus, Example shows that $D_1(0)$ is open. A set S s called a closed set if it contains all its boundary points. A set S is said to be a connected set if every pair of points z_1 and z_2 contained in S can be joined by a curve that lies entirely in S. Roughly speaking, a connected set consists of a "single piece." The unit disk $D_1(0) = \{z : |z| < 1\}$ is a connected open set.

We ask you to verify in the exercises that, if z_1 and z_2 lie in $D_1(0)$, then the straight-line segment joining them lies entirely in $D_1(0)$. The annulus $A = \{z : 1 < |z| < 2\}$ is a connected open set because any two points in A can be joined by a curve C that lies entirely in A, as shown in Figure. The set $B = \{z : |z+2| < 1 \text{ or } |z-2| < 1\}$ consists of two disjoint disks. We leave it as an exercise for you to show that the set is not connected, as shown in Figure .

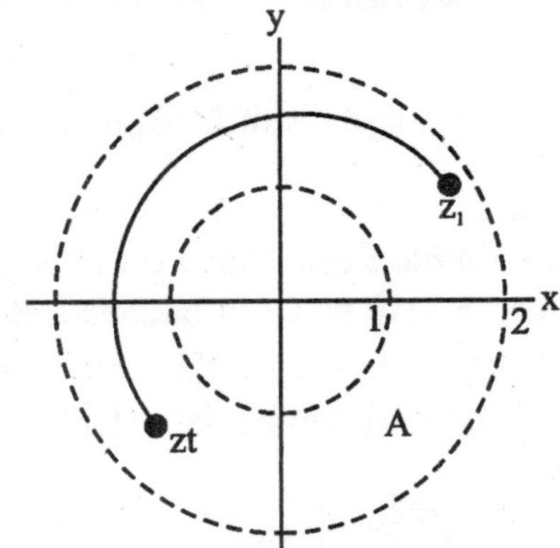

Fig. The annulus $A = \{z : 1 < | z | < 2\}$ is a connected set.

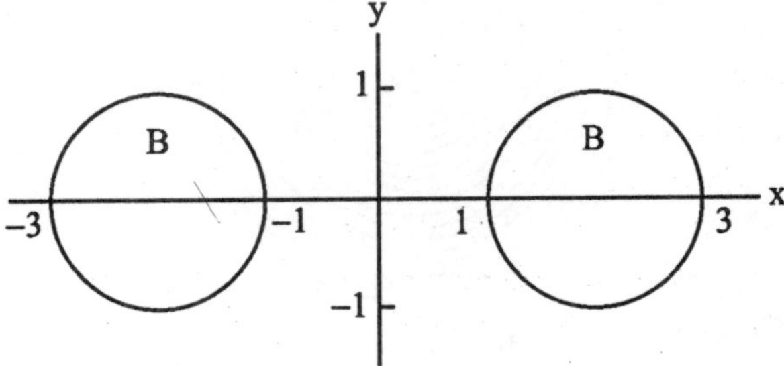

Fig. The set $B = \{z : | z + 2 | < 1 \text{ or } | z - 2 | < 1\}$ is not a connected set.

We call a connected open set a domain. In the exercises we ask you to show that the open unit disk $D_1(0) = \{z : |z|| < 1\}$ is a domain and that the closed unit disk $\overline{D_1}(0) = \{z : |z| \leq 1\}$ is not a domain. The term domain is a noun and is a type of set. In previous chapter we note that it also refers to the set of points on which a function is defined. In the latter context, it does not necessarily mean a connected open set.

A domain, together with some, none, or all its boundary points, is called a region. For example, the horizontal strip $\{z : 1 < \text{In}(z)\}$ is a region. A set formed by taking the union of a domain and its boundary is called a closed region; thus $\{z : 1 \leq \text{In}(z) \leq 2\}$ is a closed region. A set S is said to be a bounded set if it can be completely contained in some closed disk, that is, if there exists an $R > 0$ such that for each z in S we have $|z| \leq R$. The rectangle given by $\{z : |x| \leq 4 \text{ and } |y| \leq 3\}$ is bounded because it is contained inside the disk $\overline{D_5}(0) = \{z : |z| \leq 5\}$. A set that cannot be enclosed by any closed disk is called an unbounded set.

We mentioned earlier that a simple closed curve is positively oriented if its interior is on the left when the curve is traversed.

JORDAN CURVE THEOREM [MH]

If J is a simple closed curve in $\mathbb{R}^2$, then the Jordan curve theorem, also called the Jordan-Brouwer theorem states that $\mathbb{R}^2 - J$ has two components (an "inside" and "outside"), with J the boundary of each.

The Jordan curve theorem is a standard result in algebraic topology with a rich history.

Theorem. The compliment of any simple closed curve C can be partitioned into two mutually exclusive domains I and E in such a way that I is bounded, E is unbounded, and C is the boundary for both I and E. In addition $I \cup E \cup C$ is the entire complex plane. The domain I is called the interior of C, and the domain E is called the exterior of C.

Fig. Are the points z_1 (blue), and z_2 (red) in the interior or exterior of this simple closed curve?

EXERCISE

1. Let S be the open set consisting of all points z such that $|z|<1$ *or* $|z-2|<1$. State why S is not connected.
2. Show that a set S is open if and only if each point in S is an interior point.
3. Determine the accumulation points of each of the following sets:

   (a) $z_n = i^n (n=1,2,...)$; (b) $z_n = i^n / n(n=1,2,...)$;

   (c) $0 \le \arg z < \pi/2(z \ne 0)$; (d) $z_n = (-1)^n (1+i)(n-1)/2(n=1,2,...)$.

Ans. (a) None; (b) 0; (c) $\pm(1+i)$.

4. Prove that if a set contatins each of its accumulation points, then it must be a closed set.
5. Show that any point z_0 of a domain is an accumulation point of that domain.
6. Prove that a finite set of points z_1, z_2, ..., z_n cannot have any accumulation points.
7. Suppose f is analytic on a connected open set D, and $f(z) = 0$ for all zÎD. Prove that f is constant.
8. Suppose D is a bounded domain. For any e > 0, let D_e denote the larger domain containing D and every point in C whose distance from D is less than e. Give an example of a simply connected domain D such that D_e is not simply connected for any e > 0.

SEQUENCE AND SERIES IN COMPLEX NUMBER

SEQUENCE OF COMPLEX NUMBERS

A sequence is nothing more than a list of numbers written in a specific order. The list may or may not have an infinite number of terms in them General sequence terms are denoted as follows,

a_1 – first term
a_2 – second term
⋮
a_n – n^{th} term
a_n + 1 – $(n+1)^{st}$ term
⋮

There is a variety of ways of denoting a sequence. Each of the following are equivalent ways of denoting a sequence.

$$\{a_1, a_2,...,a_n, a_n+1,...\} \ \{a_n\} \ \{a_n\}_{n-1}^{\infty}$$

In the second and third notations above a_n is usually given by a formula.

Let's take a look at a couple of sequences.

(a) $\left\{\frac{n+1}{n^2}\right\}_{n-1}^{\infty}$ (b) $\left\{\frac{(-1)^{n+1}}{2^n}\right\}_{n-0}^{\infty}$

$\{b_n\}_{n-1}^{\infty}$ where $b_n = n^{\text{th}}$ digit of π

Definition: A sequence of complex number is a function g: $Z_+ \to C$ from the positive ingegers into the complex numbers. It is traditional to use subscripts to indicato the values of the function. Thus we write $g(n) \equiv z_n$ and an explicit name for the sequence is seldom used; we write simply (z_n) to stand for the sequence g which is such that $g(n) = z_n$. For example, $\left(\frac{i}{n}\right)$ is the sequence g for which $g(n) = \frac{i}{n}$

The number L is a limit of the sequence (z_n) if given an $\varepsilon > 0$, there is an integer N_ε such that $|z_n - L| < \varepsilon$ for all $n \geq N_\varepsilon$. If L is a limit of (z_n), we sometimes say that (z_n) converges to L. We frequently write $\lim(z_n) = L$. It is relatively easy to see that is the complex sequence $(z_n) = (u_n + iv_n)$ converges to L, then the two real sequences (u_n) and (v_n) each have a limit: (u_n) converges to ReL and (v_n) converges to Im L. Conversely, if the two real sequences (u_n) and (v_n) each have a limit, then so also does the complex sequence $(u_n + iv_n)$. All the usual nice properties of limits of sequences are thus true:

Now if $z_n = x_n + y_n i$ and $L = a + bi$, then $\lim_{n\to\infty} z_n = L$ if and only if

$$\lim_{n\to\infty} |z_n - L| = \lim_{n\to\infty} \sqrt{(x_n - a)^2 + (y_n - b)^2} = 0,$$

the latter of which occurs if and only if $\lim_{n\to\infty} x_n = a$ and $\lim_{n\to\infty} y_n = b$. Hence we have the following useful result.

Proposition: Let $z_n = x_n + y_{ni}$ and $L = a + bi$. Then

$$\lim_{n\to\infty} z_n = L$$

if and only if

$$\lim_{n\to\infty} x_n = a \text{ and} \lim_{n\to\infty} y_n = b.$$

if and only if

$\lim_{n\to\infty} x_n = a$ and $\lim_{n\to\infty} y_n = b$.

Thus to determine the limiting behaviour of a sequence $\{z_n\}$ of complex numbers, we need only condider the behaviour of the two sequences of real numbers, $\{\Re(z_n)\}$ and $\{\Im(z_n)\}$.

A complex sequence is a function whose domain is the positive integers and whose range is a subset of the complex numbers. The following are examples of sequences:

$$f(n) = \left(2-\frac{1}{n}\right)+\left(5+\frac{1}{n}\right)i \ (n = 1, 2, 3, ...):$$

$$g(n) = e^{i}\frac{\pi n}{4} \ (n = 1, 2, 3, ...);$$

$$h(k) = 5 + 3\,i + \left(\frac{1}{1+i}\right)^{k} \ (k = 1, 2, 3, ...);$$

$$r(n) = \left(\frac{1}{4}+\frac{i}{2}\right)^{n} \ (n = 1, 2, 3, ...).$$

For convenience, we use the term sequence rather than complex sequence.

The sequence r given by Equation (4) could be written in a variety of ways:

$$\left\{\left(\frac{1}{4}+\frac{i}{2}\right)^{n}\right\}_{n=1}^{\infty}, \left\{\left(\frac{1}{4}+\frac{i}{2}\right)^{n}\right\}_{1}^{\infty}, \left\{\left(\frac{1}{4}+\frac{i}{2}\right)^{n}\right\}_{n=1}^{\infty},$$

$$\left\{\left(\frac{1}{4}+\frac{i}{2}\right)^{n+2}\right\}_{n=-2}^{\infty}, \left\{\left(\frac{1}{4}+\frac{i}{2}\right)^{k}\right\}_{k=1}^{\infty}$$

The sequences f and g given by Equations behave differently as n gets larger. The terms in Equation approach $2 + 5\,i = (2, 5)$, but those in Equation do not approach any particular number, as they oscillate around the eight eighth roots of unity on the unit circle. Informally, the sequence $\{z_n\}_1^{\infty}$ has ζ as its limit as n approaches infinity, provided the terms z_n can be made as close as we want to ζ by making n large enough. When this happens, we write

$$\lim_{n\to\infty} z_n = \zeta \text{ or } z_n \to \zeta \text{ as } n \to \infty$$

Limit of a Sequence

$\lim_{n\to\infty} z_n = \zeta$ means that for any real number $\in > 0$ there corresponds a positive integer $N_{\in}$ (which depends on $\in$) such that $z_n \in D_{\in}(\zeta)$ whenever $n > N_{\in}$. That is $|\zeta - z_n| > \in$ whenever. Figure illustrates a convergent sequence.

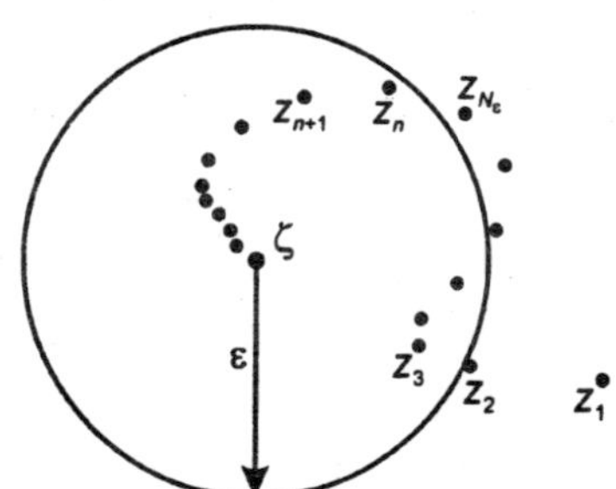

Fig. A sequence $\{z_n\}_1^\infty$ that converges to ζ. (If n > N_ϵ then $z_n \in D_\epsilon(\zeta)$)

Remark. The reason we use the notation N_ϵ is to emphasize the fact that this number depends on our choice of ϵ. Sometimes it will be convenient to drop the subscript.

Theorem. Let $z_n = x_n + i\,y_n$ and $\zeta = u + i\,v$. Then

$$\lim_{n\to\infty} z_n = \zeta, \text{ iff}$$

$$\lim_{n\to\infty} x_n = u \text{ and } \lim_{n\to\infty} y_n = v.$$

If $\lim_{n\to\infty} z_n = \zeta$, we say that the sequence $\{z_n\}_1^\infty$ converges to ζ.

Bounded Sequence

A complex sequence $\{z_n\}$ is bounded provided that there exists a positive real number R and an integer N such that $|z_n| < R$ for all $n > N$. In other words, for $n > N$, the sequence $\{z_n\}$ is contained in the disk $D_R(0)$.

Convergent sequences are always bounded.

Theorem. If $\{z_n\}$ is a convergent sequence, then $\{z_n\}$ is bounded.

Cauchy Sequence

The sequence $\{z_n\}$ is said to be a Cauchy sequence if for every $\epsilon > 0$ there exists a positive integer N_ϵ, such that if n, $m > N_\epsilon$ then $|z_n - z_m| < \epsilon$, or, equivalently, $z_n - z_m \in D_\epsilon(0)$.

Theorem. If $\{z_n\}$ is a Cauchy sequence, then $\{z_n\}$ converges.

Example. Evaluate the limit of $z_n = \dfrac{3n-1}{2n+2} + \dfrac{n+1}{n-1} i$, for $n = 1, 2, 3, \ldots$

Solution: $\lim_{n\to\infty} \Re(z_n) = \lim_{n\to\infty} \dfrac{3n-1}{2n+2} = \lim_{n\to\infty} \dfrac{3-\frac{1}{n}}{2+\frac{2}{n}} = \dfrac{3}{2}$

and $\lim_{n\to\infty} \Im(z_n) = \lim_{n\to\infty} \dfrac{n+1}{n-1} = \lim_{n\to\infty} \dfrac{1+\frac{1}{n}}{1-\frac{1}{n}} = 1,$

so $\lim_{n\to\infty} z_n = \dfrac{3}{2} + i.$

Example. Evaluate the limit of $z_n = \frac{1}{n}\left(\cos\left(\frac{n\pi}{3}\right) + \sin\left(\frac{n\pi}{3}\right)i\right)$ for $n = 1, 2, 3, \ldots$.

Solution: Now, $\lim_{n\to\infty} \Re(z_n) = \lim_{n\to\infty} \frac{\cos\left(\frac{n\pi}{3}\right)}{n} = 0$

and $\lim_{n\to\infty} \Im(z_n) = \lim_{n\to\infty} \frac{\sin\left(\frac{n\pi}{3}\right)}{n} = 0$,

so $\lim_{n\to\infty} z_n = 0$.

Geometrically, since $|z_n| = \frac{1}{n}$ and $\arg(z_n) = \frac{n\pi}{3}$, the points in this sequence are converging to 0 along a spiral path, as shown in the figure

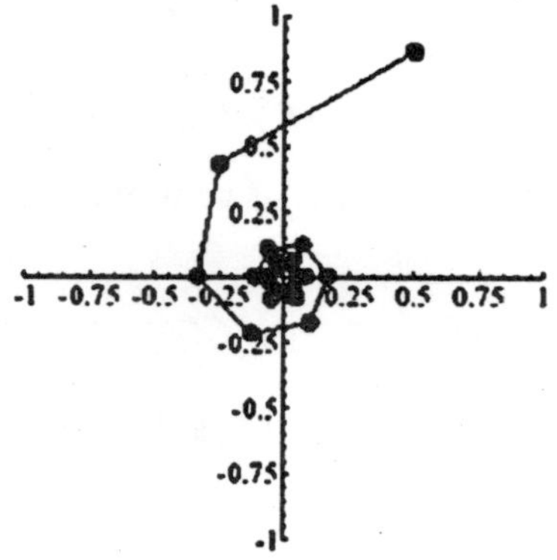

Fig. Showing the plots of the points $z_n = \frac{1}{n}\left(\cos\left(\frac{n\pi}{3}\right) + \sin\left(\frac{n\pi}{3}\right)i\right)$, $n = 1, 2, 3, \ldots 20$

THEOREM ON LIMITS

(i) $\lim(z_n \pm w_n) = \lim(z_n) \pm \lim(w_n)$;

(ii) $\lim(z_n w_n) = \lim(z_n)\lim(w_n)$; and

(iii) $\lim\left(\frac{z_n}{w_n}\right) = \frac{\lim(z_n)}{\lim(w_n)}$.

Provided that $\lim(z_n)$ and $\lim(w_n)$ exist. (And in the last equation, we must, of course, insist that $\lim(w_n) \neq 0$.)

A necessary and sufficient condition for the convergence of a sequence (a_n) is the celebrated Cauchy criterion: given $\varepsilon > 0$, there is an integer $N_\in$ so that $|a_n - a_m| < \in$ whenever $n, m > N_\in$.

CONVERGENCE OF SEQUENCE

A sequence (f_n) of functions on a domain D is the obvious thing: a function from the positive integers into the set of complex functions on D. Thus, for each $z \in D$, we have an ordinary sequence $(f_n(z))$. If each of the sequences $(f_n(z))$ converges, then we say the sequence of functions (f_n) converges to the function f defined by $f(z) = \lim(f_n(z))$.

Definition: The sequence (f_n) is said to converge to f uniformly on a set S if given an e > 0, there is an integer N_ϵ so that $|f_n(z) - f(z)| < \epsilon$ for all $n \geq N_\epsilon$ and all $z \in S$.

Note that this is possible for a sequence of continuous functions to have a limit function that is *not* continuous. This cannot happen if the convergence us uniform. To see this, suppose the sequence (f_n) of continuous functions converges uniformly to f on a domain D, let $Z_0 \in D$, and let $\varepsilon > 0$. We need to show there is a δ so that $|f(z_0) - f(z)| < \varepsilon$ whenever.

$|z_0 - z| < \delta$. Let's do it. First, choose N so that $|f_N(z) - f(z)| < \frac{\varepsilon}{3}$. We can do this because of the uniform convergence of the sequence (f_n). Next, choose δ so that $|f_N(z_0) - f_N(z)| < \frac{\varepsilon}{3}$ whenever $|z_0 - z| < \delta$, This is possible because f_N is continuous. Now then, when $|z_0 - z| < \delta$, we have

$$\begin{aligned}|f(z_0) - f(z)| &= |f(z_0) - f_N(z_0) + f_N(z_0) - f_N(z) + f_N(z) - f(z)| \\ &\leq |f(z_0) - f_N(z_0)| + |f_N(z_0) - f_N(z)| + |f_N(z) - f(z)| \\ &< \frac{\varepsilon}{3} + \frac{\varepsilon}{3} + \frac{\varepsilon}{3} = \varepsilon,\end{aligned}$$

Now suppose we have a sequence (f_n) of continuous functions which converges uniformly on a contour C to the function f. Then the sequence $\left(\int_C f_n(z)dz\right)$ converges to $\int_C f(z)dz$. This is easy to see. Let $\varepsilon > 0$. Now let N be so that $|f_n(z) - f(z)| < \frac{\varepsilon}{A}$ for $n > N$, where A is the length of C. Then,

$$\left|\int_C f_n(z)dz - \int_C f(z)dz\right| = \left|\int_C (f_n(z) - f(z))dz\right|$$

$$< \frac{\varepsilon}{A} A = \varepsilon$$

whenever $n > N$.

Now suppose (f_n) is a sequence of functions each *analytic* on some region D, and suppose the sequence converges uniformly on D to the function f. Then f is analytic.

Example. Find the limit of the function $\{Z_N\} = \left\{\dfrac{\sqrt{n}+i(n+1)}{n}\right\}$.

Solution. We write $z_n = x_n + i\,y_n = \dfrac{1}{\sqrt{n}} + i\dfrac{n+1}{n}$

Using results concerning sequences of real numbers, we find that

$$\lim_{n\to\infty} x_n = \lim_{n\to\infty}\frac{1}{\sqrt{n}} = 0 \text{ and } \lim_{n\to\infty} y_n = \lim_{n\to\infty}\frac{n+1}{n} = 1.$$

Therefore $\lim\limits_{n\to\infty} z_n = \lim\limits_{n\to\infty}\dfrac{\sqrt{n}+i(n+1)}{n} = 0 + i = i.$

Example. Show that the sequence $\{z_n\} = \{(1+i)^n\}$ diverges.

Solution. We have

$$z_n = (1+i)^n = x_n + i\,y_n$$

$$z_n = \left(\sqrt{2}\right)^n \cos\frac{n\pi}{4} + i\left(\sqrt{2}\right)^n \sin\frac{n\pi}{4}$$

The real sequences $x_n = \left(\sqrt{2}\right)^n \cos\dfrac{n\pi}{4}$ and $y_n = \left(\sqrt{2}\right)^n \sin\dfrac{n\pi}{4}$ both exhibit divergent oscillations, so we conclude that $z_n = (1+i)n$ diverges.

SERIES OF COMPLEX NUMBER

Consider a sequence $\{z_n\}$, and form a new sequence $\{S_n\}$, called the sequence of partial sums, as follows.

$$S_1 = z_1,$$
$$S_1 = z_1 + z_2,$$
$$S_2 = z_1 + z_2 + z_3,$$
$$\vdots$$
$$S_n = z_1 + z_2 + \dots + z_n = \sum_{k=1}^{n} z_k,$$
$$\vdots$$

Infinite Series

The formal expression $\sum\limits_{k=1}^{\infty} z_k = z_1 + z_2 + \dots + z_n + \dots$ is called an infinite series, and $z_1, z_2, \dots, z_n, \dots$, are called the terms of the series.

If there is a complex number S for which

$$S = \lim_{n\to\infty} S_n = \lim_{n\to\infty}\sum_{k=1}^{n} z_k$$

we will say that the infinite series $\sum_{k=1}^{\infty} z_k$ converges to S, and that S is the sum of the infinite series. When this occurs, we write

$$S = \sum_{k=1}^{\infty} z_k$$

The series $\sum_{k=1}^{\infty} z_k$ is said to be absolutely convergent provided that the (real) series of magnitudes $\sum_{k=1}^{\infty} |z_k|$ converges.

If a series does not converge, we say that it diverges.

Remark: The first finitely many terms of a series do not affect its convergence or divergence and, in this respect, the beginning index of a series is irrelevant. Thus, we will without comment conclude that if a series $\sum_{k=n+1}^{\infty} z_k$ converges, then so does $\sum_{k=1}^{\infty} z_k$, where $z_1, z_2, ..., z_n$ is any finite collection of terms. A similar remark holds for determining divergence of a series.

We now give some more standard theorems for complex series, along with examples of how they are used.

Theorem: Let $z_n = x_n + i\, y_n$ and $S = U + i\, V$.

$$S = \sum_{n=1}^{\infty} z_n = \sum_{n=1}^{\infty}(x_n + i\, y_n) \text{ (converges)}$$

if and only if both

$$U = \sum_{n=1}^{\infty} x_n \text{ and } V = \sum_{n=1}^{\infty} y_n \text{ (converge).}$$

Theorem: *If* $\sum_{n=1}^{\infty} z_n$ is a convergent complex series, then $\lim_{n\to\infty} z_n = 0$.

Example. Show that the series

$$\sum_{n=1}^{\infty}\frac{1+in(-1)^n}{n^2} = \sum_{n=1}^{\infty}\left[\frac{1}{n^{\prime}} + i\frac{(-1)^n}{n}\right] \text{ is convergent.}$$

Solution. Recall that the real series

$\sum_{n=1}^{\infty}\frac{1}{n^{\prime}}$ and $\sum_{n=1}^{\infty}\frac{(1)^n}{n}$ are convergent. Hence, Theorem implies that the given series is convergent.

Example. Show that the series

$$\sum_{n=1}^{\infty}\frac{(-1)^n+i}{n}=\sum_{n=1}^{\infty}\left[\frac{(-1)^n}{n}+i\frac{1}{n}\right] \text{ is divergent.}$$

Solution. We know that the real series $\sum_{n=1}^{\infty}\frac{1}{n}$ is divergent. Hence, theorem 4 implies that the given series is divergent.

Example. Show that the series $\sum_{n=1}^{\infty}(1+i)^n$ is divergent.

Solution. Here we set $z_n = (1 + i)^n$ and observe that

$$\lim_{n\to\infty}|z_n| = \lim_{n\to\infty}|(1+i)^n| = \lim_{n\to\infty}\left(\sqrt{2}\right)^n = \infty$$

Thus $\lim_{n\to\infty} z_n \neq 0$, and theorem 5 implies that the series is not convergent; hence it is divergent.

Theorem. Let $\sum_{n=1}^{\infty} z_n$ and $\sum_{n=1}^{\infty} w_n$ be convergent series, and let c be a complex number. Then

$$\sum_{n=1}^{\infty} cZ_n = \sum_{n=1}^{\infty} c z_n = c\sum_{n=1}^{\infty} z_n$$

and

$$\sum_{n=1}^{\infty}(z_n+w_n) = \sum_{n=1}^{\infty} z_n + \sum_{n=1}^{\infty} w_n$$

Cauchy Product of Series

Let $\sum_{n=1}^{\infty} a_n$ and $\sum_{n=1}^{\infty} b_n$ be convergent series, where a_n and b_n are complex numbers. The Cauchy product of the two series is defined to be the series $\sum_{n=1}^{\infty} c_n$, where $c_n \sum_{k=1}^{n} a_k b_{n-k}$.

Theorem. If the Cauchy product converges, then

$$\sum_{n=1}^{\infty} c_n = \left(\sum_{n=1}^{\infty} a_n\right)\left(\sum_{n=1}^{\infty} b_n\right)$$

where,

$$c_n = \sum_{k=1}^{n} a_k b_{n-k}$$

Comparison Test

Let $\sum_{n=1}^{\infty} M_n$ be a convergent series of real nonnegative terms. If $\{z_n\}$ is a sequence of complex numbers and $|z_n| \le M_n$ holds for all n, then $\sum_{n=1}^{\infty} z_n$ converges.

Corollary. If $\sum_{n=1}^{\infty} | z_n |$, then $\sum_{n=1}^{\infty} z_n$ converges. In other words, absolute convergence implies convergence for complex series.

Example. Show that the series $\sum_{n=1}^{\infty} \frac{(3+4i)^n}{5^n n^t}$ is convergent.

Solution: We calculate $|z_n|$ $\left|\frac{(3+4i)^n}{5^n n^t}\right| = \frac{1}{n^t} = M_n$. Using the comparison test and the fact that $\sum_{n=1}^{\infty} \frac{1}{n^t}$ converges, we determine that $\sum_{n=1}^{\infty} \left|\frac{(3+4i)^n}{5^n n^t}\right|$ converges and hence, by Corollary .1, so does $\sum_{n=1}^{\infty} \frac{(3+4i)^n}{5^n n^t}$

GEOMETRIC SERIES AND CONVERGENCE THEOREMS

We begin this section by presenting a series of the form $\sum_{n=0}^{\infty} z^n$, which is called a geometric series.

Geometric Series

If $| z | < 1$, the series $\sum_{n=0}^{\infty} z^n$ converges to $f(z) = \frac{1}{1-z}$. That is, if $| z | < 1$ then

$$\sum_{n=0}^{\infty} z^n = 1 + z + z^2 + \dots + z^k + \dots = \frac{1}{1-z}.$$

If $| z | \ge 1$, the series diverges.

Corollary : If $| z | > 1$, the series $\sum_{n=1}^{\infty} z^{-n}$ converges to $f(z) = \frac{1}{1-z}$. That is, if $| z | > 1$ then

$$\sum_{n=1}^{\infty} z^{-n} = z^{-1} + z^{-2}. + \ldots + z^{-4} + \ldots = \frac{1}{z-1},$$

or equivalently,

$$-\sum_{n=1}^{\infty} z^{-n} = -z^{-1} - z^{-2} - \ldots - z^{-4} - \ldots = \frac{1}{1-z}.$$

If $| z | \le 1$, the series diverges.

Corollary. If $z \neq 1$, then for all n,

$$\frac{1}{1-z} = 1 + z + z^2 + \ldots + z^{n-1} + \frac{z^n}{1-z}.$$

Example. Show that $\sum_{n=0}^{\infty} \frac{(1-i)^n}{2^n} = 1 - i.$.

Solution. If we set $z = \frac{1-i}{2}$, then $| z | = \left|\frac{1-i}{2}\right| = \frac{\sqrt{2}}{2} < 1$. By using theorem 9, you get the sum is $1/\left(1-\frac{1-i}{2}\right) = \frac{2}{2-1+i} = \frac{2}{1+i} = 1-i$

Example. Evaluate $\sum_{n=3}^{\infty} \frac{i^n}{2^n}$.

Solution. We can put this expression in the form of a geometric series:

$$\begin{aligned}\sum_{n=3}^{\infty} \frac{i^n}{2n} &= \sum_{n=3}^{\infty} \left(\frac{i}{2}\right)^n = \sum_{n=3}^{\infty} \left(\frac{i}{2}\right)^3 \left(\frac{i}{2}\right)^{n-3} \\ &= \left(\frac{i}{2}\right)^3 \sum_{n=3}^{\infty} \left(\frac{i}{2}\right)^{n-3} = \left(\frac{i}{2}\right)^3 \frac{1}{1-\frac{i}{i}} \\ &= -\frac{i}{8}\frac{1}{1-\frac{i}{i}} = \frac{-i}{8-4i} \\ &= \frac{-i}{8-4i}\frac{8+4i}{8+4i} = \frac{4-8i}{80} \\ &= \frac{1}{20} - \frac{i}{10}\end{aligned}$$

***Remark*.** If we think of z as the "ratio'' by which a given term of the series is multiplied to generate successive terms, we see that the sum of a geometric series equals $\frac{\text{first term}}{1-\text{ratio}}$, provided $| \text{ratio} | < 1$.

The geometric series is used for determining the convergence or divergence of series. The proof is similar to the one used for real series, and we leave it for you to do.

D'Alembert's Ratio Test

If $\sum_{n=0}^{\infty} \zeta_n$ is a complex series with the property that

$$\lim_{n\to\infty} \frac{|\zeta_{n+1}|}{|\zeta_n|} = L,$$

then the series converges absolutely if L < 1 and diverges if L > 1.

Example. Show that $\sum_{n=0}^{\infty} \frac{(1-i)^n}{n!}$ converges.

Solution: Using the ratio test, we find that

$$\begin{aligned}
\lim_{n\to\infty} \left| \frac{(1-i)^{n+1}}{(n+1)!} \Big/ \frac{(1-i)^n}{n!} \right| &= \lim_{n\to\infty} \left| \frac{n!(1-i)}{(n+1)!} \right| \\
&= \lim_{n\to\infty} \left| \frac{n!(1-i)}{(n+1)!} \right| \\
&= \lim_{n\to\infty} \frac{n!|1-i|}{(n+1)!} \\
&= \lim_{n\to\infty} \frac{|1-i|}{n+1} \\
&= \lim_{n\to\infty} \frac{\sqrt{2}}{n+1} \\
&= 0 = L
\end{aligned}$$

Because $L < 1$, the series converges.

Example. Show that $\sum_{n=0}^{\infty} \frac{(z-i)^n}{2^n}$ converges for all z in the disk $|z-i| < 2$.

Solution: Using the ratio test, we find that

$$\lim_{n\to\infty} \left| \frac{(z-i)^{n+1}}{2^{n+1}} \Big/ \frac{(z-i)^n}{2^n} \right| = \lim_{n\to\infty} \left| \frac{z-i}{2} \right| = \frac{|z-i|}{2} = L$$

If $|z-i| < 2$, then $\frac{|z-i|}{2} = L < 1$, and the series converges.

If $|z-i| > 2$, then $L > 1$, and the series diverges.

Limit Supremum

Let $\{t_n\}$ be a sequence of positive real numbers. The limit supremum of the sequence (denoted by $\limsup_{x\to\infty} t_n$) is the smallest real number L with the property that for any $\in > 0$ there are at most finitely many terms in the sequence that are larger than L + $\in$. If there is no such number L, then we set $\limsup_{x\to\infty} t_n = \infty$.

Root Test[sh]

Suppose that the series $\sum_{n=0}^{\infty} \zeta_n$, has $\limsup_{n\to\infty}$ has $|\zeta_n|^{\frac{1}{n}} = L \left(\text{i.e} \limsup_{n\to\infty} \sqrt[n]{|\zeta_n|} = L\right)$

Then the series is absolutely convergent if $L < 1$ and divergent if $L < 1$.

Example. Find the radius of convergence of the infinite series $S(z) = 1 + \frac{z}{2} + \frac{z^2}{3^2} + \frac{z^3}{2^3} + \frac{z^4}{3^4} + \frac{z^5}{3^5} + \frac{z^6}{3^6} + \frac{z^7}{3^7} + \frac{z^8}{3^8} + \frac{z^9}{3^9} + \frac{z^{10}}{3^{10}} + \frac{z^{11}}{3^{11}} + \ldots$

Solution: The sequence of coefficients is $\left\{1, \frac{1}{2}, \frac{1}{3^2}, \frac{1}{2^3}, \frac{1}{3^4}, \frac{1}{2^5},\right.$

$\left.\frac{1}{2^7}, \frac{1}{3^8}, \frac{1}{2^9}, \frac{1}{3^{10}}, \frac{1}{2^{11}}, \ldots\right\}$.

The coefficients with even subscripts are $a_2k\frac{1}{3^2k}$, and the coefficients with odd subscripts are $a_{2k+1} = \frac{1}{2^{2k+1}}$.

So that we have

$$|a_{2k}|^{\frac{1}{2k}} = \left(\frac{1}{3^{2k}}\right)^{\frac{1}{2k}} = \frac{1}{3} \text{ and} |a_{2h+1}|^{\frac{1}{2k+1}} = \left(\frac{1}{2^{2k+1}}\right)^{\frac{1}{2k+1}} = \frac{1}{2}.$$

It follows that $\limsup_{n\to\infty} \sqrt[n]{|a_n z^n|} = \limsup_{n\to\infty} \left\{1, \frac{1}{2}, \frac{1}{3}, \frac{1}{2}, \frac{1}{3}, \frac{1}{2}, \frac{1}{3}, \ldots\right\} = \frac{1}{2}$.

Hence $\rho = \dfrac{1}{\limsup_{n\to\infty} |c_n|^{\frac{1}{n}}} = \dfrac{1}{\frac{1}{2}} = 2$

Hence, 2 is the required radius of convergence.

POWER SERIES OF COMPLEX NUMBER

In mathematics, a power series (in one variable) is an infinite series of the form

$$f(z) = \sum_{n=0}^{\infty} a_n (z-c)^n = a_0 + a_1(z-c) + a_2(z-c)^2 + a_3(z-c)^3 + \ldots$$

where a_n represents the coefficient of the nth term, c is a constant, and x varies around c (for this reason one sometimes speaks of the series as being *centered* at c). This series usually arises as the Taylor series of some known function.

In many situations c is equal to zero, for instance when considering a Maclaurin series. In such cases, the power series takes the simpler form

$$f(z) = \sum_{n=0}^{\infty} a_n z^n = a_0 + a_1 z + a_2 z^2 + a_3 z^3 + \ldots$$

Remember: Any polynomial can be easily expressed as a power series around any center c, albeit one with most coefficients equal to zero.

Suppose that we have a series $\sum_{n=0}^{\infty} \xi_n$ where $\xi_n = c_n(z - \alpha)^n$. If α and the collection of c_n are fixed complex numbers, we will get different series by selecting different values for z. For example, if $\alpha = 2$ and $c_n = \frac{1}{n!}$ for all n, we get the series $\sum_{n=0}^{\infty} \frac{1}{n!}\left(\frac{i}{2} - 2\right)^n$ if $z = \frac{i}{2}$, and $\sum_{n=0}^{\infty} \frac{1}{n!}(2 + i)^n$ if $z = 4 + i$. Note that when $\alpha = 0$ and $c_n = 1$ for all n, we get a geometric series. The collection of points for which the series $\sum_{n=0}^{\infty} c_n (z - \alpha)^n$ converges is the domain of a function $f(z) = \sum_{n=0}^{\infty} c_n (z - \alpha)^n$, which we call a power series function. The series $\sum_{n=0}^{\infty} c_n (z - \alpha)^n$ is really compact notation for $c_0 + \sum_{n=1}^{\infty} c_n (z - \alpha)^n$.

Power Series

The function $f(z) = \sum_{n=0}^{\infty} c_n (z - \alpha)^n$ is called a power series, with center $z = \alpha$.

Theorem. Suppose $f(z) = \sum_{n=0}^{\infty} c_n (z - \alpha)^n$. Then the set of points z for which the series converges is one of the following:

(i) The single point $z = \alpha$.

(ii) The disk $D_\rho(\alpha) = \{z: | - \alpha | < r\}$, along with part (either none, or some or all) of the circle $c_r(\alpha) = \{z: | z - \alpha| = \rho\}$.

(iii) The entire complex plane.

The power series $f(z) = \sum_{n=0}^{\infty} c_n (z - \alpha)^n$ converges if $| z - \alpha | < \rho$ and diverges if $| z - \alpha | > \rho$. We call the number ρ the radius of convergence of the power series. For case (i) of Theorem, we say that the radius of convergence is zero and that the radius of convergence is infinity for case (iii).

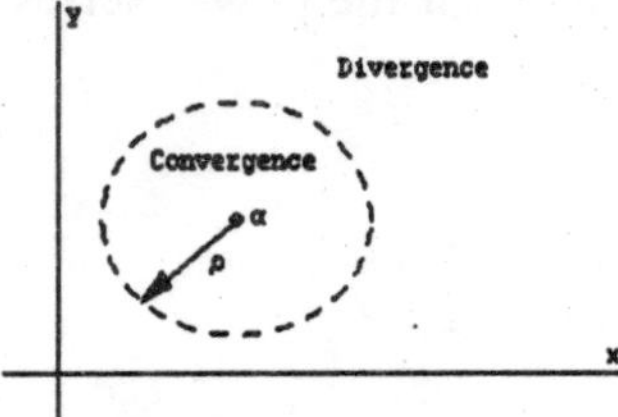

Fig. The Radius of Convergence of a Power Series. What happens on the boundary circle may be unknown.

Radius of Convergence

For the power series function $f(z) = \sum_{n=0}^{\infty} c_n (z-\alpha)^n$, we can find ρ, its radius of convergence, by any of the following methods:

(i) Cauchy's Root Test: $\rho = \dfrac{1}{\lim_{n\to\infty} |c_n|^{\frac{1}{n}}}$
(provided that the limit exists.)

(ii) Cauchy-Hadamard Formula: $\rho = \dfrac{1}{\limsup\limits_{n\to\infty} |c_n|^{\frac{1}{n}}}$
(the limit superior always exists.)

(iii) d'Alembert's Ratio Test: $\rho = \dfrac{1}{\lim_{n\to\infty} |\frac{c_n+1}{c_n}}$
(provided that the limit exists.)
We set $\rho = \infty$ if the limit equals 0, and $\rho = 0$ if the limit equals ∞.

OPERATIONS ON POWER SERIES

ADDITION AND SUBTRACTION

When two functions *f* and *g* are decomposed into power series around the same center *c*, the power series of the sum or difference of the functions can be obtained by termwise addition and subtraction. That is, if:

$$f(z) = \sum_{n=0}^{\infty} a_n z^n \text{ and } g(z) = \sum_{n=0}^{\infty} b_n z^n$$

then $(f \pm g)(z) = \sum_{n=0}^{\infty} (a_n \pm b_n) z^n$

MULTIPLICATION AND DIVISION

With the same definitions above, for the power series of the product and quotient of the functions can be obtained as follows:

$$f(z)(z)=\left[\sum_{n=0}^{\infty}a_n(z-c)^n\right]\left[\sum_{n=0}^{\infty}b_n(z-c)^n\right]$$

$$=\sum_{i=0}^{\infty}\sum_{j=0}^{\infty}a_n b_n(z-c)^{i+j}$$

$$=\sum_{i=0}^{\infty}\left[\sum_{j=0}^{n}(a_n b_n-i)(z-c)^n\right]$$

The sequence $m_n=\sum_{i=0}^{n}a_i b_n-i$ is known as the convolution of the sequence a_n and b_n.

For division, observe:

$$\frac{f(z)}{g(z)}=\frac{\sum_{n=0}^{\infty}a_n(z-c)^n}{\sum_{n=0}^{\infty}b_n(z-c)^n}=\sum_{j=0}^{n}d_n(z-c)^n$$

$$f(z)=\left[\sum_{n=0}^{\infty}a_n(z-c)^n\right]\left[\sum_{n=0}^{\infty}d_n(z-c)^n\right].$$

and then use the above, comparing coefficients.

DIFFERENTIATION AND INTEGRATION

Once a function is given as a power series, it is continuous wherever it converges and is differentiable on the interior of this set. It can be differentiated and integrated quite easily, by treating every term separately:

$$f'(z)=\sum_{n=1}^{\infty}a_n(z-c)^{n-1}$$

$$\int f(z)=\sum_{n=0}^{\infty}\frac{a_n(z-c)^{n+1}}{n+1}+C$$

Both of these series have the same radius of convergence as the original one.

ANALYTIC FUNCTIONS

A function f defined on some open subset U of R or C is called analytic if it is locally given by power series. This means that every $a–U$ has an open neighborhood $V–U$, such that there exists a power series with center a which converges to $f(x)$ for every $x –V$.

Every power series with a positive radius of convergence is analytic on the interior of its region of convergence. All holomorphic functions are complex-analytic. Sums and products of analytic functions are analytic, as are quotients as long as the denominator is non-zero.

If a function is analytic, then it is infinitely often differentiable, but in the real case the converse is not generally true. For an analytic function, the coefficients a_n can be computed as

$$a_n = \frac{f^{(n)}(c)}{n!}$$

where $f^{(n)}(c)$ denotes the nth derivative of f at c, and $f^{(0)}(c) = f(c)$. This means that every analytic function is locally represented by its Taylor series.

The power series expansion of the inverse function of an analytic function can be determined using the Lagrange inversion theorem.

Example. The infinite series $f(z) = \sum_{n=0}^{\infty}\left(\frac{n+2}{3n+1}\right)^n (z-4)^n$ has radius of convergence $\rho = 3$ Cauchy's root test because

$$\lim_{n\to\infty} |c_n|^{\frac{1}{n}} = \lim_{n\to\infty}\left(\left(\frac{n+2}{3n+1}\right)^n\right)^{\frac{1}{n}} = \lim_{n\to\infty}\frac{n+2}{3n+1} = \frac{1}{3},$$

Hence $$\rho = \frac{1}{\lim_{n\to\infty} |c_n|^{\frac{1}{n}}} = \frac{1}{\frac{1}{3}} = 3.$$

Example. By using the Cauchy-Hadamard formula we get known that the infinite series

$f(z) = \sum_{n=1}^{\infty} c_n z^n = 4z + 5^2 z^2 + 4^3 z^3 + 5^4 z^4 + 4^5 z^5 + 5^6 z^6 + \ldots$ has radius of convergence $\rho = \frac{1}{5}$. Also using property of supremum, you get

$$\left\{|c_n|^{\frac{1}{n}}\right\} = \{4, 5, 4, 5, 4, 5, \ldots\}, \text{ so}$$

$$\limsup_{n\to\infty} |c_n|^{\frac{1}{n}} = \limsup_{n\to\infty} \{4, 5, 4, 5, 5, \ldots\} = 5,$$

Hence $$\rho = \frac{1}{\limsup_{n\to\infty} |c_n|^{\frac{1}{n}}} = \frac{1}{5}.$$

Example. The infinite series $f(z) = \sum_{n=0}^{\infty} \frac{z^n}{n!}$ has radius of convergence ρ = 8 by the ratio test because

$$\lim_{n\to\infty}\left|\frac{c_n+1}{c_n}\right| = \lim_{n\to\infty}\left|\frac{n!}{(n+1)!}\right| = \lim_{n\to\infty}\frac{1}{n+1} = 0$$

Since the limit equals 0, we set ρ = 8.

Example. Find the radius of convergence of the infinite series.

$$S(z) = 1+\frac{z}{2}+\frac{z^2}{3^2}+\frac{z^3}{2^3}+\frac{z^4}{3^4}+\frac{z^5}{2^5}+\frac{z^6}{3^6}+\frac{z^7}{2^7}+\frac{z^8}{3^8}+\frac{z^9}{2^9}+\frac{z^{10}}{3^{10}}+\frac{z^{11}}{2^{11}}+\ldots$$

Solution: The sequence of coefficients is

$$\left\{1,\frac{1}{2},\frac{1}{3^2},\frac{1}{2^3},\frac{1}{3^4},\frac{1}{2^5},\frac{1}{3^6},\frac{1}{2^7},\frac{1}{3^8},\frac{1}{2^9},\frac{1}{3^{10}},\frac{1}{2^{11}},\ldots\right\}$$

The coefficients with even subscripts are $a_{2k} = \frac{1}{3^{2k}}$, and the coefficients with odd subscripts are $a_{2k+1} = \frac{1}{2^{2k+1}}$.

So that we have

$$|a_{2k}|^{\frac{1}{2k}} = \left(\frac{1}{3^{2k}}\right)^{\frac{1}{2k}} = \frac{1}{3} \text{ and } |a_{2k+1}|^{\frac{1}{2_{k+1}}} = \left(\frac{1}{2^{2k+1}}\right)^{\frac{1}{2k+1}} = \frac{1}{2}$$

It follows that

$$\limsup_{n\to\infty}\sqrt[n]{|a_n z^n|} = \limsup_{n\to\infty}\left\{1,\frac{1}{2},\frac{1}{3},\frac{1}{2},\frac{1}{3},\frac{1}{2},\frac{1}{3},\ldots\right\} = \frac{1}{2}$$

Hence $\rho = \dfrac{1}{\limsup_{n\to\infty}|c_n|^{\frac{1}{n}}} = \dfrac{1}{\frac{1}{2}} = 2$.

Hence, the required radius of convergence is 2.

Theorem. Suppose the function $f(z) = \sum_{n=0}^{\infty} c_n(z-\alpha)^n$ has radius of convergence ρ > 0. Then

(i) $f(z)$ is infinitely differentiable for all $z \in D\rho$ (α). In fact

(ii) for all k, $f(h)$ $(z) = \sum_{n=0}^{\infty} n(n-1)\ldots(n-k+1)c_n(z-\alpha)^{n-k}$; and

(iii) $c_k = \dfrac{f^{(k)}(\alpha)}{k!}$

where $f(k)$ (α) denotes the Kth derivative of f. (When $k = 0, f^{(k)}$ denotes the function f itself so that $f^{(0)}(z) = f(z)$ for all z.)

UNIFORM CONVERGENCE OF COMPLEX NUMBER

Here, we begin by discussing an important property associated with power series-uniform convergence.

Recall that, for a function f(z) defined on a set T, the sequence of functions $\{S_n(z)\}$ converges to the function $f(z)$ at the point $z_0 \in T$ provided that $\lim_{n\to\infty} S(z_0) = f(z_0)$. Thus, for the particular point z_0, we know that for each $\in > 0$, there exists a positive integer $N_{\in,z_0}$ (which depends on both $\in$ and z_0) such that

if $n \geq N_{\in,z_0}$ then $|S(z_0) - f(z_0)| < \in$

If $S_n(z)$ is the nth partial sum of the series $\sum_{k=0}^{\infty} c_k (z-\alpha)^k$, Statement becomes

If $n \geq N_{\in,z_0}$, then $\left|\sum_{k=0}^{n-1} c_k (z_0-\alpha)^k - f(z_0)\right| < \in$

For a given value of $\in$, the integer $N_{\in,z_0}$ needed to satisfy Statement often depends on our choice of z_0. This is not the case if the sequence $\{S_n(z)\}$ converges uniformly. For a uniformly convergent sequence, it is possible to find an integer $N_\in$ (depending only on $\in$) that guarantees Statement no matter what value for $z_0 \in T$ we pick. In other words, if n is large enough, the function $S_n(z)$ is uniformly close to the function $f(z)$ for all $z \in T$.

UNIFORM CONVERGENCE

The sequence $\{S_n(z)\}$ converges uniformly to $f(z)$ on the set T if for every $\in > 0$, there exists a positive integer $N_\in$ (Depending only on $\in$) such that

if $n \geq N$, then $| S_n(z) - f(z) < \in$ for all $z \in T$.

If $S_n(z)$ is the nth partial sum of the series $\sum_{k=0}^{\infty} c_k (z-\alpha)^k$, we say that the series $\sum_{k=0}^{\infty} c_k (z-\alpha)^k$ converges uniformly to $f(z)$ on the set T.

Example. The sequence $\{S_n(z)\} = \left\{e^z + \frac{1}{n}\right\}$ converges uniformly to the function $f(z) = e^z$ on the entire complex plane because for any $\in > 0$, statement is satisfied for all z for $n \geq N_\in$, where $N_\in$ is any integer greater than $\frac{1}{\in}$.

The given sequence $S_n(z)$ diverges, if there exists $\in > 0$, such that for all positive integers N, there is some $n \geq N$ and some $z_0 \in T$

such that $| S_n(s_0) - f(z_0) | \geq \in$.

The geometric series do not converge uniformly to $f(z) = \frac{1}{1-z}$ for $z \in D_1(0)$, which is shown in the figure below:

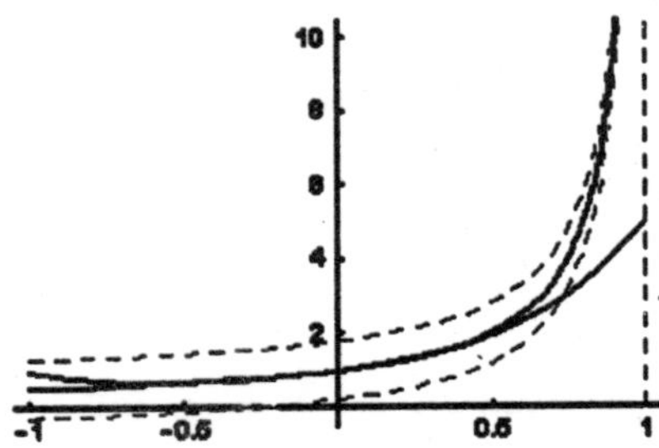

Fig. The Geometric Series does not Converge Uniformly on (-1,1).

The Weierstrass M-test can be used to determine whether an infinite series is uniformly convergent or not .

WEIERSTRASS M-TEST

Suppose the infinite series $\sum_{k=0}^{\infty} u_k(z)$ has the property that for each k, we have $| u_k(z) | \leq M_k$ for all $z \in T$. If $\sum_{k=0}^{\infty} M_k$ converges, then $\sum_{k=0}^{\infty} u_k(z)$ converges uniformly on T.

Theorem. Suppose the power series $\sum_{k=0}^{\infty} c_k (z-\alpha)^k$ has radius of convergence ρ > 0. Then for each *r*, (where $0 < r < \rho$), the series converges uniformly on the closed disk $\overline{D_r}(\alpha) = \{z : | z - \alpha | \leq r\}$.

Corollary. For each r, (where 0 < r < 1), the geometric series $\sum_{n=0}^{\infty} z^n$ converges uniformly on the closed disk $\overline{D_r}(0) = \{z : | z | \leq r\}$.

Theorem. Suppose $\{z_k(z)\}$ is a sequence of continuous functions defined on a set T containing the contour C.

If $\{S_k(z)\}$ converges uniformly to $f(z)$ on the set T, then

(i) $f(z)$ is continuous on T, and (ii) $\lim_{k\to\infty}\int_c S_k(z)dz = \int_c \lim_{k\to\infty} S_k(z)dz = \int_c f(z)dz.$

Corollary. If the series $\sum_{n=0}^{\infty} c_n(z-\alpha)^n$ converges uniformly to $f(z)$ on the set T, and C is a contour contained in T, then $\sum_{n=0}^{\infty}\int_c c_n(z-\alpha)^n dz = \int_c \sum_{n=0}^{\infty} c_n(z-\alpha)^n dz = \int_c f(z)dz$.

TAYLOR AND LAURENT SERIES

TAYLOR SERIES

Definition: If $f(z)$ is analytic at $z = \alpha$, then the series

$$f(\alpha)+f'(\alpha)(z-\alpha)+\frac{f^{(2)}(\alpha)}{2!}(z-\alpha)^2+\frac{f^{(2)}(\alpha)}{3!}(z-\alpha)^3+\ldots=\sum_{k=0}^{\infty}\frac{f^{(k)}(\alpha)}{k!}(z-\alpha)^k$$

is called the Taylor series for $f(z)$ centered at $z = \alpha$. When the center is $\alpha = 0$, the series is called the Maclaurin series for $f(z)$.

Definition: Suppose f is analytic on the open disk :$|z - z_0| < r$. Let z be any point in this disk and .choose C to be the positively oriented circle of radius r, where $|z - z_0| < \rho < r$. Then for $s \in C$we have

$$\frac{1}{s-z} = \frac{1}{(s-z_0)(z-z_0)} = \frac{1}{s-z_0}\left[\frac{1}{1-\frac{z-z_0}{s-z_0}}\right] = \sum_{j=0}^{\infty}\frac{(z-z_0)^j}{(s-z_0)^{j+1}}$$

since $\left|\frac{z-z_0}{s-z_0}\right| < 1$ The convergence is uniform, so we may integrate

$$\int_C \frac{f(s)}{s-z}ds = \sum_{j-0}^{\infty}\left(\int_C \frac{f(s)}{(s-z_0)^{j+1}}ds\right)(z-z_0)^j, \text{ or}$$

$$f(z)=\frac{1}{2\pi i}\int_C \frac{f(s)}{s-z}ds = \sum_{j-0}^{\infty}\left(\frac{1}{2\pi i}\int_C \frac{f(s)}{(s-z_0)^{j+i}}ds\right)(z-z_0)^j.$$

We have thus produced a power series having the given analytic function as a limit:

$$f(z) = \sum_{j-0}^{\infty} c_j(z-z_0)^j, |z-z_0| < r,$$

where

$$C_j = \frac{1}{2\pi i}\int_C \frac{f(s)}{(s-z_0)^{j+1}}ds.$$

We know we may differentiate the series to get

$$f(z) = \sum_{j-1}^{\infty} jc_j(z-z_0)^{j-1}$$

and this one converges uniformly where the series for f does. We can thus differentiate again and again to obtain

$$f^{(n)}(z) = \sum_{j-n}^{\infty} j(j-1)(j-2)....(j-n+1)_{cj}(z-z_0)^{j-n}.$$

Hence,

$$f^{(n)}(z_0) = n!c_n, \text{ or}$$

$$c_n = \frac{f^{(n)}(z_0)}{n!}.$$

But we also know that

$$c_n = \frac{1}{2\pi i}\int_C \frac{f(s)}{(s-z_0)^{n+1}}ds.$$

This gives us

$$f^{(n)}(z_0) = \frac{n!}{2\pi i}\int_C \frac{f(s)}{(s-z_0)^{n+1}}ds, \text{ for } n = 0, 1, 2, ...$$

which is the Cauchy Integral Formula. Suppose we have

$$f(z) = \sum_{j-0}^{\infty} c_j(z-z_0)^j,$$

and the radius of convergence is R. Then we know, of course, that the limit function f is analytic for $|z - z_0| < R$.

Lemma. If z, z_0 and α are complex numbers with $z \neq z_0$, and $z \neq \alpha$, then

$$\frac{1}{z-z_0} = \frac{1}{z-\alpha} + \frac{z_0-\alpha}{(z-\alpha)^2} + \frac{(z_0-\alpha)^2}{(z-\alpha)^3} + ... + \frac{(z_0-\alpha)^n}{(z-\alpha)^{n+1}} + \frac{1}{z-z_0}\frac{(z_0-\alpha)^{n+1}}{(z-\alpha)^{n+1}}$$

where n is a positive integer.

TAYLOR'S THEOREM

Suppose $f(z)$ is analytic in a domain G, and that $D_R(\alpha) = \{z : |z - \alpha| < R\}$ is any disk contained in G. Then the Taylor series for $f(z)$ converges to $f(z)$ for all z in $D_R(\alpha)$; that is,

$$f(z) = \sum_{k=0}^{\infty} \frac{f^{(k)}(\alpha)}{k!}(z-\alpha)^k \text{ for all } Z \in DR(\alpha).$$

Furthermore, for any r, $0 < r < R$, the convergence is uniform on the closed subdisk $\overline{Dr}$ $(\alpha) = \{z : |z - \alpha| \leq r\}$ for $0 < r < R$.

A singular point of a function is a point at which the function fails to be analytic. In other words a non-removable singular point of a function has the property that it is impossible to redefine the value of the function at that point so as to make it analytic there. For example, the function $f(z) = \dfrac{1}{1-z}$ has a nonremovable singularity at $z = 1$.

Corollary. Suppose that $f(z)$ is analytic in the domain G that contains the point $z = \alpha$. Let z_0 be a nonremovable singular point of minimum distance to the point $z = \alpha$. If $|z_0 - \alpha| = R$, then

(i) the Taylor series $\sum_{n=0}^{\infty} \frac{f^{(n)}(\alpha)}{n!}(z-\alpha)^n$ converges to $f(z)$ on all of $D_R(\alpha)$, and

(ii) if $|z_1 - \alpha| = S > R$, the series $\sum_{n=0}^{\infty} \frac{f^{(n)}(\alpha)}{n!}(z_1-\alpha)^n$ does not converges to $f(z_1)$.

Remark. In the following figure you observe that their interval of convergence is the open interval $(-1, 1)$.

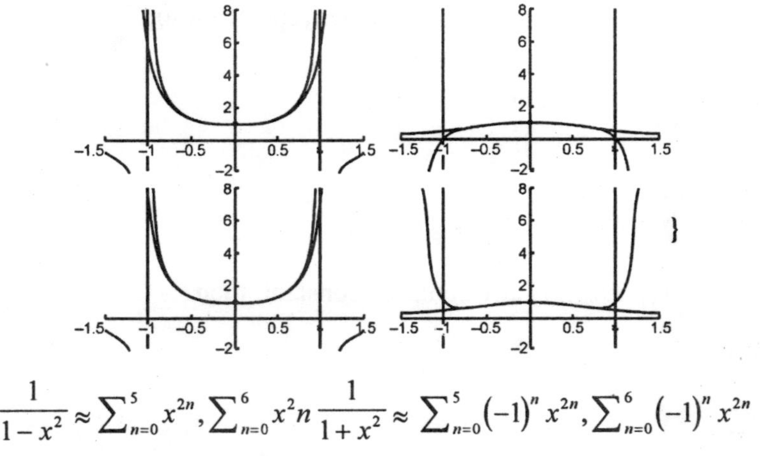

$$\frac{1}{1-x^2} \approx \sum_{n=0}^{5} x^{2n}, \sum_{n=0}^{6} x^2 n \quad \frac{1}{1+x^2} \approx \sum_{n=0}^{5} (-1)^n x^{2n}, \sum_{n=0}^{6} (-1)^n x^{2n}$$

Fig. The Maclaurin series $\sum_{n=0}^{\infty} (-1)^n x^{2n}$ does not Converge Uniformly on $(-1, 1)$.

Uniqueness of Power Series

Suppose that in some disk $D_r(a)$ we have

$$f(z) = \sum_{n=0}^{\infty} a_n (z-\alpha)^n = \sum_{n=0}^{\infty} b_n (z-\alpha)^n$$

Then $a_n = b_n$ for $n = 0, 1, 2, \ldots$

Example. Find the Maclaurin series for $f(z) = \sin^3 z$.

Solution: Computing derivatives for $f(z)$ would be an onerous task. Fortunately, we can make use of the trigonometric identity

$$\sin^3 z = \frac{3}{4}\sin z - \frac{1}{4}\sin 3z.$$

Recall that the series for sin z (valid for all z) is $\sin z = \sum_{n=0}^{\infty} \frac{(-1)^n}{(2n+1)!} z^{2n+1}$. Using the identity for $\sin^3 z$, we obtain

$$\sin^3 z = \frac{3}{4}\sum_{n=0}^{\infty} \frac{(-1)^n}{(2n+1)!} z^{2n+1} - \frac{1}{4}\sum_{N=0}^{\infty} \frac{(-1)^n}{(2n+1)!} (3z)^{2n+1}$$

$$= \frac{3}{4}\sum_{n=0}^{\infty} \frac{(-1)^n}{(2n+1)!} z^{2n+1} - \frac{3}{4}\sum_{n=0}^{\infty} \frac{(-1)^n g^n}{(2n+1)!} z^{2n+1}$$

$$= \sum_{n=0}^{\infty} (-1)^n \frac{3(1-g^n)}{4(2n+1)!} z^{2n+1}$$

By the uniqueness of power series, this last expression is the Maclaurin series for $\sin^3 z$.

Theorem. Let f(z) and g(z) have the power series representations

$$f(z) = \sum_{n=0}^{\infty} (z-\alpha)^n \text{ for } z \in D_{r1}(\alpha),$$

and

$$g(z) = \sum_{n=1}^{\infty} b_n (z-\alpha)^n \text{ for } z \in D_{r2}(\alpha)$$

If $r = \min\{r_1, r_2\}$ and β is any complex constant, then

$$\beta f(z) = \sum_{n=0}^{\infty} \beta a_n (z-\alpha)^n \text{ for } z \in D_{r1}(\alpha),$$

$$f(z) + g(z) = \sum_{n=0}^{\infty} (a_n + b_n)(z-\alpha)^n \text{ for } z \in D_r(\alpha), \text{and}$$

$$f(z)+g(z) = \sum_{n=0}^{\infty} c_n (z-\alpha)^n \text{ for } z \in D_r(\alpha), \text{ where}$$

$$c_n = \sum_{k=0}^{n} a_n b_{n-k}.$$

Example. Use the Cauchy product of series to show that

$$\frac{1}{(1-z)^2} = \sum_{n=0}^{\infty}(n+1)z^n \text{ for } z \in D_1(0).$$

Solution. We let $f(z) = g(z) = \frac{1}{1-z}$, for $z \in D_1(0)$. In terms of Theorem 20, we have $a_n = b_n = 1$, for all n, and thus Equation (52) gives

$$\frac{1}{(1-z)^2} = \text{h}(z) = f(z)\, g(z) = \sum_{n=0}^{\infty}\left(\sum_{k=0}^{n} a_k b_{n-k}\right) z^n$$

$$= \sum_{n=0}^{\infty}\left(\sum_{k=0}^{n} 1\right) z^n$$

$$= \sum_{n=0}^{\infty}(n+1)z^n$$

Example. Use the Cauchy product of series to show that

$$\frac{z}{(z-1)^4} = \sum_{n=0}^{\infty}\frac{n(1+n)(2+n)}{}z^n \text{ for } z \in D_1(0).$$

Solution: Use the following facts $\frac{z}{(z-1)^2} = \sum_{n=0}^{\infty} nz^n$ for $z \in D_1(0)$ and $\frac{z}{(z-1)^2} = \sum_{n=0}^{\infty}(n+1)z^n$ for $z \in D_1(0)$.

Take the coefficients $a_n = n$ and b_n and $b_n = n+1$ and let carry out the calculations.

$$a_n = n$$

$$bn = 1 + n$$

$$f[z] = \sum_{n=0}^{\infty} a_n z^n = \sum_{n=0}^{\infty}(k)z^n = \frac{z}{(-1+z)^2}$$

$$g[z] = \sum_{n=0}^{\infty} a_n z^n = \sum_{n=0}^{\infty}(1-k+n)z^n = \frac{1}{(-1+z)^2}$$

$$c_n = \sum_{k=0}^{n} a_k b_{n-k}$$

$$C_n = \sum_{k=0}^{n} (k)(1-k+n) = \sum_{k=0}^{n} (k(1-k+n))$$

$$\sum_{k=0}^{n} (k(1-k+n)) = \frac{1}{6} n(1+n)(2+n)$$

$$c_n = \frac{1}{6} n(1+n)(2+n)$$

$$h[z] = \sum_{n=0}^{\infty} \left(\frac{1}{6} n(1+n)(2+n) z^n\right) = \frac{z}{(-1+z)^4}$$

Example. Show that $\frac{z}{(1-z)^2} = \sum_{n=1}^{\infty} nz^n$ for $z \in D_1(0)$

Solution. Use the result of Example and equation (48) , we obtain

$$\frac{z}{(1-z)^2} = \frac{z-1+1}{(1-z)^2}$$

$$= \frac{1}{(1-z)^2} + \frac{-1}{(1-z)}$$

$$= \sum_{n=0}^{\infty} (n+1) z^n - \sum_{n=0}^{\infty} z^n = \sum_{n=0}^{\infty} nz^n$$

We could calculate the sum of the series, $\sum_{n=1}^{\infty} nz^n = \frac{z}{(-1+z)^2}$.

LAURENT SERIES REPRESENTATIONS

Suppose $f(z)$ is not analytic in $D_R(\alpha)$, but is analytic in $D_R^*(\alpha) = \{z: 0 < |z - \alpha| < R\}$.For example, the function $f(z) = \frac{1}{z^3} e^z$ is not analytic when $z = 0$ but is analytic for $|z| > 0$. Clearly, this function does not have a Maclaurin series representation. If we use the Maclaurin series for $g(z) = e^z$, however, and formally divide each term in that series by z^3, we obtain the representation

$$f(z) = \frac{1}{z^3} e^3 = \frac{1}{z^3} + \frac{1}{z^2} + \frac{1}{2!z} + \frac{1}{3!} + \frac{z}{4!} + \frac{z^2}{5!} + \frac{z^3}{5!} + \ldots$$

that is valid for all z such that $|z| > 0$.

LAURENT SERIES

Let C_n be a complex number for $n = 0, \pm 1, \pm 2, \pm 3, \ldots..$ The doubly infinite series $\sum_{n=-\infty}^{\infty} c_n (z-\alpha)^n$, called a Laurent series, is defined by

$$\sum_{n=-\infty}^{\infty} c_n (z-\alpha)^n = \sum_{n=1}^{\infty} c_{-n} (z-\alpha)^{-n} + \sum_{n=0}^{\infty} c_n (z-\alpha)^n,$$

provided the series on the right-hand side of this equation converge.

Example. Find the series for $f(z) = \frac{e^z}{z^3}$

Solution: Enter the function $f(z) = \frac{e^z}{z^3}$ and find the first few terms of the series in the punctured plane $0 < z$.

$$f[z] = \frac{e^z}{z^3}$$

$$S[z] = \left(1 + z + \frac{z^2}{2} + \frac{z^3}{6} + \frac{z^4}{24} + \frac{z^5}{120} + \frac{z^6}{720} + \frac{z^7}{5040} + \frac{z^8}{40320} + \frac{z^9}{362880} + \ldots\right) / \left(z^3\right)$$

$$L[z] = \frac{1}{6} + \frac{1}{z^3} + \frac{1}{z^2} + \frac{1}{2z} + \frac{z}{24} + \frac{z^2}{120} + \frac{z^3}{720} + \frac{z^4}{5040} + \frac{z^5}{40320} + \frac{z^6}{362880} + \ldots$$

The latter is a series is valid for $0 < |z|$. Sum up infinitely many terms to verify that we have things right.

$$S_g[z] = \frac{1}{6} + \frac{1}{z^3} + \frac{1}{2^2} + \frac{1}{2z} + \frac{z}{24} + \frac{z^2}{120} + \frac{z^3}{720} + \frac{z^4}{5040} + \frac{z^5}{40320} + \frac{z^6}{362880}$$

$$S_\infty[z] = \sum_{n=0}^{\infty} \frac{z^{n-3}}{n!} = \frac{e^z}{z^3}$$

Remark. Recall that $\sum_{n=0}^{\infty} c_n (z-\alpha)^n$ is a simplified expression for the sum $c_{n0} + \sum_{n=1}^{\infty} c_n (z-\alpha)^n$.

At times it will be convenient to write $\sum_{n=-\infty}^{\infty} c_n (z-\alpha)^n$ as

$$\sum_{n=-\infty}^{\infty} c_n (z-\alpha)^n = \sum_{n=-\infty}^{-1} c_n (z-\alpha)^n + \sum_{n=0}^{\infty} c_n (z-\alpha)^n,$$

ANNALUS

Given $0 \leq r \leq R$, we define the annulus centered at α with radii r and R by.

$A(a, r, R) = \{z : r < |z - a| < R\}$.

The closed annulus centered at with radii r and R is denoted by.

$\bar{A}(\alpha, r, R) = \{z : r \leq |z - \alpha| \leq R\}$.

Figure illustrates these terms.

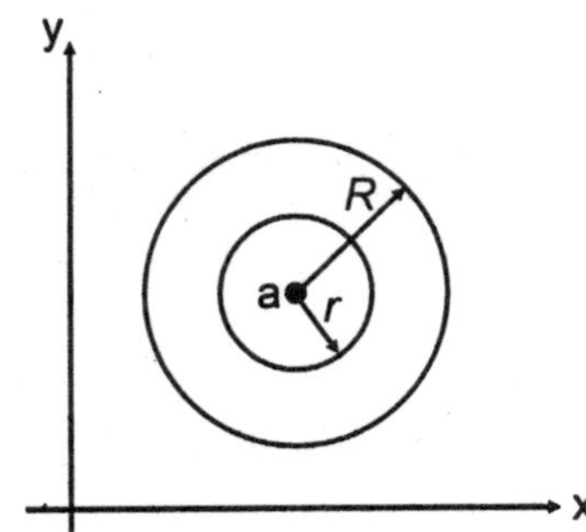

Fig. The closed annulus $\bar{A}$ (*a*, *r*, R).The shaded portion is the open annulus *A*(*a*, *r*, R).

Theorem. Suppose that the Laurent series $\sum_{n=-\infty}^{\infty} c_n (z - \alpha)^n$ converges on an annulus $A(\alpha, r, R) = \{z : r < |z - \alpha| < R\}$. Then the series converges uniformly on any closed subannulus $\bar{A}(\alpha, s, t) = \{z : s \leq |z - \alpha| \leq t\}$ where $r < s < t < R$.

In Laurent series, we will use symbols of the form $C_\rho^+(\alpha)$, which we remind you to designate the positively oriented circle with radius ρ and center α. That is, $C_\rho^+(\alpha) = \{z : |z - \alpha| < \rho\}$, oriented counterclockwise.

Laurent's Theorem

Suppose $0 \leq r \leq R$, and that $f(z)$ is analytic in the annulus $A = A(\alpha, r, R) = \{z : r < |z - \alpha| < R\}$. If ρ is any number such that $r < \rho < R$, then for all $z_0 \in A(\alpha, r, R)$ the function value $f(z_0)$ has the Laurent series representation

$$f(z_0) = \sum_{n=-\infty}^{\infty} c_n (z_0 - \alpha)^n = \sum_{n=1}^{\infty} c_{-n} (z_0 - \alpha)^{-n} + \sum_{n=0}^{\infty} c_n (z_0 - \alpha)^n,$$

where for $n = 0, 1, 2, \ldots$, the coefficients c_{-n} and c_n are given by

$$c_{-n} = \frac{1}{2\pi i} \int_{C_\rho^+(\alpha)} \frac{f(z)}{(z - \alpha)^{-n+1}} dz \text{ and}$$

$$c_n = \frac{1}{2\pi i} \int_{C_\rho^+(\alpha)} \frac{f(z)}{(z - \alpha)^{n+1}} dz.$$

Remark. If $f(z)$ is analytic in the disk $D_R(\alpha)$ then the coefficient C_n for the positive power $(z_0 - \alpha)^n$ equals $\frac{f(n)(z_0)}{n!}$ by using Cauchy's integral formula for derivatives. Hence, the series in equation (30) involving the positive powers of $(z_0 - \alpha)$ is actually the Taylor series for $f(z)$.

Uniqueness and Differentiation of Laurent Series

Suppose that $f(z)$ is analytic in the annulus $A(\alpha, r, R)$, and has the Laurent series representation

$$f(z) = \sum_{n=-\infty}^{\infty} c_n (z-\alpha)^n \text{ for all } z \in A(\alpha, r, R).$$

(i) If $f(z) = \sum_{n=-\infty}^{\infty} b_n (z-\alpha)^n$ for all $z \in A(\alpha, r, R)$, then $b_n = c_n$ for all n. (In other words, the Laurent series for $f(z)$ in a given annulus is unique.)

(ii) For all $z \in A(\alpha, r, R)$, the derivatives for $f(z)$ may be obtained by termwise differentiation of its Laurent series.

Example. Find three different Laurent series representations for the function $f(z) = \frac{3}{2+z-z^2}$ involving powers of z.

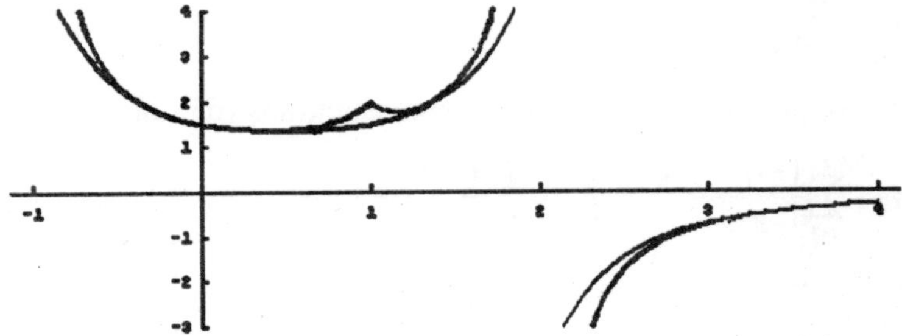

Solution: The function $f(z)$ has singularities at $z = -1, 2$ and is analytic in the disk $D : |z| < 1$, in the annulus $A : 1 < |z| < 2$, and in the region $R : 1\ |z| > 2$. We want to find a different Laurent series for $f(z)$ in each of the three domains D, A, and R. We start by writing $f(z)$ in its partial fraction form:

$$f(z) = \frac{3}{(1+z)(2-z)} = \frac{1}{1+z} + \frac{1}{2-z} = \frac{1}{1+z} + \frac{1}{2}\frac{1}{1-\frac{z}{2}}$$

We use previous Theorem and Corollary to obtain the following representations for the terms on the right side of Equation (33):

$$\frac{1}{1+z} = \sum_{n=0}^{\infty} (-1)^n z^n$$

valid for $|z| < 1$,

$$\frac{1}{1+z} = \sum_{n=1}^{\infty} \frac{(-1)^{n+1}}{z^n}$$

valid for $|z| > 1$,

$$\frac{1}{2}\frac{1}{1-\frac{z}{2}} = \sum_{n=0}^{\infty} \frac{z^n}{2^{n+1}}$$

valid for $|z| < 2$,

$$\frac{1}{2}\frac{1}{1-\frac{z}{2}} = \sum_{n=1}^{\infty} \frac{-2^{n-1}}{z^n}$$

valid for $|z| > 2$.

In the disk $D : |z| <1$, and thus we have

$$f(z) = \sum_{n=0}^{\infty} (-1)^n z^n + \sum_{n=0}^{\infty} \frac{z^n}{z^{n+1}} = \sum_{n=0}^{\infty} \left((-1)^n + \frac{1}{2^{n+1}}\right) z^n$$

valid for $|z| < 1$, which is a Laurent series that reduces to a Maclaurin series.

In the annulus $A : 1 < |z| < 2$, representations (65) and (66) are valid; hence we get

$$f(z) = \sum_{n=1}^{\infty} \frac{(-1)^{n+1}}{z^n} + \sum_{n=0}^{\infty} \frac{z^n}{2^{n+1}}$$

valid for $1 < |z| < 2$.

Finally, in the region $R : |z| > 2$ we use Representations (65) and (67) to obtain6

$$f(z) = \sum_{n=1}^{\infty} \frac{(-1)^{n+1}}{z^n} + \sum_{n=1}^{\infty} \frac{-2^{n-1}}{z^n} = \sum_{n=1}^{\infty} \frac{(-1)^{n+1} - 2^{n-1}}{z^n}$$

valid for $|z| > 2$.

EXERCISE

1. Give a sequence (f_n) of functions continuous on a set D with a limit that is not continuous.
2. Prove that a sequence cannot have more than one limit.
3. Give an example of a sequence that does not have a limit, or explain carefully why there is no such sequence.
4. Give an example of a bounded sequence that does not have a limit, or explain carefully why there is no such sequence.
5. Give a sequence of real functions differentiable on an interval which converges uniformly to a non-differentiable function.

6. Prove that a sequence cannot have more than one limit.
7. Give an example of a sequence that does not have a limit, or explain carefully why there is no such sequence.
8. Give an example of a bounded sequence that does not have a limit, or explain carefully why there is no such sequence.
9. Give a sequence (f_n) of functions continuous on a set D with a limit that is not continuous.
10. Give a sequence of real functions differentiable on an interval which converges uniformly to a non-differentiable function.
11. Find the set D of all z for which the sequence $\dfrac{z_n}{z_n - 3^n}$ has a limit. Find the limit.
12. Prove that the series $\sum_{j=1}^{n} a_j$ converges if and only if both the series $\sum_{j=1}^{n} \text{Im}\, a_j$ and $\sum_{j=1}^{n} \text{Im}\, a_j$ converge.
13. Explain how you know that the series $\sum_{j=1}^{n} \left(\dfrac{1}{z}\right)^j$ converges uniformly on the set $|z|$ ³ 5.
14. Suppose the sequence of real numbers (a_j) has a limit. Prove that lim sup(a_j) = lim(a_j).

For each of the following, find the set D of points at which the series converges:

15. $\sum_{j=0}^{n} j!z^{j}$ 11. $\sum_{j=0}^{n} jz^{j}$

16. $\sum_{j=0}^{n} \dfrac{j^2}{3^j} z^{j}$ 13. $\sum_{j=0}^{n} \dfrac{(-1)^j j^2}{2^{2j}(j!)^2} z^{2j}$

17. Find the limit of $\sum_{j=0}^{n} (j+1) z^{j}$ For what values of z does the series converge?
18. Find the limit of $\sum_{j=1}^{n} \dfrac{z^j}{j}$. For what values of z does the series converge?
19. Find a power series $\sum_{j=0}^{n} c_j (z-1)^{j}$ such that $\dfrac{1}{z} = \sum_{j=0}^{\infty} c_j (z-1)^{j}$ for $|z - 1| < 1$.
20. Find a power series such that for $|z - 1| < 1$, Log z = for $|z - 1| < 1$.

21. Show that

$$\frac{1}{1-z}=\sum_{j=0}^{\infty}\frac{(z-i)^j}{(1-i)^{j+1}} \text{ for } |z-i| < \sqrt{2}.$$

22. If $f(z)=\frac{1}{1-z}$ what is $f^{(10)}(i)$?

23. Suppose f is analytic at $z = 0$ and $f(0)=f'(0)=f''(0)$. Prove there is a function g analytic at 0 such that $f(z)=z^3 g(z)$ in a neighborhood of 0.

24. Find the Taylor series for $f(z)=\sin z$ at $z_0 = 0$.

25. Show that the function f defined by

$$f(z)=\begin{cases}\frac{\sin z}{z} & \text{for } z \neq 0\\ 1 & \text{for } z = 0\end{cases}$$

is analytic at $z = 0$, and find $f ¢(0)$.

26. Find two Laurent series in powers of z for the function f defined by

$$f(z)=\frac{1}{z^2(1-z)}$$

and specify the regions in which the series converge to $f(z)$.

27. Find two Laurent series in powers of z for the function f defined by

$$f(z)=\frac{1}{z(1+z^2)}$$

and specify the regions in which the series converge to $f(z)$.

===

which is valid for $|z| > 0$.

28. Show in two ways that the sequence

$$z_n=-2+i\frac{(-1)^n}{n^2} \qquad (n = 1, 2, ...)$$

29. Show that

if $\lim_{n\to\infty} z_n = z$, then $\lim_{n\to\infty}|z_n|=|z|$.

30. By considering the remainders ρN(z), verify that

$$\sum_{n=1}^{\infty} z_n=\frac{z}{1-z} \text{ when } |z| < 1.$$

31. Show that a limit of a convergent sequence of complex numbers is unique by appealing to the corresponding result for a sequence of real numbers.
32. Letc denote any complex number and show that

$$\text{if} \sum_{n=1}^{\infty} z_n = S, \text{ then } \sum_{n=1}^{\infty} cz_n = cS.$$

33. Obtain the Maclaurin series representation

$$z\cosh(z^2) = \sum_{n=0}^{\infty} \frac{z^{4n+1}}{(2n)!} \qquad (|z| < \infty).$$

34. Show that

$$e^z e \sum_{n=0}^{\infty} \frac{(z-1)^n}{n!} \qquad (|z| < \infty).$$

35. Find the Maclaurin series expansion of the function

$$f(z) = \frac{z}{z^4 + 9} = \frac{z}{9} \cdot \frac{1}{1 + (z^{4/9})}.$$

36. Derive the Taylor series representation

$$\frac{1}{1-z} = \sum_{n=0}^{\infty} \frac{(z-i)^n}{(1-i)^{z+1}} \qquad (|z - 1| < \sqrt{2}).$$

37. Find the Laurent series that represents the function

$$f(z) = z^2 \sin\left(\frac{1}{z^2}\right)$$

in the domain $0 < |z| < x\ \infty$.

38. Derive the Laurent series representation

$$\frac{e^2}{(z+1)^2} = \frac{1}{e}\left[\sum_{n=0}^{\infty} \frac{(z+1)^n}{(n+2)!} + \frac{1}{z+1} + \frac{1}{(z+1)^2}\right] \quad 3(0 < |z+1| < \infty).$$

39. Find a representation for 1(1 + z) in negative powers of z that is valid in the domin $1 < |z| < \infty$.
40. Give two Laurent series expansions n powers of z for the function

$$f(z) = \frac{1}{z^2(1-z)},$$

and specify the regions in which those expansions are valid.

41. By differentiating the Maclaurin series representation

$$\frac{1}{1-z}=\sum_{n=0}^{\infty} z^n \quad (|z|<1)$$

obtain the expansions

$$\frac{1}{(1-z)^2}=\sum_{n=0}^{\infty}(n+1)z^n,$$

$$\frac{2}{(1-z)^3}=\sum_{n=0}^{\infty}(n+1)(n+2)z^n \quad (|z|<1).$$

42. Find the Taylor series for the function

$$\frac{1}{z}=\frac{1}{2+(z-2)}=\frac{1}{2}\cdot\frac{1}{1+(z-2)/2}$$

about the point $z_0 = 2$. Then, by differentating that series term by term, show that

$$\frac{1}{z^2}\frac{1}{4}\sum_{n=0}^{\infty}(-1)^n(n+1)\left(\frac{z-2}{2}\right)^n \quad (|z-2|<2).$$

Chapter 8

Sequences Series

In mathematics, a sequence is an ordered list of objects (or events). Like a set, it contains members (also called *elements* or *terms*), and the number of terms (possibly infinite) is called the *length* of the sequence. Unlike a set, order matters, and the exact same elements can appear multiple times at different positions in the sequence.

For example, (C, R, Y) is a sequence of letters that differs from (Y, C, R), as the ordering matters. Sequences can be *finite*, as in this example, or *infinite*, such as the sequence of all even positive integers (2, 4, 6,...).

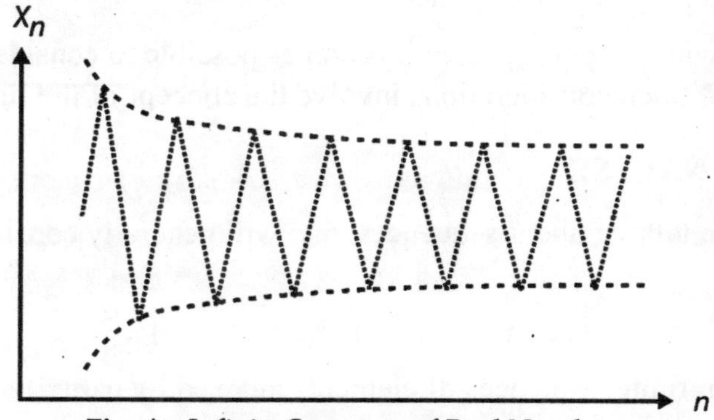

Fig. An Infinite Sequence of Real Numbers.

EXAMPLES AND NOTATION

There are various and quite different notions of sequences in mathematics, some of which are not covered by the notations introduced below.

A sequence may be denoted $(a_1, a_2,...)$. For shortness, the notation (a_n) is also used.

A more formal definition of a finite sequence with terms in a set S is a function from $\{1, 2,..., n\}$ to S for some $n \geq 0$. An infinite sequence in S is a function from $\{1, 2,...\}$ (the set of natural numbers without 0) to S.

Sequences may also start from 0, so the first term in the sequence is then a_0.

A sequence of a fixed-length n is also called an n-tuple. Finite sequences include the *empty sequence* () that has no elements.

A function from all integers into a set is sometimes called a bi-infinite sequence, since it may be thought of as a sequence indexed by negative integers grafted onto a sequence indexed by positive integers.

TYPES AND PROPERTIES OF SEQUENCES

A subsequence of a given sequence is a sequence formed from the given sequence by deleting some of the elements without disturbing the relative positions of the remaining elements.

If the terms of the sequence are a subset of an ordered set, then a *monotonically increasing* sequence is one for which each term is greater than or equal to the term before it; if each term is strictly greater than the one preceding it, the sequence is called *strictly monotonically increasing*. A monotonically decreasing sequence is defined similarly. Any sequence fulfilling the monotonicity property is called monotonic or *monotone*. This is a special case of the more general notion of monotonic function.

The terms *non-decreasing* and *non-increasing* are used in order to avoid any possible confusion with strictly increasing and strictly decreasing, respectively. If the terms of a sequence are integers, then the sequence is an integer sequence. If the terms of a sequence are polynomials, then the sequence is a polynomial sequence.

If S is endowed with a topology, then it becomes possible to consider *convergence* of an infinite sequence in S. Such considerations involve the concept of the limit of a sequence.

SEQUENCES IN ANALYSIS

In analysis, when talking about sequences, one will generally consider sequences of the form

$$(x_1, x_2, x_3, ...) \text{ or } (x_0, x_1, x_2, ...)$$

which is to say, infinite sequences of elements indexed by natural numbers.

It may be convenient to have the sequence start with an index different from 1 or 0. For example, the sequence defined by $x_n = 1/\log(n)$ would be defined only for $n \geq 2$. When talking about such infinite sequences, it is usually sufficient (and does not change much for most considerations) to assume that the members of the sequence are defined at least for all indices large enough, that is, greater than some given N.)

The most elementary type of sequences are numerical ones, that is, sequences of real or complex numbers.

This type can be generalized to sequences of elements of some vector space. In analysis, the vector spaces considered are often function spaces. Even more generally, one can study sequences with elements in some topological space.

SERIES

The sum of terms of a sequence is a series. More precisely, if $(x_1, x_2, x_3,...)$ is a sequence, one may consider the sequence of partial sums $(S_1, S_2, S_3,...)$, with

$$S_n = x_1 + x_2 + ... + x_n \quad = \quad \sum_{i=1}^{n} x_i .$$

Formally, this pair of sequences comprises the *series* with the terms $x_1, x_2, x_3,...,$ which is denoted as

$$\sum_{i=1}^{\infty} x_i .$$

If the sequence of partial sums is convergent, one also uses the infinite sum notation for its limit. For more details, see series.

INFINITE SEQUENCES IN THEORETICAL COMPUTER SCIENCE

Infinite sequences of digits (or characters) drawn from a finite alphabet are of particular interest in theoretical computer science. They are often referred to simply as *sequences* (as opposed to finite *strings*). Infinite binary sequences, for instance, are infinite sequences of bits (characters drawn from the alphabet {0,1}). The set $C = \{0, 1\}^\infty$ of all infinite, binary sequences is sometimes called the Cantor space.

An infinite binary sequence can represent a formal language (a set of strings) by setting the *n*Œth bit of the sequence to 1 if and only if the *n*Œth string (in shortlex order) is in the language. Therefore, the study of complexity classes, which are sets of languages, may be regarded as studying sets of infinite sequences.

An infinite sequence drawn from the alphabet {0, 1,..., b – 1} may also represent a real number expressed in the base-*b* positional number system. This equivalence is often used to bring the techniques of real analysis to bear on complexity classes.

SEQUENCES AS VECTORS

Sequences over a field may also be viewed as vectors in a vector space. Specifically, the set of *F*-valued sequences (where *F* is a field) is a function space (in fact, a product space) of *F*-valued functions over the set of natural numbers.

In particular, the term *sequence space* usually refers to a linear subspace of the set of all possible infinite sequences with elements in $\mathbb{C}$.

DOUBLY-INFINITE SEQUENCES

Normally, the term *infinite sequence* refers to a sequence which is infinite in one direction, and finite in the other — the sequence has a first element, but no final element (a *singly-*

infinite sequence). A *doubly-infinite sequence* is infinite in both directions — it has neither a first nor a final element. Singly-infinite sequences are functions from the natural numbers (N′) to some set, whereas doubly-infinite sequences are functions from the integers (Z) to some set.

One can interpret singly infinite sequences as element of the semigroup ring of the natural numbers $R[\mathbb{N}]$, and doubly infinite sequences as elements of the group ring of the integers $R[\mathbb{Z}]$. This perspective is used in the Cauchy product of sequences.

ORDINAL-INDEXED SEQUENCE

An ordinal-indexed sequence is a generalization of a sequence. If á is a limit ordinal and X is a set, an á -indexed sequence of elements of X is a function from á to X. In this terminology an ù-indexed sequence is an ordinary sequence.

SEQUENCES AND AUTOMATA

Automata or finite state machines can typically thought of as directed graphs, with edges labeled using some specific alphabet Ó. Most familiar types of automata transition from state to state by reading input letters from Ó, following edges with matching labels; the ordered input for such an automaton forms a sequence called a *word* (or input word).

The sequence of states encountered by the automaton when processing a word is called a *run*. A nondeterministic automaton may have unlabeled or duplicate out-edges for any state, giving more than one successor for some input letter. This is typically thought of as producing multiple possible runs for a given word, each being a sequence of single states, rather than producing a single run that is a sequence of sets of states; however, 'run' is occasionally used to mean the latter.

PROPERTIES OF SEQUENCES SERIES

Let F be a field, let V be an F-vector space, and consider an infinite sequence

$$\psi = \{\alpha_i\}_{i=0}^{\infty}$$

where $\alpha_i \in V$ for $i = 0, 1, 2$ .... We say that Ψ is linearly generated (over F) if there exist scalars $c_0, ..., c_{k-1} \in F$ such that the following recurrence relation holds:

$$\alpha_{k+i} = \sum_{j=0}^{k-1} c_j \alpha_{j+i} \text{ (for } i = 0, 1, 2, ...).$$

In this case, all of the elements of the sequence Ψ are determined by the initial segment $\alpha_0, ..., \alpha_{k-1}$, together with the coefficients $c_0, ..., c_{k-1}$ defining the recurrence relation.

The general problem we consider is this: how to determine the coefficients defining such a recurrence relation, given a sufficiently long initial segment of Ψ. To study this problem, it turns out to be very useful to rephrase the problem slightly. Let $g \in F[X]$ be a polynomial of degree, say, k, and write $g = \sum_{j=0}^{k} a_j X^j$. Next, define

$$g * \psi = \sum_{j=0}^{k} a_j \alpha_j$$

Then it is clear that Ψ is linearly generated if and only if there exists a non-zero polynomial g such that

$$(X^i g) * \Psi = 0 \text{ (for } i = 0, 1, 2, \ldots).$$

Indeed, if there is such a non-zero polynomial g, then we can take

$$c_0 := -(a'/ak),\ c_1 := -(a_1/ak),\ \ldots,\ c_{k-1} := -(a_{k-1}/ak)$$

as coe—cients defining the recurrence relation for Ψ. We call a polynomial g satisfying a generating polynomial for Ψ. The sequence Ψ will in general have many generating polynomials. Note that the zero polynomial is technically considered a generating polynomial, but is not a very interesting one.

Let $G(\Psi)$ be the set of all generating polynomials for Ψ.

Theorem. $G(\Psi)$ is an ideal of $F[X]$.

Proof: First, note that for all $g, h \in F[X]$, we have $(g + h) * \Psi = (g * \Psi) + (h * \Psi)$ – this is clear from the definitions. It is also clear that for all $c \in F$ and $g \in F[X]$, we have $(cg) * \Psi = c.(g * \Psi)$. From these two observations, it follows that $G(\Psi)$ is closed under addition and scalar multiplication. It is also easy to see from the definition that $G(\Psi)$ is closed under multiplication by X; indeed, if $(X^i g) * \Psi = 0$ for all $i \geq 0$, then certainly, $(X^i(Xg)) * \Psi = (X^{i+1}g) * \Psi = 0$ for all $i \geq 0$. But any non-empty subset of $F[X]$ that is closed under addition, multiplication by elements of F, and multiplication by X is an ideal of $F[X$.

Since all ideals of $F[X]$ are principal, it follows that $G(\Psi)$ is the ideal of $F[X]$ generated by some polynomial $\phi \in F[X]$ – we can make this polynomial unique by choosing the monic associate (if it is non-zero), and we call this polynomial the minimal polynomial of Ψ. Thus, a polynomial $g \in F[X]$ is a generating polynomial for Ψ if and only if ϕ divides g; in particular, Ψ is linearly generated if and only if $\phi \neq 0$.

We can now restate our main objective as follows: given a sufficiently long initial segment of a linearly generated sequence, determine its minimal polynomial.

Example. Linearly generated sequences can also arise in a natural way, as this example and the next illustrate. Let $E := F[X]/(f)$, where $f \in F[X]$ is a monic polynomial of degree $\ell > 0$, and let α be an element of E. Consider the sequence $\Psi := \{\alpha^i\}_{i=0}^{\infty}$ of powers of α. For every polynomial $g = \sum_{j=0}^{k} a_j X^j \in F[X]$, we have

$$g * \Psi = \sum_{j=0}^{k} a_j \alpha^j = g(\alpha).$$

Now, if $g(\alpha) = 0$, then clearly $(X^i g) * \Psi = \alpha^i g(\alpha) = 0$ for all $i \geq 0$. Conversely, if $(X^i g) * \Psi = 0$ for all $i \geq 0$, then in particular, $g(\alpha) = 0$. Thus, g is a generating polynomial for Ψ if and only if $g(\alpha) = 0$. It follows that the minimal polynomial ϕ of Ψ is the same as the minimal polynomial of α over F, as defined in §16.5. Furthermore, $\phi \neq 0$, and the degree m of ϕ may be characterized as the smallest positive integer m such that $\{\alpha^i\}_{i=0}^{\infty}$ is linearly dependent; moreover, as E has dimension ℓ over F, we must have $m \leq \ell$.

Example. One can always define a linearly generated sequence by simply choosing an initial segment $\alpha_0, \alpha_1, ..., \alpha_{k-1}$, along with scalars $c_0, ..., c_{k-1} \in F$ defining the recurrence relation. One can enumerate as many elements of the sequence as one wants by using storage for k elements of V, along with storage for the scalars $c_0, ..., c_{k-1}$, as follows:

$(\beta_0, ..., \beta_{k-1}) \leftarrow (\alpha_0, ..., \alpha_{k-1})$

repeat

output β_0

$\beta' \leftarrow \sum_{j=0}^{k-1} c_j \beta_j$

$(\beta_0, ..., \beta_{k-1}) \leftarrow (\beta_1, ..., \beta_{k-1}, \beta')$

forever

Because of the structure of the above algorithm, linearly generated sequences are sometimes also called shift register sequences. Also observe that if F is a finite field, and V is finite dimensional, the value stored in the "register" $(\beta_0, ..., \beta_{k-1})$ must repeat at some point. It follows that the linearly generated sequence must be ultimately periodic .

Example. Let V be a vector space over F of dimension $\ell > 0$, and let $\tau : V \to V$ be an F-linear map. Let $\beta \in V$, and consider the sequence $\Psi := \{\alpha_i\}_{i=0}^{\infty}$, where $i = \tau^i(\beta)$; that is, $\alpha_0 = \beta$, $\alpha_1 = \tau(\beta)$, $\alpha_2 = \tau(\tau(\beta))$, and so on. For every polynomial $g = \sum_{j=0}^{k} a_j X^j \in F[X]$, we have

$$g * \psi = \sum_{j=0}^{k} a_j \tau^j(\beta)$$

and for every $i \geq 0$, we have

$$(X^i g) * \psi = \sum_{j=0}^{k} a_j \tau^{i+j}(\beta) = \tau^i \left(\sum_{j=0}^{k} a_j \tau^t(\beta) \right) = \tau^t(g * \psi).$$

Thus, if $g * \Psi = 0$, then clearly $(X^i g) * \Psi = \tau^i(g * \Psi) = \tau^i(0) = 0$ for all $i \geq 0$. Conversely, if $(X^i g) * \Psi = 0$ for all $i \geq 0$, then in particular, $g * \Psi = 0$. Thus, g is a generating polynomial for Ψ if and only if $g * \Psi = 0$. The minimal polynomial of Ψ is non-zero and its degree m is at most ℓ; indeed, m may be characterized as the least non-negative integer such that $\{\tau^i(\beta)\}_{i=0}^{m}$ is linearly dependent, and since V has dimension ℓ over F, we must have $m \leq \ell$.

The previous example can be seen as a special case of this one, by taking V to be E, τ to be the α-multiplication map on E, and setting β to 1.

The problem of computing the minimal polynomial of a linearly generated sequence can always be solved by means of Gaussian elimination. For example, the minimal polynomial of the sequence discussed in Example can be computed using the algorithm described. The minimal polynomial of the sequence discussed in Example can be computed in a similar manner. Also, Exercise below shows how one can reformulate another special case of the problem so that it is easily solved by Gaussian elimination. However, in the following sections, we will present algorithms for computing minimal polynomials for certain types of linearly generated sequences that are much more efficient than any algorithm based on Gaussian elimination.

EXERCISE

1. Show that the only sequence for which 1 is a generating polynomial is the "all zero" sequence.
2. Let $\Psi = \{\alpha_i\}_{i=0}^{\infty}$ be a sequence of elements of an F-vector space V. Further, suppose that Ψ has non-zero minimal polynomial ϕ.
   (a) Show that for all polynomials $g, h \in F[X]$, if $g \equiv h \pmod{\phi}$, then $g * \Psi = h * \Psi$.
   (b) Let $m := \deg(\phi)$. Show that if $g \in F[X]$ and $(X^i g) * \Psi = 0$ for all $i = 0, \ldots, m - 1$, then g is a generating polynomial for Ψ.
3. Let V be a vector space over F, and consider the set $V^{\times\infty}$ of all infinite sequences $\{\alpha_i\}_{i=0}^{\infty}$, where the α_i's are in V. Let us define the scalar product of $g \in F[X]$ and $\Psi \in V^{\times\infty}$ as

   $$g \,.\, \Psi = \{(X^i g) * \psi\}_{i=0}^{\infty} \in V^{\times\infty}.$$

   Show that with this scalar product, and addition defined component-wise, $V^{\times\infty}$ is an $F[X]$-module, and that a polynomial $g \in F[X]$ is a generating polynomial for $\Psi \in V^{\times\infty}$ if and only if $g \;\Psi = 0$.
4. Let $c_0, \ldots, c_{k-1} \in F$ and $z_0, \ldots, z_{k-1} \in F$. For each $i \geq 0$, let

   $$z_{k+i} := \sum_{j=0}^{k-1} c_j z_{j+i}.$$

   Given $n \geq 0$, along with $c_0, \ldots, c_{k-1}$ and $z_0, \ldots, z_{k-1}$, show how to compute z_n using $O(\text{len}(n)k_2)$ operations in F.

MINIMAL POLYNOMIALS

FIRST CASE

We now tackle the problem of efficiently computing the minimal polynomial of a linearly generated sequence from a sufficiently long initial segment.

We shall first address a special case of this problem, namely, the case where the vector space V is just the field F. In this case, we have

$$\Psi = \{z_i\}_{i=0}^{\infty}$$

where $z_i \in F$ for $i = 0, 1, 2, ...$

Suppose that we do not know the minimal polynomial ϕ of Ψ, but we know an upper bound $M > 0$ on its degree. Then it turns out that the initial segment $z_0, z_1, ... z_{2M-1}$ completely determines ϕ, and moreover, we can very efficiently compute given this initial segment. The following theorem provides the essential ingredient.

Theorem: Let $\Psi = \{z_i\}_{i=0}^{\infty}$ be a sequence of elements of F, and define the reversed Laurent series

$$z := \sum_{i=0}^{\infty} z_i X^{-(i+1)} \in F((X^{-1})),$$

whose coefficients are the elements of the sequence Ψ. Then for every $g \in F[X]$, we have $g \in G(\Psi)$ if and only if $gz \in F[X]$. In particular, Ψ is linearly generated if and only if z is a rational function, in which case, its minimal polynomial is the denominator of z when expressed as a fraction in lowest terms.

Proof: Observe that for every polynomial $g \in F[\mathrm{X}]$ and every integer $i \geq 0$, the coefficient of $X^{-(i+1)}$ in the product gz is equal to $\mathrm{X}^i g * \Psi$ just look at the formulas defining these expressions! It follows that g is a generating polynomial for Ψ if and only if the coefficients of the negative powers of X in gz are all zero, which is the same as saying that $gz \in F[X]$. Further, if $g \neq 0$ and $h := gz \in F[X]$, then $\deg(h) < \deg(g)$ – this follows simply from the fact that $\deg(z) < 0$ (together with the fact that $\deg(h) = \deg(g) + \deg(z)$). All the statements in the theorem follow immediately from these observations.

We can compute the minimal polynomial ϕ of Ψ using the algorithm for computing the numerator and denominator of a rational function from its reversed Laurent series expansion. More precisely, we can compute ϕ given the bound M on its degree, along with the first $2M$ elements $z_0, ..., z_{2M-1}$ of Ψ, using $O(M^2)$ operations in F. Just for completeness, we write down this algorithm:

1. Run the extended Euclidean algorithm on inputs

   $f := X^{2M}$ and $h := z_0 X^{2M-1} + z_1 X^{2M-2} + ... + z_{2M-1}$, and apply Theorem with f, h, $r^* := M$, and $t^* := M$, to obtain the polynomials r', s', t'.

2. Output $\phi := t' / \mathrm{lc}(t')$.

SECOND CASE

Having dealt with the problem of finding the minimal polynomial of a linearly generated sequence Ψ, whose elements lie in F, we address the more general problem, where the elements of Ψ lie in a vector space V over F. We shall only deal with a special case of this problem, but it is one which has useful applications:

- First, we shall assume that V has finite dimension $\ell > 0$ over F.
- Second, we shall assume that the sequence $\Psi = \{\alpha_i\}_{i=0}^{\infty}$ has full rank, by which we mean the following: if the minimal polynomial ϕ of Ψ over F has degree m, then $\{\alpha_i\}_{i=0}^{m-1}$ is linearly independent. This property implies that the minimal polynomial of Ψ is the monic polynomial $\Psi \in F[X]$ of least degree such that $\phi * \Psi = 0$. The sequences considered in Examples and are of this type.
- Third, we shall assume that F is a finite field.

The dual space. Before presenting our algorithm for computing minimal polynomials, we need to discuss the dual space $D_F(V)$ of V (over F), which consists of all F-linear maps from V into F. Thus, $D_F(V) = \mathrm{Hom}_F(V, F)$, and is a vector space over F, with addition and scalar multiplication defined point-wise. We shall call elements of $D_F(V)$ projections.

Now, fix a basis $S = \{\gamma_i\}_{i=1}^{\ell}$ for V. As was discussed in §14.2, every element $\delta \in V$ has a unique coordinate vector $\mathrm{Vec}_S(\delta) = (c_1, \ldots, c_\ell) \in F^{1\times \ell}$, where $\delta = \sum_i c_i\gamma_i$. Moreover, the map $\mathrm{Vec}_S : V \to F^{1\times \ell}$ is a vector space isomorphism.

To each projection $\pi \in D_F(V)$ we may also associate the coordinate vector $(\pi(\gamma_1), \ldots, \pi(\gamma_\ell))^T \in F^{\ell\times 1}$. If T is the basis for F consisting of the single element 1_F, then the coordinate vector of π is $\mathrm{Mat}_{S,T}(\pi)$, that is, the matrix of π relative to the bases S and T. The map $\mathrm{Mat}_{S,T} : D_F(V) \to F^{\ell\times 1}$ is a vector space isomorphism.

In working with algorithms that compute with elements of V and $D_F(V)$, we shall assume that such elements are represented using coordinate vectors relative to some convenient, fixed basis for V. If $\delta \in V$ has coordinate vector $(c_1, \ldots, c_\ell) \in F^{1\times \ell}$, and $\pi \in D_F(V)$ has coordinate vector $(d_1, \ldots, d_\ell)_T \in F^{\ell\times 1}$, then $\pi(\delta)$ is easily computed, using $O(\ell)$ operations in F, as $\sum_{i=1}^{\ell} c_i d_i$.

We now return to the problem of computing the minimal polynomial ϕ of the linearly generated sequence $\Psi = \{\alpha_i\}_{i=0}^{\infty}$. Assume we have a bound $M > 0$ on the degree of ϕ. As we are assuming Ψ has full rank, we may assume that $M \le \ell$.

For each $\pi \in D_F(V)$, we may consider the projected sequence $\Psi_\pi := \{\pi(\alpha_i)\}_{i=0}^{\infty}$. Observe that ϕ is a generating polynomial for Ψ_π; indeed, for every polynomial $g \in F[X]$, we have $g * \Psi_\pi = \pi(g * \Psi)$, and hence, for all $i \geq 0$, we have $(X^i\phi) * \Psi_\pi = \pi((X^i\phi) * \Psi) = \pi(0) = 0$. Let $\Psi_\pi \in F[X]$ denote the minimal polynomial of Ψ_π. Since π divides every generating polynomial of Ψ_π, and since Ψ_π is a generating polynomial for ϕ_π, it follows that π divides ϕ.

This suggests the following algorithm for efficiently computing the minimal polynomial of Ψ, using the first $2M$ terms of Ψ:

Algorithm MP. Given the first $2M$ terms of the sequence $\Psi = \{\alpha_i\}_{i=0}^{\infty}$, do the following:

$g \leftarrow 1 \in F[X]$

repeat
choose $\pi \in D_F(V)$ at random
compute the first $2M$ terms of the projected sequence Ψ_π
use the algorithm in §18.2 to compute the minimal polynomial
ϕ_π of Ψ_π
$g \leftarrow \operatorname{lcm}(g, \phi_\pi)$
until $g * \Psi = 0$
output g

A few remarks on the above procedure are in order:

- in every iteration of the main loop, g is the least common multiple of a number of divisors of ϕ, and hence is itself a divisor of ϕ; in particular, $\deg(g) \leq M$;
- under our assumption that Ψ has full rank, and since g is a monic divisor of ϕ, if $g * \Psi = 0$, we may safely conclude that $g = \phi$;
- under our assumption that F is finite, choosing a random element π of $D_F(V)$ amounts to simply choosing at random the entries of the coordinate vector of π, relative to some basis for V;
- we also assume that elements of V are represented as coordinate vectors, so that applying a projection $\pi \in D_F(V)$ to an element of V takes $O(\ell)$ operations in F; in particular, in each loop iteration, we can compute the first $2M$ terms of the projected sequence Ψ_π using $O(M\ell)$ operations in F;
- similarly, adding two elements of V, or multiplying an element of V times a scalar, takes $O(\ell)$ operations in F; in particular, in each loop iteration, we can compute $g * \Psi$ using $O(M\ell)$ operations in F (and using the first $M + 1 \leq 2M$ terms of Ψ).

Based on the above observations, it follows that when the algorithm halts, its output is correct, and that the cost of each loop iteration is $O(M\ell)$ operations in F. The remaining question to be answered is this: what is the expected number of iterations of the main loop? The answer to this question is $O(1)$, which leads to a total expected cost of Algorithm MP of

$O(M\ell)$ operations in F. The key to establishing that the expected number of iterations of the main loop is constant is provided by the following theorem.

Theorem. If F is a finite field of cardinality q, and m and s are positive integers, then we have

$$\Lambda_F^m(s) = 1 - 1/q^{s-1} + (q-1)/q^{sm}.$$

Proof: For each positive integer n, let U_n denote the set of all tuples of polynomials $(f_1, ..., f_s) \in F[X]_{<n}^{\times s}$ with $\gcd(f_1, ..., f_s) = 1$, and let $u_n := |U_n|$. Also, for each monic polynomial $h \in F[X]$ of degree less that n, let U_n, h denote the set of all s-tuples of polynomials of degree less than n whose gcd is h. Observe that the set U_n, h is in one-to-one correspondence with U_{n-k}, where $k := \deg(h)$, via the map that sends $(f_1, ..., f_s) \in U_n$, h to $(f_1/h, ..., f_s/h) \in U_{n-k}$. As there are q^k possible choices for h of degree k, if we define V_n, k to be the set of tuples $(f_1, ..., f_s) \in F[X]_{<n}^{\times s}$ with $\deg(\gcd(f_1, ..., f_s)) = k$, we see that $|V_{n,k}| = q^k u_{n-k}$. Every non-zero tuple in $F[X]_{<n}^{\times s}$ appears in exactly one of the sets $V_{n,k}$, for $k = 0, ..., n-1$. Taking into account the zero tuple, it follows that

$$q^{sn} = 1 + \sum_{k=0}^{n-1} q^k u_{n-k},$$

which holds for all $n \geq 1$. Replacing n by $n - 1$ in , we obtain

$$q^{s(n-1)} = 1 + \sum_{k=0}^{n-2} q^k u_{n-1-k},$$

which holds for all $n \geq 2$, and indeed, holds for $n = 1$ as well. Subtracting q times from we deduce that for all $n \geq 1$,

$$q^{sn} - q^{sn-s+1} = 1 + u_n - q,$$

and rearranging terms:

$$u_n = q^{sn} - q^{sn-s+1} + q - 1.$$

Therefore,

$$\Lambda_F^m(s) = u_m/q^{sm} = 1 - 1/q^{s-1} + (q-1)/q^{sm}.$$

From the above theorem, it follows that for $s \geq 1$, the probability P_s that Algorithm MP runs for more than s loop iterations is at most $1/q^{s-1}$. If L is the total number of loop iterations, then

$$E[L] = \sum_{i\geq 1} P[L \geq i] = 1 + \sum_{s\geq 1} P_x \leq 1 + \sum_{s\geq 1} 1/q^{s-1} = 1 + \frac{q}{q-1} \leq 3.$$

Theorem. Let $\Psi = \{\alpha_i\}_{i=0}^{\infty}$ be a linearly generated sequence over the field F, where the α_i's are elements of a vector space V of finite dimension $\ell > 0$. Let ϕ be the minimal polynomial of Ψ over F, let $m := \deg(\phi)$, and assume that Ψ has full rank (i.e., $\{\alpha_i\}_{i=0}^{m-1}$ is linearly independent). Finally, let $F[X]_{<m}$ denote the vector space over F consisting of all polynomials in $F[X]$ of degree less than m.

Under the above assumptions, there exists a surjective F-linear map

$$\sigma : D_F(V) \leftarrow F[X]_{<m}$$

such that for all $\pi \in D_F(V)$, the minimal polynomial ϕ_π of the projected sequence $\Psi_\pi := \{\pi(\alpha_i)\}_{i=0}^{\infty}$ satisfies

$$\phi_\pi = \frac{\phi}{\gcd(\sigma(\phi), \phi)}$$

Proof: While the statement of this theorem looks a bit complicated, its proof is quite straightforward, given our characterization of linearly generated sequences in previous theorem in terms of rational functions. We build the linear map σ as the composition of two linear maps, σ_0 and σ_1.

Let us define the map

$$\sigma_0 : D_F(V) \to F((X^{-1}))$$

$$\pi \mapsto \sum_{i=0}^{\infty} \pi(\alpha_i) X^{-(i+1)}.$$

We also define the map σ_1 to be the -multiplication map on $F((X^{-1}))$, that is, the map that sends $z \in F((X^{-1}))$ to $\phi . z \in F((X^{-1}))$. The map σ is just the composition $\sigma = \sigma_1 \circ \sigma_0$. It is clear that both σ_0 and σ_1 are F-linear maps, and hence, so is σ.

First, observe that for $\pi \in D_F(V)$, the series $z := \sigma_0(\pi)$ is the series associated with the projected sequence Ψ_π, as in Theorem. Let ϕ_π be the minimal polynomial of Ψ_π. Since ϕ is a generating polynomial for Ψ, it is also a generating polynomial for Ψ_π. Therefore, Theorem tells us that

$$h := \sigma(\pi) = \phi . z \in F[X]_{<m},$$

and that ϕ_π is the denominator of z when expressed as a fraction in lowest terms. Now, we have $z = h/\phi$, and it follows that $\phi_\pi = \phi/\gcd(h, \phi)$ is this denominator.

Second, the hypothesis that $\{\alpha_i\}_{i=0}^{m-1}$ is linearly independent implies that $\dim_F(\mathrm{Im}\ \sigma_0) \geq m$. Also, observe that σ_1 is an injective map. Therefore, $\dim_F(\mathrm{Im}\ \sigma_0) \geq m$. In the previous paragraph, we observed that Im $\sigma \subseteq F[X]_{<m}$, and since $\dim_F (F[X]_{<m}) = m$, we may conclude that Im $\sigma = F[X]_{<m}$. That proves the theorem.

Given the above theorem, we can analyze the expected number of iterations of the main loop of Algorithm MP.

First of all, we may as well assume that the degree m of ϕ is greater than 0, as otherwise, we are sure to get in the very first iteration. Let $\pi_1, ..., \pi_s$ be the random projections chosen in the first s iterations of Algorithm MP. By Theorem each $\sigma(\pi_i)$ is uniformly distributed over $F[X]_{<m}$, and we have $g = \phi$ at the end of loop iteration s if and only if gcd $(\phi, \sigma(\pi_1), ..., (\pi_s))$ = 1.

Let us define $\Lambda_F^{\phi}(s)$ to be the probability that $\gcd(\phi, f_1, ..., f_s) = 1$, where $f_1, ..., f_s$ are randomly chosen from $F[X]_{<m}$. Thus, the probability that we have $g =$ at the end of loop iteration s is equal to $\Lambda_F^{\phi}(s)$. While one can analyze the quantity $\Lambda_F^{\phi}(s)$, it turns out to be easier, and sufficient for our purposes, to analyze a di–erent quantity. Let us define $\Lambda_F^{m}(s)$ to be the probability that $\gcd(f_1, ..., f_s) = 1$, where $f_1, ..., f_s$ are randomly chosen from $F[X]_{<m}$. Clearly, $\Lambda_F^{\phi}(s) \geq \Lambda_F^{m}(s)$.

Theorem. Let Ψ be a sequence of elements of an F-vector space V of finite dimension $\ell > 0$ over F, where F is a finite field. Assume that Ψ is linearly generated over F with minimal polynomial $\phi \in F[X]$ of degree m, and that Ψ has full rank (i.e., the first m terms of Ψ form a linearly independent family of elements). Then given an upper bound M on m, along with the first $2M$ elements of Ψ, Algorithm MP correctly computes using an expected number of $O(M\ell)$ operations in F.

We close this section with the following observation. Suppose the sequence Ψ is of the form $\{\tau^i(\beta)\}_{i=0}^{\infty}$, where $\beta \in V$ and $\tau : V \rightarrow V$ is an F-linear map. Suppose that with respect to some basis S for V, elements of V are represented by their coordinate vectors (which are elements of $F^{1\times\ell}$), and elements of $D_F(V)$ are represented by their coordinate vectors (which are elements of $F^{\ell\times 1}$). The linear map τ also has a corresponding matrix $A = \text{Mat}_{S,S}(V, V) \in F^{\ell\times\ell}$, so that evaluating τ at a point α in V corresponds to multiplying the coordinate vector of α on the right by A. Now, suppose $\beta \in V$ has coordinate vector $v \in F^{1\times\ell}$ and that $\pi \in D_F(V)$ has coordinate vector $w \in F^{\ell\times 1}$. Then if Ψ' is the sequence of coordinate vectors of the elements of Ψ, we have

$$\Psi' = \{vA^i\}_{i=0}^{\infty} \text{ and } \Psi_\pi = \{vA^i w\}_{i=0}^{\infty}.$$

This more concrete, matrix-oriented point of view is sometimes useful; in particular, it makes quite transparent the symmetry of the roles played by β and π in forming the projected sequence.

SPARSE LINEAR SYSTEMS

Let V be a vector space of finite dimension $\ell > 0$ over a finite field F, and let $\tau : V \rightarrow V$ be an F-linear map. The goal of this section is to develop time- and space-efficient algorithms for solving equations of the form

$$\tau(\gamma) = \delta;$$

that is, given δ and $\delta \in V$, find $\gamma \in V$ satisfying. The algorithms we develop will have the following properties: they will be probabilistic, and will use an expected number of $O(\ell^2)$ operations in F, an expected number of $O(\ell)$ evaluations of τ, and space for $O(\ell)$ elements of F. By an "evaluation of τ," we mean the computation of $\tau(\alpha)$ for a given $\alpha \in V$.

We shall assume that elements of V are represented as coordinate vectors with respect to some fixed basis for V. This means that a single element of V is represented as a vector of ℓ elements of F. Now, if the matrix of τ with respect to the given basis is sparse, having, say, $\ell^{1+o(1)}$ non-zero entries, then the space required to represent τ is $\ell^{1+o(1)}$ elements of F, and the time required to evaluate τ is $\ell^{1+o(1)}$ operations in F. Under these assumptions, our algorithms to solve use an expected number of $\ell^{2+o(1)}$ operations in F, and space for $\ell^{1+o(1)}$ elements of F. This is to be compared with standard Gaussian elimination: even if the original matrix is sparse, during the execution of the algorithm, most of the entries in the matrix may eventually be "filled in" with non-zero field elements, leading to a running time of $\Omega(\ell^3)$ operations in F, and a space requirement of $\Omega(\ell^2)$ elements of F. Thus, the algorithms presented here will be much more efficient than Gaussian elimination when the matrix of τ is sparse.

We hasten to point out that the algorithms presented here may be more efficient than Gaussian elimination in other cases, as well. All that matters is that τ can be evaluated using $o(\ell^2)$ operations in F and/or represented using space for $o(\ell^2)$ elements of F – in either case, we obtain a time and/or space improvement over Gaussian elimination. Indeed, there are applications where the matrix of the linear map "may not be sparse, but nevertheless has special structure that allows it to be represented and evaluated in subquadratic time and/or space.

We shall only present algorithms that work in two special, but important, cases:

- the first case is where τ is bijective,
- the second case is where τ is not bijective, $\delta = 0$, and a non-zero solution to is required (i.e., we are looking for a non-zero element of Ker τ).

In both cases, the key will be to use Algorithm MP to find the minimal polynomial ϕ of the linearly generated sequence

$$\Psi := \{\alpha_i\}_{i=0}^{\infty} \ (\alpha_i := \tau^i(\beta),\ i = 0, 1, \ldots),$$

where β is a suitably chosen element of V. From the discussion in Example, this sequence has full rank, and so we may use Algorithm MP. We may use $M := \ell$ as an upper bound on the degree of ϕ (assuming we know nothing more about τ and that would allow us to use a smaller upper bound). In using Algorithm MP in this application, note that we do not want to store α_0, ..., $\tau_{2\ell-1}$ – if we did, we would not satisfy our stated space bound. Instead of storing the α_i's in a "warehouse," we use a "just in time" strategy for computing them, as follows:

- In the body of the main loop of Algorithm MP, where we calculate the projections $z_i := \pi(\tau_i)$, for $i = 0 \ldots 2\ell - 1$, we perform the computation as follows:

  $\alpha \leftarrow \beta$

  for $i \leftarrow 0$ to $2\ell - 1$ do

  $z_i \leftarrow \pi(\alpha)$, $\alpha \leftarrow \tau(\alpha)$

- In the test at the bottom of the main loop of Algorithm MP, if $g = \sum_{j=0}^{k} a_j X^j$, we compute $v := g * \Psi \in V$ using the following Horner-like scheme:

$v \leftarrow 0$

for $j \leftarrow k$ down to 0 do

$v \leftarrow \tau(v) + a_j.\beta$

With this implementation, Algorithm MP uses an expected number of $O(\ell 2)$ operations in F, an expected number of $O(\ell)$ evaluations of τ, and space for $O(\ell)$ elements of F. Of course, the "warehouse" strategy is faster than the "just in time" strategy by a constant factor, but it uses about ℓ times as much space; thus, for large ℓ, using the "just in time" strategy is a very good time/space trade-off.

THE BIJECTIVE CASE

Now consider the case where τ is bijective, and we want to solve for a given $\delta \in V$. We may as well assume that $\delta \leftarrow 0$, since otherwise, $\gamma = 0$ is the unique solution to. We proceed as follows. First, using Algorithm MP as discussed above, compute the minimal polynomial of the sequence Ψ defined in , using $\beta := \delta$. Let $\phi = \sum_{j=0}^{m} c_j X^j$, where $c_m = 1$ and $m > 0$. Then we have

$$c_0\delta + c_1\tau(\delta) + \ldots + c_m \tau^m(\delta) = 0.$$

We claim that $c_0 \neq 0$. To prove the claim, suppose that $c_0 = 0$. Then applying τ^{-1}, we would obtain

$$c_1\delta + \ldots + c_m \tau^{m-1}(\delta) = 0,$$

which would imply that ϕ/X is a generating polynomial for Ψ, contradicting the minimality of ϕ. That proves the claim.

Since $c_0 \neq 0$, we can apply τ^{-1}, and solve for $\gamma = \tau^{-1}(\delta)$ as follows:

$$\gamma = -c_0^{-1}(c_1\delta + \ldots + c_m\tau^{m-1}(\delta)).$$

To actually compute γ, we use the same "just in time" strategy as was used in the implementation of the computation of $g * \Psi$ in Algorithm MP, which costs $O(\ell^2)$ operations in F, $O(\ell)$ evaluations of τ, and space for $O(\ell)$ elements of F.

THE NON-BIJECTIVE CASE

Now consider the case where τ is not bijective, and we want to find non-zero $\gamma \in V$ such that $\tau(\gamma) = 0$. The idea is this. Suppose we choose an arbitrary, non-zero element β of V, and use Algorithm MP to compute the minimal polynomial ϕ of the sequence Ψ defined , using this value of β. Let $= \sum_{j=0}^{m} c_j X^j$, where $m > 0$ and $c_m = 1$. Then we have

$$c_0\beta + c_1\tau(\beta) + \ldots + c_m\tau^m(\beta) = 0.$$

Let

$$\gamma := c_1\beta + \ldots + c_m \Psi^{m-1}(\beta).$$

We must have $\gamma \neq 0$, since $\gamma = 0$ would imply that $\lfloor \phi / X \rfloor$ is a non-zero generating polynomial for Ψ, contradicting the minimality of ϕ. If it happens that $c_0 = 0$, then equation implies that $\tau(\gamma) = 0$, and we are done. As before, to actually compute γ, we use the same "just in time" strategy as was used in the implementation of the computation of $g * \Psi$ in Algorithm MP, which costs $O(\ell^2)$ operations in F, $O(\ell)$ evaluations of Ψ, and space for $O(\ell)$ elements of F.

The above approach fails if $c_0 \neq 0$. However, in this "bad" case, equation implies that $\beta = -c_0^{-1}\tau(\gamma)$; in particular, $\beta \in \operatorname{Im} \tau$. One way to avoid such a "bad" β is to randomize: as τ is not surjective, the image of τ is a subspace of V of dimension strictly less than ℓ, and therefore, a randomly chosen β lies in the image of τ with probability at most $1/|F|$. So a simple technique is to choose repeatedly β at random until we get a "good" β. The overall complexity of the resulting algorithm will be as required: $O(\ell^2)$ expected operations in F, $O(\ell)$ expected evaluations of τ, and space for $O(\ell)$ elements of F.

As a special case of this situation, consider the problem that arose in Previous chapter in connection with algorithms for computing discrete logarithms and factoring. We had to solve the following problem: given an $\ell \times (\ell - 1)$ matrix A with entries in a finite field F, containing $\ell^{1+o(1)}$ non-zero entries, find non-zero $v \in F^{1\times\ell}$ such that $vA = 0$. To solve this problem, we can augment the matrix A, adding an extra column of zeros, to get an $\ell \times \ell$ matrix A'. Now, let $V = F^{1\times\ell}$ and let τ be the F-linear map on V that sends $\gamma \in V$ to A'. A non-zero solution to the equation $\tau(\gamma) = 0$ will provide us with the solution to our original problem; thus, we can apply the above technique directly, solving this problem using $\ell^{2+o(1)}$ expected operations in F, and space for $\ell^{1+o(1)}$ elements of F. As a side remark, in this particular application, we can choose a "good" in the above algorithm without randomization: just choose $\beta := (0, \ldots, 0, 1)$, which is clearly not in the image of τ.

THE ALGEBRA OF LINEAR TRANSFORMATIONS

Throughout this chapter, one could hear the whispers of the algebra of linear transformations. We develop some of the aspects of this theory here, leaving a number of details as exercises. It will not play a role in any material that follows, but it serves to provide the reader with a "bigger picture."

Let F be a field and V be an F-vector space. We denote by $L_F(V)$ the set of all F-linear maps from V into V. Thus, $L_F(V) = \text{Hom}_F(V, V)$, and is a vector space over F, with addition and scalar multiplication defined point-wise. Elements of $L_F(V)$ are called linear transformations.

For $\tau, \tau' \in L_F(V)$, the composed map, $\tau \circ \tau'$, which sends $\alpha \in V$ to $\tau(\tau'(\alpha))$ is also an element of $L_F(V)$. As always, function composition is associative (i.e., for $\tau, \tau', \in L_F(V)$, we have $\tau \circ (\tau' \circ \tau'') = (\tau \circ \tau') \circ \tau''$); however, function composition is not in general commutative (i.e., we may have $\tau \circ \tau'' \neq \tau' \circ \tau$ for some $\tau, \tau' \in L_F(V)$). The following theorem considers the interaction between composition, addition, and scalar multiplication.

Theorem. For all $\tau, \tau', \tau'', \in L_F(V)$, and for all $c \in F$, we have:

(i) $\tau \circ (\tau' + \tau'') = \tau \circ \tau' + \tau \circ \tau''$;

(ii) $(\tau' + \tau'')\ \tau = \tau' \circ \tau + \tau'' \circ \tau$;

(iii) $c(\tau \circ \tau') = (c\tau) \circ \tau' = \tau \circ (c\tau')$.

Proof: Exercise.

Under the addition operation and scalar multiplication of the vector space $L_F(V)$, and defining multiplication on $L_F(V)$ using the "$\circ$" operation, the previous theorem implies that $L_F(V)$ satisfies all the properties of an F-algebra, except for the fact that multiplication is not commutative (the identity map acts as the multiplicative identity). Thus, we can think of $L_F(V)$ as a non-commutative F-algebra.

Let $\tau \in L_F(V)$ be a linear transformation. For each integer $i \geq 0$, the map τ^i (i.e., the i-fold composition of τ) is also an element of $L_F(V)$. Note that τ^0 is by definition just the identity map on V. For each polynomial $g \in F[X]$, with $g = \sum_i a_i X^i$, we denote by $g(\tau)$ the linear transformation

$$g(\tau) := \sum_i a_i \tau^i \in L_F(V).$$

Thus, for $\alpha \in V$, the value of $g(\tau)$ at is $\sum_i a_i \tau^i(\alpha)$.

MINIMAL POLYNOMIALS IN $F[X] / (F)$ (II)

Let us consider F is a field, $f \in F[X]$ is a monic polynomial of degree $\ell > 0$, and $E := F[X] / (f)$; we are given an element $\alpha \in E$, and want to compute the minimal polynomial $\phi \in$

$F[X]$ of over F. As discussed in Example, this problem is equivalent to the problem of computing the minimal polynomial of the sequence

$$\Psi := \{\alpha_i\}_{i=0}^{\infty} \ (\alpha_i := \tau^i, i = 0, 1, ...),$$

and the sequence has full rank; therefore, we can use Algorithm MP directly to solve this problem, assuming F is a finite field. If we use the "just in time" strategy in the implementation of Algorithm MP, we get an algorithm that computes the minimal polynomial of using $O(\ell^3)$ expected operations in F, but space for just $O(\ell^2)$ elements of F. Thus, in terms of space, based on Gaussian elimination. In terms of time complexity, the algorithm based on linearly generated sequences is a bit slower than the one based on Gaussian elimination (but only by a constant factor).

However, if we use any subquadratictime algorithm for polynomial arithmetic, we immediately get an algorithm that runs in subcubic time, while still using linear space. In the exercises below, you are asked to develop an algorithm that computes the minimal polynomial of α using just $O(1^{1.5})$ operations in F, at the expense of requiring space for $O(\ell^{1.5})$ elements of F – this algorithm does not rely on fast polynomial arithmetic, and can be made even faster if such arithmetic is used.

EXERCISE

1. Let $f \in F[X]$ be a monic polynomial of degree $\ell > 0$ over a field F (not necessarily finite), and let $E := F[X] / (f)$. Further, suppose that f is irreducible, so that E is itself a field. Show how to compute the minimal polynomial of $\alpha \in E$ over F deterministically, using algorithms that satisfy the following complexity bounds:
   (a) $O(\ell^{2.5})$ operations in F and space for $O(\ell^{1.5})$ elements of F.
   (b) $O(\ell^3)$ operations in F and space for $O(\ell)$ elements of F;
2. Let $f \in F[X]$ be a monic polynomial of degree $\ell > 0$ over a field F, and let $E := F[X] / (f)$. Also, let $\xi := [X]_f \in E$. For computational purposes, we assume that elements of E and $D_F(E)$ are represented as coordinate vectors with respect to the usual "polynomial" basis $\{\xi^{i-1}\}_{i=1}^{\ell}$. For $\beta \in E$, let M_β denote the β-multiplication map on E that sends $\alpha \in E$ to $\alpha\beta \in E$, which is an F-linear map from E into E.
   (a) Given as input the polynomial f defining E, along with a projection $\pi \in D_F(E)$, an element $\alpha \in E$, and a parameter $k > 0$, show how to compute $(\pi(1), \pi(\alpha), ..., \pi(\alpha^{k-1}))$ using just $O(k\ell + k^{1/2}\ell^2)$ operations in F, and space for $O(k^{1/2}\ell)$ elements of F.
   (b) Given as input the polynomial f defining E, along with a projection $\pi \in D_F(E)$ and an element $\beta \in E$, show how to compute the projection $\pi \circ M_\beta \in D_F(E)$, using $O(\ell^2)$ operations in F.

Theorem. The scalar multiplication $\odot$, together with the usual addition operation on V, makes V into an $F[X]$-module; that is, for all $g, h \in F[X]$ and $\alpha, \beta \in V$, we have

$$g \odot (h \odot \alpha) = (gh) \odot \alpha,\ (g + h) \odot \alpha = g \odot \alpha + h \odot \alpha,$$

$$g \odot (\alpha + \beta) = g \odot \alpha + g \odot \beta,\ 1 \odot \alpha = \alpha.$$

Proof: Exercise.

Note that each choice of τ gives rise to a different $F[X]$-module structure, but all of these structures are extensions of the usual vector space structure, in the sense that for all $c \in F$ and $\alpha \in V$, we have $c \odot \alpha = c\alpha$.

Now, for fixed $\tau \in L_F(V)$ and $\alpha \in V$, consider the $F[X]$-linear map $\rho_{\tau,\alpha} : F[X] \to V$ that sends $g \in F[X]$ to $g \odot \alpha = g(\tau)(\alpha)$. The kernel of this map must be a submodule, and hence an ideal, of $F[X]$; since every ideal of $F[X]$ is principal, it follows that Ker $\rho_{\tau,\alpha}$, is the ideal of $F[X]$ generated by some polynomial $\phi_{\tau,\alpha}$, which we can make unique by insisting that it is monic or zero. We call $\phi_{\tau,\alpha}$, the minimal polynomial of α under τ. We can also characterize $\phi_{\tau,\alpha}$ as follows:

if there exists a non-zero polynomial $g \in F[X]$ such that $g(\tau)(\alpha) = 0$, then $\phi_{\tau,\alpha}$ the monic polynomial of least degree with this property; otherwise, $\phi_{\tau,\alpha} = 0$.

Another way to characterize $\phi_{\tau,\alpha}$ is as follows:

$\phi_{\tau,\alpha}$, is the minimal polynomial of the sequence $\{\tau^i(\alpha)\}_{i=0}^{\infty}$.

Note that since $\phi_\tau(\tau)$ is the zero map, we have

$$\phi_\tau \odot (\alpha) = \phi_\tau(\tau)\alpha = 0,$$

and hence $\phi_\tau \in$ Ker $\rho_{\tau,\alpha}$, which means that $\phi_{\tau,\alpha} \mid \phi_\tau$.

Now consider the image of $\rho_{\tau,\alpha}$, which we shall denote by $\langle\alpha\rangle_\tau$. As an $F[X]$-module, $\langle\alpha\rangle_\tau$ is isomorphic to $F[X] / (\phi_{\tau,\alpha})$. In particular, if $\phi_{\tau,\alpha}$ is non-zero and has degree m, then $\langle\alpha\rangle_\tau$ is a vector space of dimension m over F; indeed, the elements α, $\tau(\alpha)$, ..., $\tau^{m-1}(\alpha)$ form a basis for $\langle\alpha\rangle_\tau$ over F; moreover, m is the smallest non-negative integer such that $\{\tau^i(\alpha)\}_{i=0}^{m}$ is linearly dependent.

Observe that for every $\beta \in \langle\alpha\rangle_\tau$, we have $\phi_{\tau,\alpha} = 0$; indeed, if $\beta = g \odot \alpha$, then

$$\phi_{\tau,\alpha} \odot (g \odot \alpha) = (\phi_{\tau,\alpha} g) \odot \alpha = g \odot (\phi_{\tau,\alpha} \odot \alpha) = g \odot 0 = 0.$$

The following three theorems develop some simple facts; the proofs of these are straightforward, and left as exercises. In each theorem, τ is an element of $L_F(V)$, and $\odot$ is the associated scalar multiplication that makes V into an $F[X]$-module.

Theorem. For all $\tau \in L_F(V)$, for all $c \in F$, and for all $g, h \in F[X]$, we have:

(i) $g(\tau) + h(\tau) = (g + h)(\tau)$;

(ii) $c \,.\, g(\tau) = (cg)(\tau)$;

(iii) $g(\tau) \circ h(\tau) = (gh)(\tau) = h(\tau) \circ g(\tau)$.

Proof: Exercise.

Let $\tau \circ L_F(V)$ be a linear transformation. We define

$$F[\tau] := \{g(\tau) : g \in F[X]\},$$

which is a subset of $L_F(V)$. By the previous theorem, it is clear that $F[\tau]$ is closed under addition, multiplication (i.e., composition), and scalar multiplication, and that $F[\tau]$ is in fact an F-algebra in the usual sense (i.e., multiplication is commutative).

Let ϕ_τ be the minimal polynomial of τ over F, so that $F[\tau]$ is isomorphic as an F-algebra to $F[X] / (\phi_\tau)$. We can also characterize ϕ_τ as follows:

if there exists a non-zero polynomial $g \in F[X]$ such that $g(\tau) = 0$, then ϕ_τ is the monic polynomial of least degree with this property; otherwise, $\phi_\tau = 0$.

Another way to characterize ϕ_τ is as follows:

ϕ_τ is the minimal polynomial of the sequence $\{\tau^i\}_{i=0}^{\infty}$.

If V has finite dimension $\ell > 0$, then by Theorem, $L_F(V)$ is isomorphic as an F-vector space to $F^{\ell \times \ell}$, and so in particular, has dimension ℓ^2. Therefore, there must be a linear dependence among $1, \tau, ..., \tau^{\ell^2}$, which implies that the minimal polynomial of τ is non-zero with degree at most ℓ^2 (and at least 1). We shall show below that in this case, the minimal polynomial of τ actually has degree at most ℓ.

For a fixed $\tau \in L_F(V)$, we can define a "scalar multiplication" operation $\odot$, that maps $g \in F[X]$ and $\alpha \in V$ to

$$g \odot \alpha = \sum_i a_i \tau^i(\alpha).$$

that is, if $g = \sum_i a_i X^i$, then

$$g \odot \alpha = \sum_i a_i \tau^i(\alpha).$$

Theorem. Let $\alpha \in V$. Let $f \in F[X]$ be a monic irreducible polynomial such that $f^e \odot \alpha = 0$ but $f^{e-1} \odot \alpha \neq 0$ for some integer $e \geq 1$. Then f^e is the minimal polynomial of α under τ.

Theorem. Let $\alpha \in V$ have minimal polynomial $f \in F[X]$ under τ, and let $\beta \in V$ have minimal polynomial $g \in F[X]$ under τ. If $\gcd(f, g) = 1$, then $\langle\alpha\rangle_\tau \cap \langle\beta\rangle_\tau = \{0\}$, and $\alpha + \beta$ has minimal polynomial $f \,.\, g$ under τ.

Theorem. If V has finite dimension $\ell > 0$, then for every $\tau \in L_F(V)$, the minimal polynomial of τ is non-zero of degree at most ℓ.

We close this section with a simple observation. Let V be an arbitrary $F[X]$-module with scalar multiplication $\odot$. Restricting the scalar multiplication from $F[X]$ to F, we can naturally view V as an F-vector space. Let $\tau : V \rightarrow V$ be the map that sends $\alpha \in V$ to $X \odot \alpha$. It is easy to see that $\tau \in L_F(V)$, and that for all polynomials $g \in F[X]$, and all $\alpha \in V$, we have $g \odot \alpha = g(\tau)(\alpha)$. Thus, instead of starting with a vector space and defining an $F[X]$-module structure in terms of a given linear map, we can go the other direction, starting from an $F[X]$-module and obtaining a corresponding linear map.

Furthermore, using the language introduced in Examples, we see that the $F[X]$-exponent of V is the ideal of $F[X]$ generated by the minimal polynomial of τ, and the $F[X]$-order of any element $\alpha \in V$ is the ideal of $F[X]$ generated by the minimal polynomial of α under τ. Theorem says that there exists an element in V whose $F[X]$-order is equal to the $F[X]$-exponent of V, assuming the latter is non-zero.

So depending on one's mood, one can place emphasis either on the linear map τ, or just talk about $F[X]$-modules without mentioning any linear maps.

Theorem. Let $\alpha \in L_F(V)$, and suppose that τ has non-zero minimal polynomial ϕ. Then there exists $\beta \in V$ such that the minimal polynomial of β under τ is ϕ.

Proof. Let $\odot$ be the scalar multiplication associated with τ. Let $\phi = \phi_1^{e_1} \ldots \phi_r^{e_r}$ be the factorization of ϕ into monic irreducible polynomials in $F[X]$.

First, we claim that for each $i = 1, \ldots, r$, there exists $\alpha_i \in V$ such that $\phi/\phi_i \odot \alpha_i \neq 0$. Suppose the claim were false: then for some i, we would have $\phi/\phi_i \odot \alpha = 0$ for all $\alpha \in V$; however, this means that $(\phi/\phi_i)(\tau) = 0$, contradicting the minimality property in the definition of the minimal polynomial ϕ. That proves the claim.

Let $\alpha_1, \ldots, \alpha_r$ be as in the above claim. Then by Theorem, each $\phi/\phi_i^{e_i} \odot \alpha_i$ has minimal polynomial $\phi_i^{e_i}$ under τ. Finally, by Theorem,

$$\beta := \phi/\phi_1^{e_1} \odot \alpha_1 + \ldots + \phi/\phi_r^{e_r} \odot \alpha_r$$

has minimal polynomial ϕ under τ.

Theorem says that if τ has minimal polynomial ϕ of degree $m \geq 0$, then there exists $\beta \in V$ such that $\{\tau^i(\beta)\}_{i=0}^{m-1}$ is linearly independent.

EXERCISE

1. Let F be a finite field, and let V have finite dimension $\ell > 0$ over F. Let $\tau \in L_F(V)$ have minimal polynomial ϕ, with $\deg(\phi) = m$. Suppose that $\alpha_1, ..., \alpha_s$ are randomly chosen elements of V. Let g_j be the minimal polynomial of α_j under τ, for $j = 1, ..., s$. Let Q be the probability that $\operatorname{lcm}(g_1, ..., g_s) = \phi$. The goal of this exercise is to show that $Q \geq \Lambda_F^{\phi}(s)$, where $\Lambda_F^{\phi}(s)$ is as defined.

   (a) Show that $h_j \mid g_j$, for $j = 1, ..., s$, and conclude that $Q \geq \Lambda_F^{\phi}(s)$.

   (b) Without the assumption that $m = \ell$, things are a bit more challenging. Adopting the matrix-oriented point of view discussed at the end of §18.3, and transposing everything, show that

   – There exists $\pi \in D_F(V)$ such that the sequence $\{\pi \circ \tau^i\}_{i=0}^{\infty}$ has minimal polynomial ϕ, and

   – If, for $j = 1, ..., s$, we define h_j to be the minimal polynomial of the sequence $\{\pi(\tau^i(\alpha_j))\}_{i=0}^{\infty}$, then the probability that $\operatorname{lcm}(h_1, ..., h_s) = \phi$ is equal to $\Lambda_F^{\phi}(s)$.

2. Let $\tau \in L_F(V)$ have non-zero minimal polynomial ϕ of degree m, and let $\phi = \phi_1^{e_1} ... \phi_r^{e_r}$ be the factorization of ϕ into monic irreducible polynomials in $F[X]$. Let $\odot$ be the scalar multiplication associated with τ. Show that $\beta \in V$ has minimal polynomial ϕ under τ if and only if $\phi/\phi_i \odot \beta \neq 0$ for $i = 1, ..., r$.

3. In this exercise, you are to derive the fundamental theorem of finite dimensional $F[X]$-modules, which is completely analogous to the fundamental theorem of finite abelian groups. Both of these results are really special cases of a more general decomposition theorem for modules over a principal ideal domain. Let V be an $F[X]$-module. Assume that as an F-vector space, V has finite dimension $\ell > 0$, and that the $F[X]$-exponent of V is generated by the monic polynomial $\phi \in F[X]$ (note that $1 \leq \deg(\phi) \leq \ell$). Show that there exist monic, non-constant polynomials $\phi_1, ..., \phi_t \in F[X]$ such that

   1. $\phi_i \mid \phi_{i+1}$ for $i = 1, ..., t - 1$, and
   2. V is isomorphic, as an $F[X]$-module, to the direct product of $F[X]$-modules $V' := F[X] / (\phi_1) \times ... \times F[X] / (\phi_t)$.

   Moreover, show that the polynomials $\phi_1, ..., \phi_t$ satisfying these conditions are uniquely determined, and that $\phi_t = \phi$.

4. Let $\tau \in L_F(V)$ have non-zero minimal polynomial ϕ. Show that τ is bijective if and only if $X \mid \beta$.

5. Let $f, g \in F[X]$ with $f \neq 0$, and let $h := f / \gcd(f, g)$. Show that $g.F[X] / (f)$ and $F[X] / (h)$ are isomorphic as $F[X]$-modules.

Chapter 9

Logarithmic Functions

In mathematics, the logarithm of a given number to a given base is the power or exponent to which the base must be raised in order to produce the given number.

For example, the logarithm of 1000 to the common base 10 is 3, because 10 raised to the power of 3 is 1000; the base 2 logarithm of 32 is 5 because 2 to the power 5 is 32.

The logarithm of x to the base b is written $\log_b(x)$ or, if the base is implicit, as $\log(x)$. So, for a number x, a base b and an exponent y,

if $by = x$, then $\log_b(x) = y$.

An important feature of logarithms is that they reduce multiplication to addition, by the formula:

$\log (x \times y) = \log x + \log y$.

That is, the logarithm of the product of two numbers is the sum of the logarithms of those numbers. The use of logarithms to facilitate complex calculations was a significant motivation in their original development.

BASIC PROPERTIES OF THE LOGARITHM

When x and b are restricted to positive real numbers, $\log_b(x)$ is a unique real number. The magnitude of the base b must be neither 0 nor 1; the base used is typically 10, e, or 2. Logarithms are defined for real numbers and for complex numbers.

The major property of logarithms is that they map multiplication to addition. This ability stems from the following identity:

$$b^x \times b^y = b^{x+y},$$

which by taking logarithms becomes

$$\log_b (b^x \times b^y) = \log_b (b^{x+y}) = x + y = \log_b (b^x) + \log_b (b^y)$$

A related property is reduction of exponentiation to multiplication. Using the identity:

$$c = b^{\log_b(c)},$$

it follows that c to the power p (exponentiation) is:

$$c^p = (b^{\log_b(c)})p = b^{p\log_b(c)},$$

or, taking logarithms:

$$\log_b(c^p) = p^{\log_b}(c).$$

In words, to raise a number to a power p, find the logarithm of the number and multiply it by p. The exponentiated value is then the inverse logarithm of this product; that is, *number to power* = b^{product}.

Besides reducing multiplication operations to addition, and exponentiation to multiplication, logarithms reduce division to subtraction, and roots to division. Logarithms make lengthy numerical operations easier to perform. The whole process is made easy by using tables of logarithms, or a slide rule, antiquated now that calculators are available. Although the above practical advantages are not important for numerical work today, they are used in graphical analysis.

THE LOGARITHM AS A FUNCTION

Though logarithms have been traditionally thought of as arithmetic sequences of numbers corresponding to geometric sequences of other (positive real) numbers, as in the 1797 Britannica definition, they are also the result of applying an analytic function. The function can therefore be meaningfully extended to complex numbers.

The function $\log_b(x)$ depends on both b and x, but the term logarithm function (or logarithmic function) in standard usage refers to a function of the form $\log_b(x)$ in which the base b is fixed and so the only argument is x. Thus there is one logarithm function for each value of the base b (which must be positive and must differ from 1). Viewed in this way, the base-b logarithm function is the inverse function of the exponential function b^x. The word "logarithm" is often used to refer to a logarithm function itself as well as to particular values of this function.

The base can also be a complex number; the evaluation of the log is just slightly more complicated in this case. See imaginary base.

LOGARITHM OF A COMPLEX NUMBER

When the base b is real and z is a complex number, say $z = x + i\,y$, the logarithm of z is found easily by putting z in polar form that is, $z = (x^2 + y^2)^{1/2} \exp\,(i \tan^{"1} (y / x))$. If the base of the logarithm is chosen as e, that is, using $\log_e$ (denoted by *ln* and called the natural logarithm), the logarithm becomes:

$$\begin{aligned} \ln(z) &= \ln\left[(x^2 + y^2)^{1/2}\, e^{i\tan^{-1} y/x}\right] \\ &= \ln\left[(x^2 + y^2)\right] + \ln\left[e^{i\tan^{-1} y/x}\right] \end{aligned}$$

$$= \frac{1}{2}\ln(x^2 + y^2) + i\tan^{-1}\left(\frac{y}{x}\right).$$

This evaluation uses the properties of all logarithms (see above), regardless of choice of base: $\log_b (c\ d) = \log_b (c) + \log_b (d)$ and its generalization to arbitrary products $\log_b b^z = z$. Because the inverse tangent is a multiple valued function of its argument, the logarithm of a complex number is not unique either. See article on complex logarithm.

GROUP THEORY

From the pure mathematical perspective, the identity

$$\log (cd) = \log (c) + \log (d)$$

is fundamental in two senses. First, the remaining three arithmetic properties can be derived from it. Furthermore, it expresses an isomorphism between the multiplicative group of the positive real numbers and the additive group of all the reals.

Logarithmic functions are the only continuous isomorphisms from the multiplicative group of positive real numbers to the additive group of real numbers.

BASES

The most widely used bases for logarithms are 10, the mathematical constant $e \approx 2.71828...$ and 2. When "log" is written without a base (b missing from $\log_b$), the intent can usually be determined from context:

- Natural logarithm ($\log_e$, ln, log, or Ln) in mathematical analysis, statistics, economics and some engineering fields. The reasons to consider e the *natural* base for logarithms, though perhaps not obvious, are numerous and compelling.
- Common logarithm ($\log_{10}$ or simply log; sometimes lg) in various engineering fields, especially for power levels and power ratios, such as acoustical sound pressure, and in logarithm tables to be used to simplify hand calculations
- Binary logarithm ($\log_2$; sometimes lg, lb, or ld) in information theory and computer-related fields
- Indefinite logarithm when the base is irrelevant, e.g. in complexity theory when describing the asymptotic behavior of algorithms in big O notation.

To avoid confusion, it is best to specify the base if there is any chance of misinterpretation.

Other Notations

The notation "ln(x)" invariably means $\log_e(x)$, i.e., the natural logarithm of x, but the implied base for "log(x)" varies by discipline:

- Mathematicians generally understand "ln(x)" to mean $\log_e(x)$ and "log(x)" to mean "$\log_{10}(x)$". Calculus textbooks will occasionally write "lg(x)" to represent "$\log_{10}(x)$".

- Many engineers, biologists, astronomers, and some others write only "ln(x)" or "$\log_e(x)$" when they mean the natural logarithm of x, and take "log(x)" to mean $\log_{10}(x)$ or, sometimes in the context of computing, $\log_2(x)$.
- On most calculators, the LOG button is $\log_{10}(x)$ and LN is $\log_e(x)$.
- In most commonly used computer programming languages, including C, C++, Java, Fortran, Ruby, and BASIC, the "log" function returns the natural logarithm. The base-10 function, if it is available, is generally "log10."
- Some people use Log(x) (capital L) to mean $\log_{10}(x)$, and use log(x) with a lowercase l to mean $\log_e(x)$.
- The notation Log(x) is also used by mathematicians to denote the principal branch of the (natural) logarithm function.
- A notation frequently used in some European countries is the notation ${}^b\log(x)$ instead of $\log_b(x)$.

This chaos, historically, originates from the fact that the natural logarithm has nice mathematical properties (such as its derivative being $1/x$, and having a simple definition), while the base 10 logarithms, or decimal logarithms, were more convenient for speeding calculations (back when they were used for that purpose). Thus natural logarithms were only extensively used in fields like calculus while decimal logarithms were widely used elsewhere.

As recently as 1984, Paul Halmos in his "automathography" *I Want to Be a Mathematician* heaped contempt on what he considered the childish "ln" notation, which he said no mathematician had ever used. The notation was in fact invented in 1893 by Irving Stringham, professor of mathematics at Berkeley. As of 2005, many mathematicians have adopted the "ln" notation, but most use "log".

In computer science, the base 2 logarithm is sometimes written as lg(x), as suggested by Edward Reingold and popularized by Donald Knuth. However, lg(x) is also sometimes used for the common log, and lb(x) for the binary log. In Russian literature, the notation lg(x) is also generally used for the base 10 logarithm. In German, lg(x) also denotes the base 10 logarithm, while sometimes ld(x) or lb(x) is used for the base 2 logarithm.

The clear advice of the United States Department of Commerce National Institute of Standards and Technology is to follow the ISO standard *Mathematical signs and symbols for use in physical sciences and technology, ISO 31-11:1992*, which suggests these notations:

- The notation "ln(x)" means $\log_e(x)$;
- The notation "lg(x)" means $\log_{10}(x)$;
- The notation "lb(x)" means $\log_2(x)$.

As the difference between logarithms to different bases is one of *scale*, it is possible to consider all logarithm functions to be the same, merely giving the answer in different *units*, such as dB, neper, bits, decades, etc.; see the section Science and engineering below. Logarithms to a base less than 1 have a negative scale, or a flip about the x axis, relative to logarithms of base greater than 1.

Change of Base

While there are several useful identities, the most important for calculator use lets one find logarithms with bases other than those built into the calculator (usually $\log_e$ and $\log_{10}$). To find a logarithm with base *b*, using any other base *k*:

$$\log_b(x) = \frac{\log_k(x)}{\log_k(b)}$$

Moreover, this result implies that all logarithm functions (whatever the base) are similar to each other. So to calculate the log with base 2 of the number 16 with a calculator:

$$\log_2(16) = \frac{\log(16)}{\log(2)}$$

USES OF LOGARITHMS

Logarithms are useful in solving equations in which exponents are unknown. They have simple derivatives, so they are often used in the solution of integrals. The logarithm is one of three closely related functions. In the equation $b^n = x$, *b* can be determined with radicals, *n* with logarithms, and *x* with exponentials. See logarithmic identities for several rules governing the logarithm functions.

EXPONENTIAL FUNCTIONS

One way of defining the exponential function e^x, also written as $\exp(x)$, is as the inverse of the natural logarithm. It is positive for every real argument *x*.

The operation of "raising *b* to a power *p*" for positive arguments *b* and all real exponents *p* is defined by

$$bp = (e\ln b)p = e^{p\ln b}.$$

The antilogarithm function is another name for the inverse of the logarithmic function. It is written $\text{antilog}_b(n)$ and means the same as b^n.

EASIER COMPUTATIONS

Logarithms can be used to replace difficult operations on numbers by easier operations on their logs (in any base), as the following table summarizes. In the table, upper-case variables represent logs of corresponding lower-case variables:

Operation with numbers	Operation with exponents	Logarithmic identity
cd	$C + D$	$\log(cd) + \log(d)$
c/d	$C - D$	$\log(c/d) = \log(c) - \log(d)$
c^d	Cd	$\log(c^d) = d\log(c)$
$\sqrt[d]{c}$	C/d	$\log(\sqrt[d]{c}) = \dfrac{\log(c)}{d}$

These arithmetic properties of logarithms make such calculations much faster. The use of logarithms was an essential skill until electronic computers and calculators became available. Indeed the discovery of logarithms, just before Newton's era, had an impact in the scientific world that can be compared with that of the advent of computers in the 20th century because it made feasible many calculations that had previously been too laborious.

As an example, to approximate the product of two numbers one can look up their logarithms in a table, add them, and, using the table again, proceed from that sum to its antilogarithm, which is the desired product. The precision of the approximation can be increased by interpolating between table entries. For manual calculations that demand any appreciable precision, this process, requiring three lookups and a sum, is much faster than performing the multiplication. To achieve seven decimal places of accuracy requires a table that fills a single large volume; a table for nine-decimal accuracy occupies a few shelves. Similarly, to approximate a power c^d one can look up log c in the table, look up the log of *that*, and add to it the log of d; roots can be approximated in much the same way.

One key application of these techniques was celestial navigation. Once the invention of the chronometer made possible the accurate measurement of longitude at sea, mariners had everything necessary to reduce their navigational computations to mere additions. A five-digit table of logarithms and a table of the logarithms of trigonometric functions sufficed for most purposes, and those tables could fit in a small book.

Another critical application with even broader impact was the slide rule, an essential calculating tool for engineers. Many of the powerful capabilities of the slide rule derive from a clever but simple design that relies on the arithmetic properties of logarithms. The slide rule allows computation much faster still than the techniques based on tables, but provides much less precision, although slide rule operations can be chained to calculate answers to any arbitrary precision.

RELATED OPERATIONS

Cologarithms

The cologarithm of a number is the logarithm of the inverse of said number, meaning $\operatorname{colog}_b(x) = \log_b(1/x) = -\log_b(x)$.

Antilogarithms

The antilogarithm is the logarithmic inverse of the logarithm, meaning that the $\text{antilog}_b(\log_b(x)) = x$. Thus, setting $b^y = x$ implies that $\log_b(x) = y$. By taking the antilog_b of both sides, $\text{antilog}_b(\log_b(x)) = \text{antilog}_b y$, thus $x = \text{antilog}_b y$. Therefore, $b^y = \text{antilog}_b y$.

Calculus

The natural logarithm of a positive number x can be defined as

$$\ln(x) = \int_1^x \frac{dt}{t}.$$

The derivative of the natural logarithm function is

$$\frac{d}{dx}\ln(x) = \frac{1}{x}.$$

By applying the change-of-base rule, the derivative for other bases is

$$\frac{d}{dx}\log_b(x) = \frac{d}{dx}\frac{\ln(x)}{\ln(b)} = \frac{1}{x\ln(b)} = \frac{\log_b(e)}{x}.$$

The antiderivative of the natural logarithm $\ln(x)$ is

$$\int \ln(x)dx = x\ln(x) - x + C,$$

and so the antiderivative of the logarithm for other bases is

$$\int \log_b(x)dx = x\log_b(x) - \frac{x}{\ln(b)} + C = x\log_b\left(\frac{x}{e}\right) + C$$

Series for Calculating the Natural Logarithm

There are several series for calculating natural logarithms. The simplest, though inefficient, is:

$$\ln(z) = \sum_{n=1}^{\infty} \frac{-(-1)^n}{n}(z-1)^n \quad \text{when } |Z-1| < 1.$$

To derive this series, start with ($|x| < 1$)

$$\frac{1}{1-x} = 1 + x + x^2 + x^3 + \dots .$$

Integrate both sides to obtain

$$-\ln(1-x) = x + \frac{x^2}{2} + \frac{x^3}{3} + \dots$$

$$\ln(1-x) = -x - \frac{x^2}{2} - \frac{x^3}{3} - \frac{x^4}{4} - \ldots$$

Letting $z = 1 - x$ and thus $x = -(z - 1)$, we get

A more efficient series is

$$\ln z = (z-1) - \frac{(z-1)^2}{2} + \frac{(z-1)^3}{3} - \frac{(z-1)^4}{4} + \ldots$$

for z with positive real part.

$$\ln(z) = 2\sum_{n=0}^{\infty} \frac{1}{2n+1}\left(\frac{z-1}{z+1}\right)^{2n+1}$$

To derive this series, we begin by substituting $-x$ for x and get

$$\ln(1+x) = x - \frac{x^2}{2} - \frac{x^3}{3} - \frac{x^4}{4} - \ldots.$$

Subtracting, we get

$$\ln\frac{1+x}{1-x} = \ln(1+x) - \ln(1-x) = 2x + 2\frac{x^2}{3} + 2\frac{x^5}{5} + \ldots.$$

Letting $z = \dfrac{1+x}{1-x}$ and thus $z = \dfrac{1-x}{1+x}$, we get

$$\ln z = 2\left(\frac{z-1}{z+1} + \frac{1}{3}\left(\frac{z-1}{z+1}\right)^2 + \frac{1}{5}\left(\frac{z-1}{z+1}\right)^5 + \ldots\right).$$

For example, applying this series to

$$z = \frac{11}{9}$$

we get

$$\frac{z-1}{z+1} = \frac{\frac{11}{9}-1}{\frac{11}{9}+1} = \frac{1}{10},$$

and thus

$$\ln(1.2222222\ldots) = \frac{2}{10}\left(1 + \frac{1}{3\cdot 100} + \frac{1}{5\cdot 10000} + \frac{1}{7\cdot 1000000} + \ldots\right)$$

$$= 0.20 \times (1.000000... + 0.0033333\ ... + 0.0000200...+ 0.0000001... + ...)$$
$$= 0.2 \times 1.0033535... + = 0.2006707...$$

where we factored 1/10 out of the sum in the first line.

For any other base b, we use

$$\log_b(x) = \frac{\ln(x)}{\ln(b)}$$

SMOOTH NUMBERS

If y is a non-negative real number and m is a positive integer, then we say that m is y-smooth if all prime divisors of m are at most y.

For $0 \leq y \leq x$, let us define $\Psi(y, x)$ to be the number of y-smooth integers up to x. The following theorem gives us a lower bound on $\Psi(y, x)$, which will be crucial in the analysis of our discrete logarithm and factoring algorithms.

Theorem. Let y be a function of x such that

$$\frac{y}{\log x} \to \infty \text{ and } u := \frac{\log x}{\log y} \to \infty$$

as $x \to \infty$. Then

$$\Psi(y, x) \geq x \,.\, \exp[(-1 + o(1))u \log \log x].$$

Proof: Let us write $u = \lfloor u \rfloor + \delta$, where $0 \leq \delta < 1$. Let us split the primes up to y into two sets: the set V of "very small" primes that are at most $y^\delta/2$, and the set W of other primes that are greater than $y^\delta/2$ but at most y. To simplify matters, let us also include the integer 1 in the set V.

By Bertrand's postulate, there exists a constant $C > 0$ such that $|W| \geq Cy/\log y$ for sufficiently large y. By the assumption that $y/\log x \to \infty$ as $x \to \infty$, we also have $|W| \geq 2\lfloor u \rfloor$ for sufficiently large x.

To derive the lower bound, we shall count those integers that can be built up by multiplying together b*u*c distinct elements of W, together with one element of V.

These products are clearly distinct, y-smooth numbers, and each is bounded by x, since each is at most $y^{\lfloor u \rfloor} y^\delta = y^u = x$.

If S denotes the set of all of these products, then for x su—ciently large, we have

$$|S| = \binom{|W|}{\lfloor u \rfloor} . |V|$$

$$= \frac{|W|(|W|-1)...(|W|-\lfloor u \rfloor+1)}{\lfloor u \rfloor!}.|V|$$

$$\geq \left(\frac{|W|}{2u}\right)^{\lfloor u \rfloor}.|V|$$

$$\geq \left(\frac{Cy}{2u\log y}\right)^{\lfloor u \rfloor}.|V|$$

$$= \left(\frac{Cy}{2\log x}\right)^{u-\delta}.|V|.$$

Taking logarithms, we have

$$\log |S| \geq (u-\delta)(\log y - \log\log x + \log(C/2)) + \log |V|$$

$$= \log x - u \log\log x + (\log |V| - \delta \log y) +$$

$$O(u + \log\log x).$$

To prove the theorem, it suffices to show that

$$\log |S| \geq \log x - (1 + o(1))u \log\log x.$$

Under our assumption that $u \to \infty$, the term $O(u + \log\log x)$ in is clearly $o(u \log\log x)$, and so it will suffice to show that the term $(\log |V| - \delta \log y)$ is also $o(u \log\log x)$. But by Chebyshev's theorem, for some positive constant D, we have

$$Dy^{\delta}/\log y \leq |V| \leq y^{\delta},$$

and taking logarithms, and again using the fact that $u \to \infty$, we have

$$\log |V| - \delta \log y = O(\log\log y) = o(u \log\log x).$$

AN ALGORITHM FOR FACTORING INTEGERS

We now present a probabilistic, subexponential-time algorithm for factoring integers. The algorithm uses techniques very similar to those used in Algorithm SEDL.

Let $n > 1$ be the integer we want to factor. We make a few simplifying assumptions. First, we assume that n is odd – this is not a real restriction, since we can always pull out any factors of 2 in a pre-processing step. Second, we assume that n is not a perfect power, that is, not of the form a^b for integers $a > 1$ and $b > 1$ – this is also not a real restriction, since we can always partially factor n using the algorithm, if n is a perfect power. Third, we assume that n is not prime – this may be efficiently checked using, say, the Miller–Rabin test. Fourth, we assume that n is not divisible by any primes up to a "smoothness parameter" y – we can ensure this using trial division, and it will be clear that the running time of this pre-computation is dominated by that of the algorithm itself.

With these assumptions, the prime factorization of n is of the form

$$n = q_1^{f_1} \dots q_w^{f_w},$$

where $w > 1$, the q_i's are distinct, odd primes, each greater than y, and the f_i's are positive integers.

The main goal of our factoring algorithm is to find a random square root of 1 in $\mathbb{Z}_n^*$. Let

$$\theta : \mathbb{Z}_n \to \mathbb{Z}_{q_1^{f_1}} \times \dots \times \mathbb{Z}_{q_w^{f_w}}$$

$$[a]_n \mapsto \left([a]_{q_1^{f_1}}, \dots, [a]_{q_w^{f_w}}\right)$$

be the ring isomorphism of the Chinese remainder theorem. The square roots of 1 in $\mathbb{Z}_n^*$ are precisely those elements $\gamma \in \mathbb{Z}_n^*$ such that $\theta(\gamma) = (\pm 1, \dots, \pm 1)$. If γ is a random square root of 1, then with probability $1 - 2^{-w+1} \geq 1/2$, we have $\geq \theta(\gamma) = (\gamma_1, \dots, \gamma_w)$, where the γ_i's are neither all 1 nor all -1 (i.e., $\gamma \neq \pm 1$). If this happens, then $(\gamma_i - 1) = (1 - 1, \dots, \gamma_w - 1)$, and so we see that some, but not all, of the values $\gamma_i - 1$ will be zero. The value of gcd(rep($\gamma_i - 1$), n) is precisely the product of the prime powers $q_i^{f_i}$ such that $\gamma_i - 1 = 0$, and hence this gcd will yield a non-trivial factorization of n, unless $\gamma = \pm 1$.

Let $p_1, \dots, p_k$ be the primes up to the smoothness parameter y mentioned above. Let $\pi_i := [p_i]_n \in \mathbb{Z}_n^*$ for $i = 1, \dots, k$.

We first describe a simplified version of the algorithm, after which we modify the algorithm slightly to deal with a technical problem. Like Algorithm SEDL, this algorithm proceeds in two stages. In the first stage, we find relations of the form

$$\alpha_i^2 = \pi_1^{e_{i1}} \dots \pi_k^{e_{ik}},$$

for $i = 1, \dots, k + 1$, where $e_{i1}, \dots, e_{ik} \in \mathbb{Z}$ and $i \in \mathbb{Z}_n^*$ for each i.

We can obtain each such relation by randomized search, as follows: we select $i \in \mathbb{Z}_n^*$ at random, square it, and try to factor $mi :=$ rep (α_i^2) by trial division, trying all the primes $p_1, \dots, p_k$ up to y. If we are lucky, we obtain a factorization

$$m_i = p_1^{e_{i1}} \dots p_k^{e_{ik}},$$

for some exponents $e_{i1}, \dots, e_{ik}$, yielding the relation ; if not, we just keep trying.

For $i = 1, ..., k + 1$, let $v_i := (e_{i1}, ..., e_{ik}) \in \mathbb{Z}^{\times k}$, and let $\overline{v}_i$ denote the image of v_i in $\mathbb{Z}_2^{\times k}$ (i.e., $\overline{v}_i := ([e_{i1}]_2, ..., [e_{ik}]_2)$). Since $\mathbb{Z}_2^{\times k}$ is a vector space over the field $\mathbb{Z}_2$ of dimension k, the family of vectors $v_1, ..., v_{k+1}$ must be linearly dependent. The second stage of the algorithm uses Gaussian elimination over $\mathbb{Z}_2$ to find a linear dependence among the vectors $\overline{v}_1, ..., \overline{v}_{k+1}$, that is, to find integers $c_1, ..., c_{k+1} \in \{0, 1\}$, not all zero, such that

$$(e_1, ..., e_k) := c_1 v_1 + ... + c_{k+1} v_{k+1} \in 2\mathbb{Z}^{\times k}.$$

Raising each equation to the corresponding power c_i, and multiplying them all together, we obtain

$$\alpha^2 = \pi_1^{e_1} ... \pi_k^{e_k},$$

where

$$\alpha := \prod_{i=1}^{k+1} \alpha_i^{c_i}.$$

Since each e_i is even, we can compute

$$\beta := \pi_1^{e_1/2} ... \pi_k^{e_k/2},$$

and we see that $\alpha^2 = \beta^2$, and hence $(\alpha/\beta)^2 = 1$. Thus, $\gamma := \alpha/\beta$ is a square root of 1 in $\mathbb{Z}_n^*$. A more careful analysis shows that in fact, γ is uniformly distributed over all square roots of 1, and hence, with probability at least 1/2, if we compute gcd(rep($\gamma - 1$), n), we get a non-trivial factor of n.

That is the basic idea of the algorithm. There is, however, a technical problem. Namely, in the method outlined above for generating a relation, we attempt to factor $m_i :=$ rep(α_i^2). Thus, the running time of the algorithm will depend in a crucial way on the probability that a random square modulo n is y-smooth. Unfortunately for us, Theorem does not say anything about this situation – it only applies to the situation where a number is chosen at random from an interval $[1, x]$. There are (at least) three different ways to address this problem:

1. Ignore it, and just assume that the bounds in Theorem apply to random squares modulo n (taking $x := n$ in the theorem).
2. Prove a version of Theorem that applies to random squares modulo n.
3. Modify the factoring algorithm, so that Theorem applies.

The first choice, while not unreasonable from a practical point of view, is not very satisfying mathematically. It turns out that the second choice is indeed a viable option (i.e., the theorem is true and is not so difficult to prove), but we opt for the third choice, as it is somewhat easier to carry out, and illustrates a probabilistic technique that is more generally useful.

So here is how we modify the basic algorithm. Instead of generating relations of the form, we generate relations of the form

$$\alpha_i^2\delta = \pi_1^{e_{i1}} \ldots \pi_k^{e_{ik}},$$

for $i = 1, \ldots, k+2$, where $e_{i1}, \ldots, e_{ik} \in \mathbb{Z}$ and $\alpha_i \in \mathbb{Z}_n^*$ for each i, and $\delta \in \mathbb{Z}_n^*$. Note that the value δ is the same in all relations.

We generate these relations as follows. For the very first relation (i.e., $i = 1$), we repeatedly choose δ_1 and δ in $\mathbb{Z}_n^*$ at random, until rep($\alpha_1^2\delta$) is y-smooth. Then, after having found the first relation, we find each subsequent relation (i.e., for $i > 1$) by repeatedly choosing α_i in $\mathbb{Z}_n^*$ at random until rep($\alpha_i^2\delta$) is y-smooth, where δ is the same value that was used in the first relation. Now, Theorem will apply directly to determine the success probability of each attempt to generate the first relation. When we have found this relation, the value $\alpha_1^2\delta$ will be uniformly distributed over all y-smooth elements of $\mathbb{Z}_n^*$ (i.e., elements whose integer representations are y-smooth). Consider the various cosets of $(\mathbb{Z}_n^*)^2$ in $\mathbb{Z}_n^*$. Intuitively, it is much more likely that a random y-smooth element of $\mathbb{Z}_n^*$ lies in a coset that contains many y-smooth elements than in a coset with very few, and indeed, it is reasonably likely that the fraction of y-smooth elements in the coset containing δ is not much less than the overall fraction of y-smooth elements in $\mathbb{Z}_n^*$. Therefore, for $i > 1$, each attempt to find a relation should succeed with reasonably high probability. This intuitive argument will be made rigorous in the analysis to follow.

The second stage is then modified as follows. For $i = 1, \ldots, k+2$, let $v_i := (e_{i1}, \ldots, e_{ik}, 1) \in \mathbb{Z}^{\times(k+1)}$, and let $\bar{v}_i$ denote the image of v_i in $\mathbb{Z}_2^{\times(k+1)}$. Since $\mathbb{Z}_2^{\times(k+1)}$ is a vector space over the field $\mathbb{Z}_2$ of dimension $k+1$, the family of vectors $\bar{v}_1, \ldots, \bar{v}_{k+2}$ must be linearly dependent. Therefore, we use Gaussian elimination over $\mathbb{Z}_2$ to find a linear dependence among the vectors $\bar{v}_1, \ldots, \bar{v}_{k+2}$, that is, to find integers $c_1, \ldots, c_{k+2} \in \{0, 1\}$, not all zero, such that

$$(e_1, \ldots, e_{k+1}) := c_1 v_1 + \ldots + c_{k+2} v_{k+2} \in 2\mathbb{Z}^{\times(k+1)}.$$

Raising each equation to the corresponding power *ci*, and multiplying them all together, we obtain

$$\alpha^2\delta^{e_{k+1}} = \pi_1^{e_1} \ldots \pi_k^{e_k},$$

where

$$\alpha := \prod_{i=1}^{k+2} \alpha_i^{c_i}.$$

Since each e_i is even, we can compute

$$\beta := \pi_1^{e_1/2} ... \pi_k^{e_k/2} \delta^{-e_{k+1}/2},$$

so that $\alpha^2 = \beta^2$ and $\gamma := \alpha/\beta$ is a square root of 1 in $\mathbb{Z}_n^*$.

Now the analysis. From the discussion above, it is clear that Algorithm SEF either outputs "failure," or outputs a non-trivial factor of n. So we have the same three questions to answer as we did in the analysis of Algorithm SEDL:

1. What is the expected running time of Algorithm SEF?
2. How should the smoothness parameter y be chosen so as to minimize the expected running time?
3. What is the probability that Algorithm SEF outputs "failure"?

To answer the first question, let σ denote the probability that (the canonical representative of) a random element of $\mathbb{Z}_n^*$ is y-smooth. For $i = 1, ..., k + 2$, let L_i denote the number of iterations of the inner loop in the ith iteration of the main loop in stage 1; that is, L_i is the number of attempts made in finding the ith relation.

Lemma: For $i = 1, ..., k + 2$, we have $E[L_i] \leq \sigma -1$.

Proof. We first compute $E[L_1]$. As is chosen uniformly from $\mathbb{Z}_n^*$ and independent of α_1, at each attempt to find a relation, $\alpha_1^2\delta$ is uniformly distributed over $\mathbb{Z}_n^*$,

$i \leftarrow 0$

repeat

$i \leftarrow i + 1$

repeat

choose $\alpha_i \in \mathbb{Z}_n^*$ at random

if $i = 1$ then choose $\delta \in \mathbb{Z}_n^*$ at random

$m_i \leftarrow \text{rep}(\alpha_1^2\delta)$

test if m_i is y-smooth (trial division)

until $m_i = p_1^{e_{i1}} ... p_k^{e_{ik}}$ for some integers $e_{i1}, ..., e_{ik}$

until $i = k + 2$

set $v_i \rightarrow (e_{i1}, ..., e_{ik}, 1) \in \mathbb{Z}^{\times(k+1)}$ for $i = 1, ..., k+2$

apply Gaussian elimination over $\mathbb{Z}_2$ to find integers $c_1, ..., c_{k+2} \in \{0, 1\}$, not all zero, such that

$(e_1, ..., e_{k+1}) := c_1 v_1 + ... + c_{k+2}\, vk_{+2} \in 2\mathbb{Z}^{\times(k+1)}$.

$\alpha \leftarrow \prod_{i=1}^{k+2} \alpha_i^{c_i}, \beta \leftarrow \pi_1^{e_1/2} ... \pi_k^{e_k/2} \delta^{-e_{k+1}/2}, \gamma \leftarrow \alpha/\beta$

if $\gamma = \pm 1$

then output "failure"

else output gcd(rep($\gamma - 1$), n)

and hence the probability that the attempt succeeds is precisely σ. This means $E[L_1] = \sigma_{-1}$.

We next compute $E[L_i]$ for $i > 1$. To this end, let us denote the cosets of $(\mathbb{Z}_n^*)^2$ by $\mathbb{Z}_n^*$ as $C_1, ..., C_t$. As it happens, $t = 2^w$, but this fact plays no role in the analysis. For $j = 1, ..., t$, let σ_j denote the probability that a random element of C_j is y-smooth, and let σ_j denote the probability that the final value of δ belongs to C_j.

We claim that for $j = 1, ..., t$, we have $\tau_j = \sigma_j \sigma^{-1} t^{-1}$. To see this, note that each coset C_j has the same number of elements, namely, $|\mathbb{Z}_n^*|^{t-1}$, and so the number of y-smooth elements in C_j is equal to $\sigma_j |\mathbb{Z}_n^*|^{t-1}$. Moreover, the final value of $\alpha_1^2\delta$ is equally likely to be any one of the y-smooth numbers in $\mathbb{Z}_n^*$, of which there are $\sigma|\mathbb{Z}_n^*|$, and hence

$$\tau_j = \frac{\sigma_j |\mathbb{Z}_n^*| t^{-1}}{\sigma |\mathbb{Z}_n^*|} = \sigma_j \sigma^{-1} t^{-1},$$

which proves the claim.

Now, for a fixed value of and a random choice of $\alpha_i \in \mathbb{Z}_n^*$, one sees that $\alpha_i^2\delta$ is uniformly distributed over the coset containing δ. Therefore, for $j = 1, ..., t$, if $\tau_j > 0$, we have

$E[L_i \mid \delta \in C_j] = -1$

Summing over all $j = 1, ..., t$ with $\tau_j > 0$, it follows that

$$E[L_i] = \sum_{\tau_j > 0} E[L_i \mid \delta \in C_j].P[\delta \in C_j]$$

$$= \sum_{\tau_j > 0} \sigma_j^{-1}.\tau_j = \sum_{\tau_j > 0} \sigma_j^{-1}.\sigma_j \sigma^{-1} t^{-1} \leq \sigma^{-1},$$

which proves the lemma.

So in first stage, the expected number of attempts made in generating a single relation is σ^{-1}, each such attempt takes time $k.\text{len}(n)^{O(1)}$, and we have to generate $k + 2$ relations, leading to a total expected running time in stage 1 of $-1k2$ len$(n)O(1)$.

Second stage is dominated by the cost of performing Gaussian elimination, which takes time $k^3 . \text{len}(n)^{O(1)}$. Thus, if Z is the total running time of the algorithm, we have

$$\text{E}[Z] \leq (\sigma^{-1} k^2 + k^3) . \text{len}(n)^{O(1)}.$$

By our assumption that n is not divisible by any primes up to y, all y-smooth integers up to $n - 1$ are in fact relatively prime to n. Therefore, the number of y-smooth elements of $\mathbb{Z}_n^*$ is equal to $\Psi(y, n - 1)$, and since n itself is not y-smooth, this is equal to $\Psi(y, n)$. From this, it follows that

$$\sigma = \psi(y,n)/|\mathbb{Z}_n^*| \geq \psi(y,n)/n.$$

The rest of the running time analysis is essentially the same as in the analysis of Algorithm SEDL; that is, assuming $y = \exp[(\log n)^{\lambda+o(1)}]$ for some constant $0 < \lambda < 1$, we obtain

$$\text{E}[Z] \leq \exp[(1 + o(1)) \max \{(\log n/ \log y) \log \log n + 2 \log y, 3 \log y\}].$$

Setting $y = \exp[(1/\sqrt{2})(\log n \log \log n)^{1/2}]$, we obtain

$$\text{E}[Z] \leq \exp[(2\sqrt{2} + o(1))(\log n \log \log n)^{1/2}].$$

That basically takes care of the first two questions. As for the third, we have:

Lemma: Algorithm SEF outputs "failure" with probability $2^{-w+1} \leq 1/2$.

Proof. Let F be the event that the algorithm outputs "failure." We may view the final values assigned to δ and $\alpha_1, ..., \alpha_{k+2}$ as random variables, which we shall denote by these same names. Let $\delta' \in \mathbb{Z}_n^*$ and $\alpha_1', ..., \alpha_{k+2}' \in (\mathbb{Z}_n^*)^2$ be arbitrary, fixed values such that rep$(\alpha_i'\delta')$ is y-smooth for $i = 1, ..., k + 2$. Let B be the event that $\delta = \delta'$ and $\alpha_i^2 = \alpha_i'$ for $i = 1, ..., k + 2$. We shall show that $P[F | B] = 2^{-w+1}$, and since this holds for all relevant B, it follows by total probability that $P[F] = 2^{-w+1}$.

For the rest of the argument, we focus on the conditional distribution given B. With respect to this conditional distribution, the distribution of each random variable α_i is (essentially) the uniform distribution on $\rho^{-1}(\{\alpha_i'\})$, where ρ is the squaring map on $\mathbb{Z}_n^*$. Moreover, the family of random variables $\{\alpha_i\}_{i=1}^{k+2}$ is mutually independent. Also, the values and $c_1, ..., c_{k+2}$ computed by the algorithm are fixed. It follows that the distribution of is (essentially) the uniform distribution on $\rho^{-1}(\{\beta^2\})$, and hence $\gamma := \alpha/\beta$ is a random square root of 1 in $\mathbb{Z}_n^*$. Thus, $\gamma = \pm 1$ with probability 2^{-w+1}.

Let us summarize the above discussion in the following theorem.

Theorem. With the smoothness parameter set as

$$y := \exp[(1/\sqrt{2})(\log n \log \log n)^{1/2}],$$

the expected running time of Algorithm SEF is at most

$$\exp[(2\sqrt{2} + o(1))(\log n \log \log n)^{1/2}].$$

The probability that Algorithm SEF outputs "failure" is at most 1/2.

ALGORITHM OF LOGARITHMS

We now present a probabilistic, subexponential-time algorithm for computing discrete logarithms. The input to the algorithm is p, q, γ, α, where p and q are primes, with $q \mid (p - 1)$, γ is an element of $\mathbb{Z}^*_p$ generating a subgroup G of $\mathbb{Z}^*_p$ of order q, and $\alpha \in G$.

We shall make the simplifying assumption that $q^2 \mid (p - 1)$, which is equivalent to saying that $q \mid m := (p - 1)/q$. Although not strictly necessary, this assumption simplifies the design and analysis of the algorithm, and moreover, for cryptographic applications, this assumption is almost always satisfied. At a high level, the main goal of our discrete logarithm algorithm is to find a random representation of 1 with respect to γ and α, this allows us to compute $\log_\gamma \alpha$ (with high probability). More precisely, our main goal is to compute integers r and s in a probabilistic fashion, such that $\gamma^r\alpha^s = 1$ and $[s]_q$ is uniformly distributed over $\mathbb{Z}_q$. Having accomplished this, then with probability $1 - 1/q$, we shall have $s \not\equiv 0 \pmod q$, which allows us to compute $\log_\gamma \alpha$ as $-rs^{-1} \bmod q$.

Let H be the subgroup of $\mathbb{Z}^*_p$ of order m. Our assumption that $q \mid m$ implies that $G \cap H = \{1\}$, since the multiplicative order of any element in the intersection must divide both q and m, and so the only possibility is that the multiplicative order is 1. Therefore, the map $\rho : G \times H \to \mathbb{Z}^*_p$ that sends (β, δ) to $\beta\delta$ is injective, and since $|\mathbb{Z}^*_p| = qm$, it must be surjective as well.

We shall use this fact in the following way: if β is chosen uniformly at random from G, and δ is chosen uniformly at random from H (and independent of β), then $\beta\delta$ is uniformly distributed over $\mathbb{Z}^*_p$. Furthermore, since H is the image of the q-power map on $\mathbb{Z}^*_p$, we may generate a random $\delta \in H$ simply by choosing $\hat{\delta} \in Zp$ at random, and setting $\delta = \hat{\delta}^q$.

The discrete logarithm algorithm uses a "smoothness parameter" y. We will discuss choice of y below, when we analyze the running time of the algorithm; for now, we only assume that $y < p$. Let $p_1, ..., p_k$ be an enumeration of the primes up to y. Let $p_i := [p_i]_p \in \mathbb{Z}^*_p$ for $i = 1, ..., k$.

The algorithm has two stages.

In the first stage, we find relations of the form

$$^{r_i}\alpha^{s_i}\delta_i = \pi_1^{e_{i1}} \ldots \pi_k^{e_{ik}},$$

for $i = 1, ..., k + 1$, where $r_i, s_i, e_{i1}, ..., e_{ik} \in Z$ and $\delta_i \in H$ for each i.

We obtain each such relation by a randomized search, as follows: we choose $r_i, s_i \in \{0, ..., q - 1]$ at random, as well as $\hat{\delta}_i \in \mathbb{Z}_p^*$ at random; we then compute $\delta_i := \hat{\delta}_i^q, \beta_i := \gamma^{r_i}\alpha^{s_i}$, and $m_i :=$ rep($\beta_i\delta_i$). Now, the value β_i is uniformly distributed over G, while δ_i is uniformly distributed over H; therefore, the product $\beta_i\beta_i$ is uniformly distributed over $\mathbb{Z}_p^*$, and hence m_i is uniformly distributed over $\{1, ..., p - 1\}$. Next, we simply try to factor *mi* by trial division, trying all the primes $p_1, ..., p_k$ up to y. If we are lucky, we completely factor m_i in this way, obtaining a factorization

$$m_i = p_1^{e_{i1}} .....p_k^{e_{ik}},$$

for some exponents $e_{i1}, ..., e_{ik}$, and we get the relation . If we are unlucky, then we simply keep trying until we are lucky.

For $i = 1, ..., k + 1$, let $v_i := (e_{i1}, ..., e_{ik}) \in \mathbb{Z}^{\times k}$, and let $\bar{v}_i$ denote the image of v_i in $\mathbb{Z}_q^{\times k}$ (i.e., $\bar{v}_i := ([e_{i1}]_q, ..., [e_{ik}]_q)$). Since $\mathbb{Z}_q^{\times k}$ is a vector space over the field $\mathbb{Z}_q$ of dimension k, the family of vectors $v_1, ..., v_{k+1}$ must be linearly dependent. The second stage of the algorithm uses Gaussian elimination over $\mathbb{Z}_q$ to find a linear dependence among the vectors $\bar{v}_1, ..., \bar{v}_{k+1}$, that is, to find integers $c_1, ..., c_{k+1} \in \{0, ..., q - 1\}$, not all zero, such that

$$(e_1, ..., e_k) := c_1v_1 + ... + c_{k+1} v_{k+1} \in q\mathbb{Z}^{\times k}.$$

Raising each equation to the corresponding power c_i, and multiplying them all together, we obtain

$$\gamma^r\alpha^s\beta = \pi_1^{e_1}...\pi_k^{e_k},$$

where

$$r := \sum_{i=1}^{k+1} c_i r_i,\ s := \sum_{i=1}^{k+1} c_i s_i, \text{ and } \delta := \sum_{i=1}^{k+1} \delta_i^{c_i}$$

Now, $\delta \in H$, and since each e_j is a multiple of q, we also have $\pi_j^{e_j} \in H$ for $j = 1, ..., k$. It follows that $\gamma^r\alpha^s \in H$. But since $\gamma^r\alpha^s \in G$ as well, and $G \cap H = \{1\}$, it follows that $\gamma^r\alpha^s = 1$. If we are lucky (and we will be with overwhelming probability, as we discuss below), we will have $s \not\equiv 0 \pmod q$, in which case, we can compute $s' := s^{-1} \bmod q$, obtaining

$$= \gamma^{-rs'},$$

and hence $-rs' \bmod q$ is the discrete logarithm of γ to the base γ. If we are very unlucky, we will have $s \equiv 0 \pmod q$, at which point the algorithm simply quits, reporting "failure."

If Algorithm SEDL does not output "failure," then its output is indeed the discrete logarithm of α to the base γ. There remain three questions to answer:

1. What is the expected running time of Algorithm SEDL?
2. How should the smoothness parameter y be chosen so as to minimize the expected running time?
3. What is the probability that Algorithm SEDL outputs "failure"?

Let us address these questions in turn. As for the expected running time, let be the probability that a random element of $\{1, ..., p - 1\}$ is y-smooth. Then

$i \leftarrow 0$

repeat

$i \leftarrow i + 1$

repeat

choose $r_i, s_i \in \{0, ..., q - 1\}$ at random

choose $\hat{\delta}_i \in \mathbb{Z}^*_p$ at random $\beta_i \leftarrow \gamma^{r_i}\alpha^{s_i}, \delta_i \leftarrow \hat{\delta}_i^q, m_i \leftarrow rep(\beta_i \delta_i)$

test if m_i is y-smooth (trial division)

until $m_i = \hat{\delta}_i \in \mathbb{Z}^*_p$ for some integers $e_{i1}, ..., e_{ik}$

until $i = k + 1$

set $v_i \rightarrow (e_{i1}, ..., e_{ik}) \in \mathbb{Z}^{\times k}$ for $i = 1, ..., k + 1$

apply Gaussian elimination over $\mathbb{Z}_q$ to find integers $c_1, ..., c_{k+1} \in \{0, ..., q - 1\}$, not all zero, such that

$c_1 v_1 + ... + c_{k+1} v_{k+1} \in q\mathbb{Z}^{\times k}$.

$r \in \sum_{i=1}^{k+1} c_i r_i, s \leftarrow \sum_{i=1}^{k+1} c_i s_i$

if $s \equiv 0 \pmod{q}$

then output "failure"

else output $-rs^{-1} \bmod q$

the expected number of attempts needed to produce a single relation is σ^{-1}, and so the expected number of attempts to produce $k + 1$ relations is $(k + 1)\sigma^{-1}$. In each attempt, we perform trial division using $p_1, ..., p_k$, along with a few other minor computations, leading to a total expected running time in stage 1 of $k^2\sigma^{-1} \cdot \text{len}(p)^{O(1)}$. The running time in stage 2 is dominated by the Gaussian elimination step, which takes time $k^3 \cdot \text{len}(p)^{O(1)}$. Thus, if Z is the total running time of the algorithm, then we have

$$E[Z]\ \sigma\ (k^2\sigma^{-1} + k^3) \cdot \text{len}(p)^{O(1)}.$$

Let us assume for the moment that

$$y = \exp[(\log p)^{\lambda+o(1)}]$$

for some constant λ with $0 < \lambda < 1$. Our final choice of y will indeed satisfy this assumption. Consider the probability σ. We have

$$\sigma = \Psi(y, p-1)/(p-1) = \Psi(y, p)/(p-1) \geq \Psi(y, p)/p,$$

where for the second equality we use the assumption that $y < p$, so p is not y-smooth. With our assumption, we may apply Theorem (with the given value of y and $x := p$), obtaining

$$\sigma \geq \exp[(-1 + o(1))(\log p/ \log y) \log \log p].$$

By Chebyshev's theorem, we know that $k = \Theta(y/ \log y)$, and so $\log k = (1 + o(1)) \log y$. Moreover, assumption implies that the factor $\text{len}(p)O(1)$ is of the form $\exp[o(\min(\log y, \log p/\log y))]$, and so we have

$$E[Z] \leq \exp[(1 + o(1)) \max f(\log p/ \log y) \log \log p + 2 \log y, 3 \log y\}].$$

Let us find the value of y that minimizes the right-hand side of, ignoring the "o(1)" terms. Let $\mu := \log y$, $A := \log p \log \log p$, $S_1 := A/\mu + 2$, and $S_2 := 3\mu$. We want to find μ that minimizes $\max \{S_1, S_2\}$. Using a little calculus, one sees that S_1 is minimized at $\mu = (A/2)^{1/2}$. With this choice of μ, we have $S_1 = (2\sqrt{2})A^{1/2}$ and $S2 = (3/\sqrt{2})A^{1/2} < S_1$. Thus, choosing

$$y = \exp[(1/\sqrt{2})(\log p \log \log p)^{1/2}],$$

we obtain

$$E[Z] \leq \exp[(2\sqrt{2} + o(1))(\log p \log \log p)^{1/2}].$$

That takes care of the first two questions, although strictly speaking, we have only obtained an upper bound for the expected running time, and we have not shown that the choice of y is actually optimal, but we shall nevertheless content ourselves (for now) with these results. Finally, we deal with the third question, on the probability that the algorithm outputs "failure."

Lemma. The probability that Algorithm SEDL outputs "failure" is $1/q$.

Proof. Let F be the event that the algorithm outputs "failure." For $i = 1, ..., k+1$, we may view the final values assigned to r_i, s_i, δ_i, and m_i as random variables, which we shall denote by these same names (to avoid additional notation). Similarly, we may view s as a random variable.

Let $m_1', ..., m_{k+1}'$ be arbitrary, fixed y-smooth numbers, and let B be the event that $m_1 = m_1', ..., m_{k+1} = m_{k+1}'$. We shall show that $P[F \mid B] = 1/q$, and since this holds for all relevant B, it follows by total probability that $P[F] = 1/q$.

For the rest of the argument, we focus on the conditional distribution given B. With respect to this conditional distribution, the distribution of each random variable (r_i, s_i, δ_i) is the uniform distribution on the set

$$P_i := \{(r', s'; \delta') \in I_q \times I_q \times H : \gamma^{r'}\alpha^{s'}\delta' = [m_i']p\},$$

where $I_q := \{0, ..., q-1\}$; also, the family of random variables $\{(r_i, s_i, \delta_i)\}_{i=1}^{k+1}$ is mutually independent. It is easy to see that for $i = 1, ..., k+1$, and for each $s' \in I_q$, there exist unique values $r' \in I_q$ and $\delta' \in H$ such that $(r', s', \delta') \in P_i$. From this, it easily follows that each s_i is uniformly distributed over I_q, and the family of random variables $\{s_i\}_{i=1}^{k+1}$ is mutually independent. Also, the values $c_1, ..., c_{k+1}$ computed by the algorithm are fixed (as they are determined by $m_1', ..., m_{k+1}'$), and since $s = c_1 s_1 + ... + c_{k+1} s_{k+1}$, and not all the *ci*'s are zero modulo q, it follows that $s \bmod q$ is uniformly distributed over I_q, and so is equal to zero with probability $1/q$.

Let us summarize the above discussion in the following theorem.

Theorem. With the smoothness parameter set as

$$y := \exp[(1/\sqrt{2})(\log p \log\log p)^{1/2}],$$

the expected running time of Algorithm SEDL is at most

$$\exp[(2\sqrt{2} + o(1))(\log p \log\log p)^{1/2}].$$

The probability that Algorithm SEDL outputs "failure" is $1/q$.

In the description and analysis of Algorithm SEDL, we have assumed that the primes $p_1, ..., p_k$ were pre-computed. Of course, we can construct this list of primes using, for example, the sieve of Eratosthenes, and the running time of this pre-computation will be dominated by the running time of Algorithm SEDL.

In the analysis of Algorithm SEDL, we relied crucially on the fact that in generating a relation, each candidate element $\gamma^{r_i}\alpha^{s_i}\delta_i$ was uniformly distributed over $\mathbb{Z}_p^*$. If we simply left out the δ_i's, then the candidate element would be uniformly distributed over the subgroup G, and Theorem simply would not apply. Although the algorithm might anyway work as expected, we would not be able to prove this.

EXERCISE

1. Using the algorithm of the previous exercise as a subroutine, design and analyze an algorithm for the following problem. The input is p, q, γ, α, where p is a prime, q is a prime dividing $p-1$, γ generates the subgroup G of $\mathbb{Z}_p^*$ of order q, and $\alpha \in G$; note that we may have $q^2 \mid (p-1)$. The output is $\log \alpha$. Your algorithm should always succeed in computing this discrete logarithm, and its expected running time

should be bounded by a constant times the expected running time of the algorithm of the previous exercise.

2. Show how to modify Algorithm SEDL to solve the following problem: given a prime p, a generator γ for $\mathbb{Z}^*_p$, and an element $\alpha \in \mathbb{Z}^*_p$, compute $\log_\gamma \alpha$. Your algorithm should work without knowledge of the factorization of $p - 1$; its expected running time should be roughly the same as that of Algorithm SEDL, but its success probability may be lower. In addition, explain how the success probability may be significantly increased at almost no cost by collecting a few extra relations.

PRACTICAL IMPROVEMENTS

Our presentation and analysis of algorithms for discrete logarithms and factoring were geared towards simplicity and mathematical rigor. However, if one really wants to compute discrete logarithms or factor numbers, then a number of important practical improvements should be considered. In this section, we briefly sketch some of these improvements, focusing our attention on algorithms for factoring numbers (although some of the techniques apply to discrete logarithms as well).

THE QUADRATIC SIEVE ALGORITHM

We now describe a practical improvement to Algorithm SEF. This algorithm, known as the quadratic sieve, is faster in practice than Algorithm SEF; however, its analysis is somewhat heuristic.

First, let us return to the simplified version of Algorithm SEF, where we collect relations of the form. Furthermore, instead of choosing the values α_i at random, we will choose them in a special way, as we now describe. Let

$$\tilde{n} := \lfloor \sqrt{n} \rfloor,$$

and define the polynomial

$$F := (X + \tilde{n})^2 - n \in \mathbb{Z}[X].$$

In addition to the usual "smoothness parameter" y, we need a "sieving parameter" z, whose choice will be discussed below. We shall assume that both y and z are of the form $\exp[(\log n)^{1/2+o(1)}]$, and our ultimate choices of y and z will indeed satisfy this assumption.

For all $s = 1, 2, \ldots, \lfloor x \rfloor$, we shall determine which values of s are "good," in the sense that the corresponding value $F(s)$ is y-smooth. For each good s, since we have $F(s) \equiv (s + \tilde{n})^2 \pmod{n}$, we obtain one relation of the form , with $\alpha_i := [s + \tilde{n}]_n$. If we find at least $k + 1$ good values of s, then we can apply Gaussian elimination as usual to find a square root γ of 1 in $\mathbb{Z}^*_n$. Hopefully, we will have $\gamma \neq \pm 1$, allowing us to split n.

Observe that for $1 \leq s \leq z$, we have

$$1 \leq F(s) \leq z^2 + 2zn^{1/2} \leq n^{1/2+o(1)}.$$

Now, although the values $F(s)$ are not at all random, we might expect heuristically that the number of good s up to z is roughly equal to $\hat{\sigma} z$, where $\hat{\sigma}$ is the probability that a random integer in the interval $[1, n^{1/2}]$ is y-smooth, and by Theorem 15.7, we have

$$\hat{\sigma} = \exp[(-\frac{1}{4} + o(1))(\log n/\log y)\log\log n].$$

If our heuristics are valid, this already yields an improvement over Algorithm SEF, since now we are looking for y-smooth numbers near $n^{1/2}$, which are much more common than y-smooth numbers near n. But there is another improvement possible; namely, instead of testing each individual number $F(s)$ for smoothness using trial division, we can test them all at once using the following "sieving procedure."

The sieving procedure works as follows. First, we create an array $v[1 \ldots \lfloor z \rfloor]$, and initialize $v[s]$ to $F(s)$, for $1 \leq s \leq z$. Then, for each prime p up to y, we do the following:

1. Compute the roots of the polynomial F modulo p.

   This can be done quite *efficiently*, as follows. For $p = 2$, F has exactly one root modulo p, which is determined by the parity of $\tilde{n}$. For $p > 2$, we may use the familiar quadratic formula together with an algorithm for computing square roots modulo p, as discussed in Exercise 12.7. A quick calculation shows that the discriminant of F is $4n$, and thus, F has a root modulo p if and only if n is a quadratic residue modulo p, in which case it will have two roots (under our usual assumptions, we cannot have $p \mid n$).

2. Assume that F has v_p distinct roots modulo p lying in the interval $[1, p]$; call them $r_1, \ldots, r_{v_p}$.

Note that $v_p = 1$ for $p =\in$ and $v_p \in \{0, 2\}$ for $p > 2$. Also note that $F(s) \equiv 0 \pmod{p}$ if and only if $s \equiv r_i \pmod{p}$ for some $i = 1, \ldots, v_p$.

For $i = 1, \ldots, v_p$, do the following:

$s \leftarrow r_i$

while $s \leq z$ do

repeat $v[s] \leftarrow v[s]/p$ until $p \mid v[s]$

$s \leftarrow s + p$

At the end of this sieving procedure, the good values of s may be identified as precisely those such that $v[s] = 1$. The running time of this sieving procedure is at most $\text{len}(n)^{O(1)}$ times

$$\sum_{p \leq y} \frac{z}{p} = z \sum_{p \leq y} \frac{1}{p} = O(z \log\log y) = z^{1+o(1)}.$$

Here, we have made use of Theorem , although this is not really necessary – for our purposes, the bound $\sum_{p \le y} 1/p = O(\log y)$ would suffice. Note that this sieving procedure is a factor of $k^{1+o(1)}$ faster than the method for finding smooth numbers based on trial division. With just a little extra book-keeping, we can not only identify the good values of s but also compute the factorization of $F(s)$ into primes, at essentially no extra cost.

Now, let us put together all the pieces. We have to choose z just large enough so as to find at least $k + 1$ good values of s up to z. So we should choose z so that $z \approx k / \hat{\sigma}$ – in practice, we could choose an initial estimate for z, and if this choice of z does not yield enough relations, we could keep doubling z until we do get enough relations. Assuming that $z \approx k / \hat{\sigma}$, the cost of sieving is $(z / \hat{\sigma})^{1+o(1)}$, or

$$\exp[(1 + o(1))(\frac{1}{4} (\log n/ \log y) \log \log n + \log y)].$$

The cost of Gaussian elimination is still $O(k^3)$, or

$$\exp[(3 + o(1)) \log y].$$

Thus, the total running time is bounded by

$$\exp[(1 + o(1)) \max\{\frac{1}{4} (\log n/ \log y) \log \log n + \log y, 3 \log y\}].$$

Let $\mu := \log y$, $A := (1/4) \log n \log \log n$, $S_1 := A/\mu + \mu$ and $S_2 := 3\mu$, and let us find the value of μ that minimizes max $\{S_1, S_2\}$. Using a little calculus, one finds that S_1 is minimized at $\mu = A^{1/2}$. For this value of μ, we have $S_1 = 2A^{1/2}$ and $S_2 = 3A^{1/2} > S_1$, and so this choice of μ is a bit larger than optimal. For $\mu < A^{1/2}$, S_1 is decreasing (as a function of μ), while S_2 is always increasing. It follows that the optimal value of μ is obtained by setting

$$A/\mu + \mu = 3\mu,$$

and solving for μ. This yields $\mu = (A/2)^{1/2}$. So setting

$$y := \exp[(1/2\sqrt{2})(\log n \log \log n)^{1/2}],$$

the total running time of the quadratic sieve factoring algorithm is bounded by

$$\exp[(3/2\sqrt{2} + o(1))(\log n \log \log n)^{1/2}].$$

Thus, we have reduced the constant in the exponent from∈(for Algorithm SEF with the more accurate smoothness density estimates) to $3/2\sqrt{2} \approx 1.061$.

We mention one final improvement. The matrix to which we apply Gaussian elimination in stage 2 is "sparse"; indeed, since any integer less than n has $O(\log n)$ prime factors, the total number of non-zero entries in the matrix is $k^{1+o(1)}$. There are special algorithms for working with such sparse matrices, which allow us to perform stage 2 of the factoring algorithm in time $k^{2+o(1)}$, or

$$\exp[(2 + o(1)) \log y].$$

Setting

$$y := \exp\left[\frac{1}{2}(\log n \log\log n)^{1/2}\right],$$

the total running time is bounded by

$$\exp[(1 + o(1))(\log n \log\log n)^{1/2}].$$

Thus, this improvement reduces the constant in the exponent from $3/2\sqrt{2} \approx 1.061$ to 1. Moreover, the special algorithms designed to work with sparse matrices typically use much less space than ordinary Gaussian elimination (even if the input to Gaussian elimination is sparse, the intermediate matrices will not be). We shall discuss in detail later, in §18.4, one such algorithm for solving sparse systems of linear equations.

The quadratic sieve may fail to factor n, for one of two reasons: first, it may fail to find $k + 1$ relations; second, it may find these relations, but in stage 2, it finds only a trivial square root of 1. There is no rigorous theory to say why the algorithm should not fail for one of these two reasons, but experience shows that the algorithm does indeed work as expected.

BETTER SMOOTHNESS DENSITY ESTIMATES

From an algorithmic point of view, the simplest way to improve the running times of both Algorithms SEDL and SEF is to use a more accurate smoothness density estimate, which dictates a different choice of the smoothness bound y in those algorithms, speeding them up significantly. While our Theorem is a valid lower bound on the density of smooth numbers, it is not "tight," in the sense that the actual density of smooth numbers is somewhat higher. We quote from the literature the following result:

Theorem. Let y be a function of x such that for some $\varepsilon > 0$, we have

$$y = \Omega((\log x)^{1+\varepsilon}) \text{ and } u := \frac{\log x}{\log y} \to \infty$$

as $x \to \infty$ Then

$$\Psi(y, x) = x \,.\, \exp[(-1 + o(1))u \log u].$$

Proof. Let us apply this result to the analysis of Algorithm SEF. Assume that

$$y = \exp[(\log n)^{1/2+o(1)}].$$

Our choice of y will in fact be of this form. With this assumption, we have log log y = (1/2 + o(1)) log log n, and using Theorem, we can improve the inequality, obtaining instead (as the reader may verify)

$$E[Z] \leq \exp[(1 + o(1)) \max \{\frac{1}{2}(\log n/ \log y) \log \log n + 2 \log y, 3 \log y\}].$$

From this, if we set

$$y := \exp[\frac{1}{2} (\log n \log \log n)^{1/2})],$$

we obtain

$$E[Z] \leq \exp[(2 + o(1))(\log n \log \log n)^{1/2}].$$

An analogous improvement can be obtained for Algorithm SEDL.

Although this improvement only reduces the constant $2\sqrt{2} \approx 2.828$ to 2, the constant is in the exponent, and so this improvement is not to be sco–ed at!

LOGARITHMS IN $\mathbb{Z}^*_p$

As we have seen in Theorem , for a prime p, $\mathbb{Z}^*_p$ is a cyclic group of order $p - 1$. This means that there exists a generator $\gamma \in \mathbb{Z}^*_p$, such that each $\alpha \in \mathbb{Z}^*_p$ can be written uniquely as $\alpha = \gamma^x$, where x is an integer with $0 \leq x < p - 1$; the integer x is called the discrete logarithm of to the base γ, and is denoted $\log_\gamma \alpha$.

This chapter discusses some computational problems in this setting; namely, how to efficiently find a generator γ, and given γ and α, how to compute $\log_\gamma \alpha$.

More generally, if γ generates a subgroup G of $\mathbb{Z}^*_p$ of order q, where $q \mid (p - 1)$, and $\alpha \in G$, then $\log_\gamma \alpha$ is defined to be the unique integer x with $0 \leq x < q$ and $\alpha = \gamma^x$. In some situations it is more convenient to view $\log_\gamma \alpha$ as an element of $\mathbb{Z}_q$. Also for $x \in \mathbb{Z}_q$, with $x = [a]_q$, one may write γ^x to denote γ_a. There can be no confusion, since if $x = [a']_q$, then $\gamma^{a'} = \gamma^a$. However, in this chapter, we shall view $\log_\gamma \alpha$ as an integer.

Although we work in the group $\mathbb{Z}^*_p$, all of the algorithms discussed in this chapter trivially generalize to any finite cyclic group that has a suitably compact representation of group elements and an e—cient algorithm for performing the group operation on these representations.

LOGARITHMS IN $\mathbb{Z}^*_p$

Let us consider algorithms for computing the discrete logarithm of $\alpha \in \mathbb{Z}^*_p$ to a given base γ. The algorithms we present here are, in the worst case, exponential-time algorithms, and are by no means the best possible; however, in some special cases, these algorithms are not so bad.

Baby Step/giant Step Method

As above, suppose that $\gamma \in \mathbb{Z}_p^*$ generates a subgroup G of $\mathbb{Z}_p^*$ of order $q > 1$ (not necessarily prime), and we are given p, q, γ, and $\alpha \in G$, and wish to compute $\log_\gamma \alpha$.

A faster algorithm than brute-force search is the baby step/giant step method. It works as follows.

Let us choose an approximation m to $q^{1/2}$. It does not have to be a very good approximation we just need $m = \Theta(q^{1/2})$. Also, let $m' = \lfloor q/m \rfloor$, so that $m' = \Theta(q^{1/2})$ as well.

The idea is to compute all the values γ^i for $i = 0, ..., m - 1$ (the "baby steps") and to build an "associative array" (or "lookup table") T that maps the key γ^i to the value i. For $\beta \in \mathbb{Z}_p^*$, we shall write $T[\beta]$ to denote the value associated with the key, writing $T[\beta] = \perp$ if there is no such value. We shall assume that T is implemented so that accessing $T[\beta]$ is fast. Using an appropriate data structure, T can be implemented so that accessing individual elements takes time $O(\text{len}(p))$. One such data structure is a radix tree (also called a search trie). Other data structures may be used (for example, a hash table or a binary search tree), but these may have somewhat different access times.

We can build the associative array T using the following algorithm:

initialize T // $T[\beta] = \perp$ for all $\beta \in \mathbb{Z}_p^*$

$\beta \leftarrow 1$

for $i \leftarrow 0$ to $m - 1$ do

$T[\beta] \leftarrow i$

$\beta \leftarrow \beta \, . \, \gamma$

Clearly, this algorithm takes time $O(q^{1/2} \text{ len}(p)^2)$.

After building the lookup table, we execute the following procedure (the "giant steps"):

$\gamma' \leftarrow \gamma^{-m}$

$\beta \leftarrow \alpha, j \leftarrow 0, i \leftarrow T[\beta]$

while $i = \perp$ do

$\beta \leftarrow 0, j \leftarrow j + 1, i \leftarrow T[\beta]$

$x \leftarrow jm + i$

output x

To analyze this procedure, suppose that $\alpha = x$ with $0 \leq x < q$. Now, x can be written in a unique way as $x = vm + u$, where u and v are integers with $0 \leq u < m$ and $0 \leq v \leq m'$. In the jth loop iteration, for $j = 0, 1, \ldots$, we have

$$\beta = \alpha\gamma^{-mj} = \gamma^{(v-j)m+u}.$$

So we will detect $i \neq \perp$ precisely when $j = v$, in which case $i = u$. Thus, the output will be correct, and the total running time of the algorithm (for both the "baby steps" and "giant steps" parts) is easily seen to be $O(q^{1/2}\,\text{len}(p)^2)$.

While this algorithm is much faster than brute-force search, it has the drawback that it requires space for about $q^{1/2}$ elements of $\mathbb{Z}_p$. Of course, there is a "time/space trade-off" here: by choosing m smaller, we get a table of size $O(m)$, but the running time will be proportional to $O(q/m)$. We discuss an algorithm that runs (at least heuristically) in time $O(q^{1/2}\,\text{len}(q)\,\text{len}(p)^2)$, but which requires space for only a constant number of elements of $\mathbb{Z}_p$.

GROUPS OF ORDER Q^E

Suppose that $\gamma \in \mathbb{Z}_p^*$ generates a subgroup G of $\mathbb{Z}_p^*$ of order q^e, where $q > 1$ and $e \geq 1$, and we are given p, q, e, γ, and $\alpha \in G$, and wish to compute $\log_\gamma \alpha$.

There is a simple algorithm that allows one to reduce this problem to the problem of computing discrete logarithms in the subgroup of $\mathbb{Z}_p^*$ of order q.

It is perhaps easiest to describe the algorithm recursively. The base case is when $e = 1$, in which case, we use an algorithm for the subgroup of $\mathbb{Z}_p^*$ of order q.

Suppose now that $e > 1$. We choose an integer f with $0 < f < e$. Different strategies for choosing f yield different algorithms – we discuss this below. Suppose $\alpha = \gamma^x$, where $0 \leq x < q^e$. Then we can write $x = q^f v + u$, where u and v are integers with $0 \leq u < q^f$ and $0 \leq v < q^{e-f}$. Therefore,

$$\alpha^{q^{e-f}} = \gamma^{q^{e-f}u}.$$

Note that $\gamma^{q^{e-f}}$ has multiplicative order q^f, and so if we recursively compute the discrete logarithm of $\alpha^{q^{e-f}}$ to the base $\gamma^{q^{e-f}}$, we obtain u.

Having obtained u, observe that

$$\alpha / \gamma^u = \gamma^{q^f v}.$$

Note also that γ^{q^f} has multiplicative order q^{e-f}, and so if we recursively compute the discrete logarithm of α/γ^u to the base γ^{q^f}, we obtain v, from which we then compute $x = q^f v + u$.

Let us put together the above ideas succinctly in a recursive procedure:

Algorithm RDL. On input p, q, e, γ, α as above, do the following:

if $e = 1$ then

return $\log_\gamma \alpha$ // base case: use a different algorithm

else

select $f \in \{1, ..., e-1\}$

$u \leftarrow \mathrm{RDL}(p, q, f, \gamma^{q^{e-f}}, \alpha^{q^{e-f}})$ // $0 \leq u < q^f$

$v \leftarrow \mathrm{RDL}(p, q, e-f, \gamma^{q^f}, \alpha/\gamma^u)$ // $0 \leq v < q^{e-f}$

return $q^f v + u$

To analyze the running time of this recursive algorithm, note that the running time of the body of one recursive invocation (not counting the running time of the recursive calls it makes) is $O(e \operatorname{len}(q) \operatorname{len}(p)^2)$. To calculate the total running time, we have to sum up the running times of all the recursive calls plus the running times of all the base cases.

Regardless of the strategy for choosing f, the total number of base case invocations is e. Note that all the base cases compute discrete logarithms to the base $\gamma^{q^{e-1}}$. Assuming we implement the base case using the baby step/giant step algorithm, the total running time for all the base cases is therefore $O(eq^{1/2} \operatorname{len}(p)^2)$.

The total running time for the recursion (not including the base case computations) depends on the strategy used to choose the split f. It is helpful to represent the behavior of the algorithm using a recursion tree. This is a binary tree, where every node represents one recursive invocation of the algorithm; the root of the tree represents the initial invocation of the algorithm; for every node N in the tree, if N represents the recursive invocation I, then N's children (if any) represent the recursive invocations made by I. We can naturally organize the nodes of the recursion tree by levels: the root of the recursion tree is at level 0, its children are at level 1, its grandchildren at level 2, and so on. The depth of the recursion tree is defined to be the maximum level of any node.

We consider two different strategies for choosing the split f:

- If we always choose $f = 1$ or $f = e - 1$, then the depth of the recursion tree is $O(e)$. The running time contributed by each level of the recursion tree is $O(e \operatorname{len}(q) \operatorname{len}(p)^2)$, and so the total running time for the recursion is $O(e^2 \operatorname{len}(q) \operatorname{len}(p)^2)$. Note that if $f = 1$, then the algorithm is essentially tail recursive, and so may be easily converted to an iterative algorithm without the need for a stack.
- If we use a "balanced" divide-and-conquer strategy, choosing $f \approx e/2$, then the depth of the recursion tree is $O(\operatorname{len}(e))$, while the running time contributed by each level of the recursion tree is still $O(e \operatorname{len}(q) \operatorname{len}(p)^2)$. Thus, the total running time for the recursion is $O(e \operatorname{len}(e) \operatorname{len}(q) \operatorname{len}(p)^2)$.

Assuming we use the faster, balanced recursion strategy, and that we use the baby step/giant step algorithm for the base case, the total running time of Algorithm RDL is:

$$O((eq^{1/2} + e\ \text{len}(e)\ \text{len}(q))\ .\ \text{len}(p)^2).$$

Brute-force Search

Suppose that $\gamma \in \mathbb{Z}_p^*$ generates a subgroup G of $\mathbb{Z}_p^*$ of order $q > 1$ (not necessarily prime), and we are given p, q, γ, and $\alpha \in G$, and wish to compute $\log_\gamma \alpha$.

The simplest algorithm to solve this problem is brute-force search:

$\beta \leftarrow 1$

$i \leftarrow 0$

while $\beta \neq \alpha$ do

$\beta \leftarrow \beta.\gamma$

$i \leftarrow i + 1$

output i

This algorithm is clearly correct, and the main loop will always halt after at most q iterations (assuming, as we are, that $\alpha \in G$). So the total running time is $O(q\ \text{len}(p)^2)$.

A Space-efficient Square-root Time Algorithm

We present a more space-efficient alternative to the algorithm, the analysis of which we leave as a series of exercises for the reader.

The algorithm makes a somewhat heuristic assumption that we have a function that "behaves" for all practical purposes like a random function. Such functions can indeed be constructed using cryptographic techniques under reasonable intractability assumptions; however, for the particular application here, one can get by in practice with much simpler constructions.

Let p be a prime, q a prime dividing $p - 1$, γ an element of $\mathbb{Z}_p^*$ that generates a subgroup G of $\mathbb{Z}_p^*$ of order q, and $\alpha \in G$. Let F be a function mapping elements of G to $\{0, ..., q - 1\}$. Define $H : G \to G$ to be the function that sends to β to $\beta\alpha\gamma^{F(\beta)}$.

The algorithm runs as follows:

$i \leftarrow 1$

$x \leftarrow 0$, $\beta \leftarrow \alpha$,

$x' \leftarrow F(\beta)$, $\beta' \leftarrow H(\beta)$

while $\beta \neq \beta'$ do

$x \leftarrow (x + F(\beta)) \bmod q$, $\beta \leftarrow H(\beta)$

repeat 2 times

$x' \leftarrow (x' + F(\beta')) \bmod q$, $\beta' \leftarrow H(\beta')$

$i \leftarrow i + 1$

if $i < q$ then

output $(x - x')i^{-1} \bmod q$

else

output "fail"

To analyze this algorithm, let us define $\beta_1, \beta_2, \ldots$, as follows: $\beta_1 := \alpha$ and for $i > 1$, $\beta_i := H(\beta_{i-1})$.

Discrete Logarithms in $\mathbb{Z}_p^*$

Suppose that we are given a prime p, along with the prime factorization

$$p-1=\prod_{i=1}^{r} q_i^{e_i},$$

a generator γ for $\mathbb{Z}_p^*$, and $\alpha \in \mathbb{Z}_p^*$. We wish to compute $\log_\gamma \alpha$.

Suppose that $\alpha = \gamma^x$, where $0 \leq x < p - 1$. Then for $i = 1, \ldots, r$, we have

$$\alpha^{(p-1)/q_i^{e_i}} = \left(\gamma^{(p-1)/q_i^{e_i}}\right)^x.$$

Note that $\gamma^{(p-1)/q_i^{e_i}}$ has multiplicative order $q_i^{e_i}$, and if x_i is the discrete logarithm of $\alpha^{(p-1)/q_i^{e_i}}$ to the base $\gamma^{(p-1)/q_i^{e_i}}$, then we have $0 \leq x_i < q_i^{e_i}$ and $x \equiv x_i \pmod{q_i^{e_i}}$.

Thus, if we compute the values $x_1, \ldots, x_r$, using Algorithm RDL, we can obtain x using the algorithm of the Chinese remainder theorem. If we define $q := \max\{q_1, \ldots, q_r\}$, then the running time of this algorithm will be bounded by $q^{1/2}\,\mathrm{len}(p)\,O(1)$.

We conclude that

the difficulty of computing discrete logarithms in $\mathbb{Z}_p^*$ is determined by the size of the largest prime dividing $p - 1$.

EXERCISE

1. Let j be the smallest index such that $\beta_j = \beta_k$ for some index $k < j$. Show that $j \leq q + 1$ and that the loop terminates with $i < j$ (and in particular, $i \leq q$).
2. Show that each time the main loop of the algorithm is entered, we have $\beta = \beta_i = \gamma^x\alpha^i$, and $\beta' = \beta_{2i} = \gamma^{x'}\alpha_{2i}$.
3. Assume that F is a random function, meaning that it is chosen at random, uniformly from among all functions from G into $\{0, ..., q-1\}$. Show that this implies that H is a random function, meaning that it is uniformly distributed over all functions from G into G.
4. Show that if the loop terminates with $i < q$, the value output is equal to $\log_\gamma \alpha$.

FINDING A GENERATOR FOR $\mathbb{Z}_p^*$

In this section, we consider the problem of how to find a generator for $\mathbb{Z}_p^*$. There is no efficient algorithm known for this problem, unless the prime factorization of $p - 1$ is given, and even then, we must resort to the use of a probabilistic algorithm. Of course, factoring in general is believed to be a very difficult problem, so it may not be easy to get the prime factorization of $p - 1$. However, if our goal is to construct a large prime p, together with a generator for $\mathbb{Z}_p^*$, then we may use Algorithm RFN in §9.6 to generate a random factored number n in some range, test $n + 1$ for primality, and then repeat until we get a factored number n such that $p = n + 1$ is prime. In this way, we can generate a random prime p in a given range along with the factorization of $p - 1$.

We now present an efficient probabilistic algorithm that takes as input an odd prime p, along with the prime factorization

$$p-1=\prod_{i=1}^{r} q_i^{e_i},$$

and outputs a generator for $\mathbb{Z}_p^*$. It runs as follows:

for $i \leftarrow 1$ to r do

repeat

choose $\alpha \in \mathbb{Z}_p^*$ at random

compute $\beta \leftarrow \alpha^{(p-1)/q_i}$

until $\beta \neq 1$

$\gamma_i \leftarrow \alpha^{(p-1)/q_i^{e_i}}$ $\gamma \leftarrow \prod_{i=1}^{r} \gamma_i$

output γ

First, let us analyze the correctness of this algorithm. When the ith loop iteration terminates, by construction, we have

$$\gamma_i^{q_i^{e_i}} = 1 \text{ but } \gamma_i^{q_i^{e_i - 1}} \neq 1$$

It follows that γ_i has multiplicative order $q_i^{e_i}$. From this, it follows that has multiplicative order $p - 1$. Thus, we have shown that if the algorithm terminates, its output is always correct.

Let us now analyze the running time of this algorithm. Fix $i = 1, ..., r$, and consider the repeat/until loop in the ith iteration of the outer loop. Let L_i be the random variable whose value is the number of iterations of this repeat/until loop. Since α is chosen at random from $\mathbb{Z}^*_p$, the value of β is uniformly distributed over the image of the $(p - 1)/q_i$-power map, and since the latter is a subgroup of $\mathbb{Z}^*_p$ of order q_i, we see that $\beta = 1$ with probability $1/q_i$. Thus, L_i has a geometric distribution with associated success probability $1 - 1/q_i$, and $E[L_i] = 1/(1 - 1/q_i) \leq 2$.

Now set $L := L_1 + ... + L_r$. By linearity of expectation, we have $E[L] = E[L_1] + ... + E[L_r] \leq 2r$. The running time Z of the entire algorithm is $O(L.\text{len}(p)^3)$, and hence the expected running time is $E[Z] = O(r \text{ len}(p)^3)$, and since $r \leq \log_2 p$, we have $E[Z] = O(\text{len}(p)^4)$.

Although this algorithm is quite practical, there are asymptotically faster algorithms for this problem.

EXERCISE

1. Suppose we are given a positive integer n, along with its prime factorization $n = p_1^{e_1} ... p_r^{e_r}$, and that for each $i = 1, ..., r$, we are also given the prime factorization of $p_i - 1$. Show how to efficiently compute the multiplicative order of any element $\alpha \in \mathbb{Z}^*_n$.
2. Suppose we are not given the prime factorization of $p - 1$, but rather, just a prime q dividing $p - 1$, and we want to find an element of multiplicative order q in $\mathbb{Z}^*_p$. Design and analyze an efficient algorithm to do this.
3. Suppose there is an efficient algorithm that takes as input a positive integer n and an element $\alpha \in \mathbb{Z}^*_n$, and computes the multiplicative order of α.

   Show how to use this algorithm to build an efficient integer factoring algorithm.
4. Suppose we are given a prime p, along with the prime factorization $p - 1 = \prod_{i=1}^{r} q_i^{e_i}$.

   (a) Improve the running time bound to $O(\text{len}(r) \text{ len}(p)_3)$.

(b) If, in addition, we are given $\alpha \in \mathbb{Z}_p^*$, show how to compute the multiplicative order of α in time $O(r \operatorname{len}(p)^3)$.

(c) Modifying the algorithm you developed for part (b), show how to construct a generator for $\mathbb{Z}_p^*$ in expected time $O(\operatorname{len}(r) \operatorname{len}(p)^3)$.

The Diffie–Hellman key establishment protocol [mh]

One of the main motivations for studying algorithms for computing discrete logarithms is the relation between this problem and the problem of breaking a protocol called the Diffie–Hellman key establishment protocol, named after its inventors.

In this protocol, Alice and Bob need never to have talked to each other before, but nevertheless, can establish a shared secret key that nobody else can easily compute. To use this protocol, a third party must provide a "telephone book," which contains the following information:

- p, q, and γ, where p and q are primes with $q \mid (p-1)$, and γ is an element generating a subgroup G of $\mathbb{Z}_p^*$ of order q;
- an entry for each user, such as Alice or Bob, that contains the user's name, along with a "public key" for that user, which is an element of the group G.

To use this system, Alice posts her public key in the telephone book, which is of the form $\alpha = \gamma^x$, where $x \in \{0, ..., q-1\}$ is chosen by Alice at random. The value x is Alice's "secret key," which Alice never divulges to anybody. Likewise, Bob posts his public key, which is of the form $\beta = \gamma^y$, where $y \in \{0, ..., q-1\}$ is chosen by Bob at random, and is his secret key.

To establish a shared key known only between them, Alice retrieves Bob's public key from the telephone book, and computes $\kappa_A := \beta^x$. Likewise, Bob retrieves Alice's public key α, and computes $\kappa_B := \alpha^y$. It is easy to see that

$$\kappa_A = \beta^x = (\gamma^y)^x = \gamma^{xy} = (\gamma^x)^y = \alpha^y = \kappa_B,$$

and hence Alice and Bob share the same secret key $\kappa := \kappa_A = \kappa_B$.

Using this shared secret key, they can then use standard methods for encryption and message authentication to hold a secure conversation. We shall not go any further into how this is done; rather, we briefly (and only superficially) discuss some aspects of the security of the key establishment protocol itself. Clearly, if an attacker obtains α and β from the telephone book, and computes $x = \log_\gamma \alpha$, then he can compute Alice and Bob's shared key as $\kappa = \kappa^x$ – in fact, given x, an attacker can efficiently compute any key shared between Alice and another user.

Thus, if this system is to be secure, it should be very difficult to compute discrete logarithms. However, the assumption that computing discrete logarithms is hard is not enough to guarantee security. Indeed, it is not entirely inconceivable that the discrete logarithm problem is hard, and yet the problem of computing κ from α and is easy. The latter problem – computing κ from α and β– is called the Diffie–Hellman problem.

As in the discussion of the RSA cryptosystem in §4.7, the reader is warned that the above discussion about security is a bit of an oversimplification. A complete discussion of all the security issues related to the above protocol is beyond the scope of this text.

Note that in our presentation of the Diffie–Hellman protocol, we work with a generator of a subgroup G of $\mathbb{Z}_p^*$ of prime order, rather than a generator for $\mathbb{Z}_p^*$. There are several reasons for doing this: one is that there are no known discrete logarithm algorithms that are any more practical in G than in $\mathbb{Z}_p^*$, provided the order q of G is sufficiently large; another is that by working in G, the protocol becomes substantially more efficient. In typical implementations, p is 1024 bits long, so as to protect against subexponential-time algorithms, while q is 160 bits long, which is enough to protect against the square-roottime algorithms discussed. The modular exponentiations in the protocol will run several times faster using "short," 160-bit exponents rather than "long," 1024-bit exponents.

For the following exercise, we need the following notions from complexity theory.

- We say problem A is deterministic poly-time reducible to problem B if there exists a deterministic algorithm R for solving problem A that makes calls to a subroutine for problem B, where the running time of R (not including the running time for the subroutine for B) is polynomial in the input length.
- We say that problems A and B are deterministic poly-time equivalent if A is deterministic poly-time reducible to B and B is deterministic poly-time reducible to A.

EXERCISE

1. Let p be a prime, q a prime that divides $p - 1$, $\gamma \in \mathbb{Z}_p^*$ an element that generates a subgroup G of $\mathbb{Z}_p^*$ of order q, and $\alpha \in G$. For $\delta \in G$, a representation of δ with respect to γ and α is a pair of integers (r, s), with $0 \leq r < q$ and $0 \leq s < q$, such that $\gamma^r\alpha^s = \delta$.
   (a) Show that given any $\delta \in G$, along with any two distinct representations of δ with respect to γ and α, we can efficiently compute $\log_\gamma \alpha$.
   (b) Show that for every $\delta \in G$, there are precisely q representations (r, s) of δ with respect to γ and α, and among these, there is precisely one with $s = 0$.
   (c) Suppose we are given access to an "oracle" that, when presented with any $\delta \in G$, tells us some representation of with respect to γ and α. Show how to use this oracle to efficiently compute $\log_\gamma \alpha$.

(d) Show that given a representation (r, s) of 1 with respect to γ and α such that $s \neq 0$, we can efficiently compute $\log_\gamma \alpha$

The following two exercises examine the danger of the use of "short" exponents in discrete logarithm based cryptographic schemes that do not work with a group of prime order.

2. Let p be a prime, q a prime dividing $p - 1$, and g an element of $\mathbb{Z}_p^*$ that generates a subgroup G of order q. Let $\alpha \in G$, and let H be the subgroup of $G \times G$ generated by (γ, α). Let $\tilde{\gamma}, \tilde{\alpha}$ be arbitrary elements of G, and define the map

$\rho : \mathbb{Z}_q \times \mathbb{Z}_q \to G \times G$

$([r]_q, [s]_q) \mapsto \left(\gamma^r \tilde{\gamma}^s, \alpha^r \tilde{\alpha}^s\right)$.

Show that the definition of ρ is unambiguous, that ρ is a group homomorphism, and that

- if $\left(\tilde{\gamma}, \tilde{\alpha}\right) \in H$, then $\operatorname{Im} \rho = H$, and
- if $\left(\tilde{\gamma}, \tilde{\alpha}\right) \notin H$, then $\operatorname{Im} \rho = G \times G$.

3. For p, q, γ as in the previous exercise, let $D_{p,q,\gamma}$ be the set of all triples of the form $(\gamma^x, \gamma^y, \gamma^{xy})$, and let $R_{p,q,\gamma}$ be the set of all triples of the form $(\gamma^x, \gamma^y, \gamma^z)$. Using the result from the previous exercise, design a probabilistic algorithm that runs in expected polynomial time, and that on input p, q, γ, along with a triple $\Gamma \in R_{p,q,\gamma}$, outputs a triple $\Gamma^* \in R_{p,q,\gamma}$ such that

- if $\Gamma \in D_{p,q,\gamma}$, then Γ^* is uniformly distributed over $D_{p,q,\gamma}$, and
- if $\Gamma \notin D_{p,q,\gamma}$, then Γ^* is uniformly distributed over $R_{p,q,\gamma}$.
- Consider the following problems.
- Given a prime p, a prime q that divides $p - 1$, an element $\gamma \in \mathbb{Z}_p^*$ generating a subgroup G of $\mathbb{Z}_p^*$ of order q, and two elements $\alpha \in G$, with $\beta \neq 1$, compute $\gamma^{xy'}$, where $x := \log_\gamma \alpha$, $y' := y^{-1} \bmod q$, and $y := \log_\gamma \beta$.
- Given a prime p, a prime q that divides $p - 1$, an element $\gamma \in \mathbb{Z}_p^*$ generating a subgroup G of $\mathbb{Z}_p^*$ of order q, and two elements $\alpha, \beta \in G$, compute γ^{xy}, where $x := \log_\gamma \alpha$ and $y := \log_\gamma \beta$. (This is just the Diffie–Hellman problem.)
- Given a prime p, a prime q that divides $p - 1$, an element $\gamma \in \mathbb{Z}_p^*$ generating

a subgroup G of $\mathbb{Z}^*_p$ of order q, and an element $\alpha \in G$, with $\alpha \neq 1$, compute $\gamma^{x'}$, where $x' := x^{-1} \bmod q$ and $x := \log_\gamma \alpha$.

- Given a prime p, a prime q that divides $p - 1$, an element $\gamma \in \mathbb{Z}^*_p$ generating a subgroup G of $\mathbb{Z}^*_p$ of order q, and an element $\alpha \in G$, compute γ^{x^2}, where $x' := \log_\gamma \alpha$.

  Show that these problems are deterministic poly-time equivalent. Moreover, your reductions should preserve the values of p, q, and γ; that is, if the algorithm that reduces one problem to another takes as input an instance of the former problem of the form $(p, q, \gamma, ...)$, it should invoke the subroutine for the latter problem with inputs of the form $(p, q, \gamma, ...)$.

4. Assume the notation of the previous exercise, but let us drop the restriction that q is prime. Design and analyze a deterministic algorithm A that takes inputs p, q, γ and $\Gamma^* \in R_{p,q,\gamma}$, that outputs 0 or 1, and that satisfies the following property: if t is the smallest prime dividing q, then A runs in time $(t + \text{len}(p))^{O(1)}$, and the "distinguishing advantage" $\varepsilon(p, q, \gamma)$ for A on inputs p, q, γ is at least $1/t$.

5. Suppose there is a probabilistic algorithm A that takes as input a prime p, a prime q that divides $p - 1$, and an element $\gamma \in \mathbb{Z}^*_p$ generating a subgroup G of of order q. The algorithm also takes as input $\alpha \in G$. It outputs either "failure," or $\log_\gamma \alpha$. Furthermore, assume that A runs in expected polynomial time, and that for all p, q, and γ of the above form, and for randomly chosen $\alpha \in G$, A succeeds in computing $\log_\gamma \alpha$ with probability $\varepsilon(p, q, \gamma)$. Here, the probability is taken over the random choice of α, as well as the random choices made during the execution of A. Show how to use A to construct another probabilistic algorithm A' that takes as input p, q, and γ as above, as well as $\alpha \in G$, runs in expected polynomial time, and that satisfies the following property:

   if $\varepsilon(p, q,) \geq 0.001$, then for all $\alpha \in G$, A' computes log with probability at least 0.999.

   The algorithm A' in the previous exercise is an example of a random selfreduction, which means an algorithm that reduces the task of solving an arbitrary instance of a given problem to that of solving a random instance of the same problem. Intuitively, the existence of such a reduction means that the problem is no harder in the worst case than on average.

6. Suppose that A is a probabilistic algorithm that takes as input p, q, γ as in the previous exercise, along with a triple $\Gamma^* \in R_{p,q,\gamma}$, and outputs either 0 or 1. Furthermore, assume that A runs in expected polynomial time. Define two random variables, $X_{p,q,\gamma}$ and $Y_{p,q,\gamma}$, as follows:

- $X_{p,q,\gamma}$ is defined to be the output of A on input p, q, γ, and Γ *, where Γ * is uniformly distributed over $D_{p,q,\gamma}$, and
- $Y_{p,q,\gamma}$ is defined to be the output of A on input p, q, γ, and Γ *, where Γ * is uniformly distributed over $R_{p,q,\gamma}$.

  In both cases, the value of the random variable is determined by the random choice of Γ *, as well as the random choices made by the algorithm. Define

  $\varepsilon(p, q, \gamma) := \left|P[X_{p,q,\gamma} = 1] - P[Y_{p,q,\gamma} = 1]\right|.$

  Using the result of the previous exercise, show how to use A to design a probabilistic, expected polynomial-time algorithm that takes as input p, q, γ as above, along with $\Gamma \in R_{p,q,\gamma}$, and outputs either "yes" or "no," so that

  if $\varepsilon(p, q, \gamma)$ 0.001, then for all $\Gamma \in' R_{p,q,\gamma}$, the probability that A' correctly determines whether $\Gamma \in D_{p,q,\gamma}$ is at least 0.999.

.7. Let p be a prime and let $p - 1 = q_1^{e_1} \ldots q_r^{e_r}$ be the prime factorization of $p - 1$. Let γ be a generator for $\mathbb{Z}_p^*$. Let y be a positive number, and let $Q_p(y)$ be the product of all the prime powers $q_i^{e_i}$ with $q_i \leq y$. Suppose you are given p, y, the primes q_i dividing $p - 1$ with $q_i \leq y$, along with γ, an element α of $\mathbb{Z}_p^*$, and a bound $\hat{x}$, where $x := \log \alpha < \hat{x}$. Show how to compute x in time

$(y^{1/2} + (\hat{x}/Q_p(y))^{1/2}).\text{len}(p)^{O(1)}$.

Chapter 10

Operations of Vector

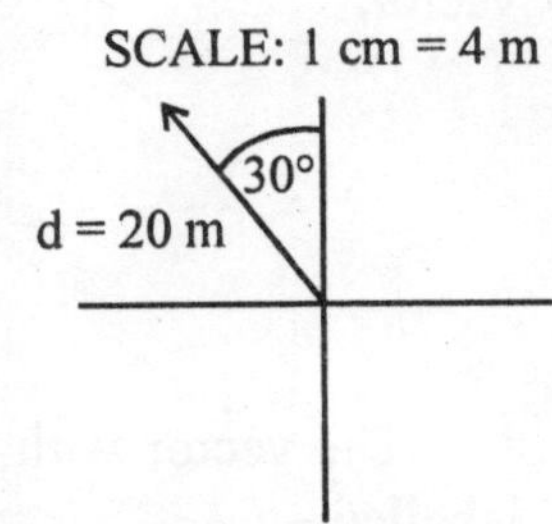

A study of motion will involve the introduction of a variety of quantities which are used to describe the physical world. Examples of such quantities include distance, displacement, speed, velocity, acceleration, force, mass, momentum, energy, work, power, etc. All these quantities can by divided into two categories-vectors and scalars. In this chapter, we will study about vectors and operations on them.

Scalars are quantities that only have a magnitude like mass and speed. Many times it is often useful to have a quantity that has not only a magnitude but also a direction; such a quantity is called a vector.

As an example of when vectors are necessary, suppose a plane traveling at 300 mph to the north with no wind present encounters a westerly crosswind of 50 mph. The resultant velocity of the plane is the sum of the velocities of the wind and the plane. To find this resultant velocity, we must use vectors.

Vectors alone have limited use other than providing a simple, yet effective, means of displaying quantities possessing both a magnitude and direction. The real power in vectors resides in the ability to perform mathematical operations on them.

SCALARS AND VECTORS

Scalars: A scalar is a real number.

Examples. Mass, volume, temperature, work, etc.

Vectors : A vector is a quantity with magnitude and a direction. The magnitude of the vector is also called its length.

Examples. Velocity, acceleration, momentum, force, electric and magnetic field intensities.

Note that all of these quantities not only have a magnitude (such as speed - the magnitude of the velocity vector) but also occur or act in a given direction. A vector is represented by a directed line segment. The direction of the vector is indicated by an arrow pointing from the tail to the head.

If the tail is at point A and the head is at point B, the vector from A to B is written $\overrightarrow{AB}$.

The length (magnitude) of a vector $\vec{v}$ is written $|\vec{v}|$. The length is always a non-negative real number.

VECTORS AS DIRECTED LINE SEGMENTS

Vectors are represented in a natural way by directed line segments, the length of the segment measuring the magnitude of the vector and the arrowhead indicating the direction of the vector. The arrowhead is usually placed at one end (or near the end) of the line segment; this end is called the head of the vector and the other end is called the tail of the vector.

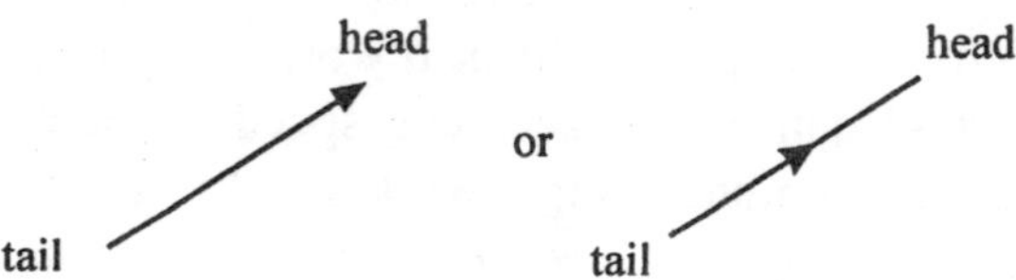

The vector with the same magnitude as u but opposite direction, the negative of u, is labelled $-u$.

The zero vector (a vector with zero length and no direction) is written 0.

REPRESENTATION OF A VECTOR

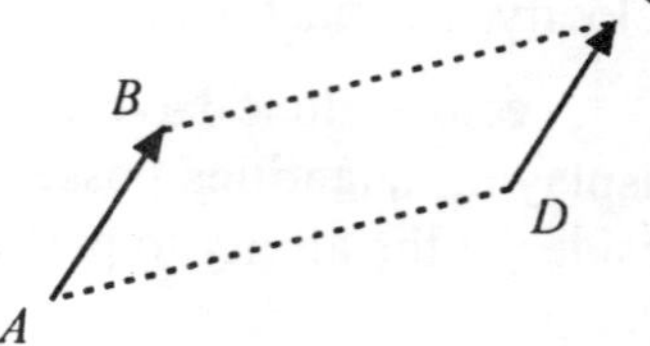

Given points A and B, we may think of the position vector $v = \overrightarrow{AB}$ as the vector that acts on the point A to get B.

If the line segments AB and DC have the same direction and the same length, then $ABCD$ is a parallelogram and the position vectors $\overrightarrow{AB}$ and $\overrightarrow{DC}$ are equal; we write this as $\overrightarrow{AB} = \overrightarrow{DC}$.

The vector v is a free vector because although it has a definite direction and length it does not have any particular position in space.

When using vectors to model physical phenomena it is not always appropriate to use free vectors. For example, in order to completely describe the effect of a force we need to give a vector representing the magnitude and direction of the force as well as a point of application of the force.

A geometric vector is a quantity that can be represented by the directed line segment with initial point *A* and terminal point *B* as shown in the figure. Its magnitude is the length of the directed line segment proportional to the magnitude of the vector. The magnitude is sometimes called the norm.

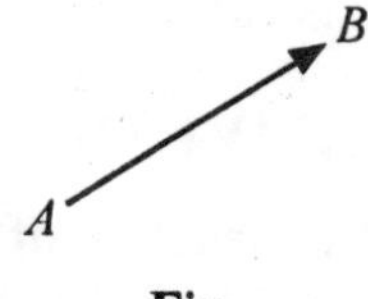

Fig.

Analytically a vector is represented by a symbol $\overrightarrow{AB}$ or a bold face letter *U*. The magnitude of $\overrightarrow{AB}$AB is written as $|\overrightarrow{AB}|$ or $|AB|$ or simply *AB*. The magnitude of *U* is written as $|U|$ or simply *U*.

ZERO OR NULL VECTOR [H]

Even though we have said that all vectors have length and direction, there is one exception, the zero vector.

The zero vector $\hat{0}$ is the vector from a point A to the same point A. We can write $0 = \overrightarrow{AA}$ no matter where the point *A* is.

The zero vector does not have a direction but it does have a length; not surprisingly, its length is 0. Thus we can write $|0| = 0$, i.e., it is a vector of magnitude zero. It is written as $\hat{0}$.

Thus zero vector is a vector, having no specific direction. Graphically it is a point. If $\overrightarrow{AB}= 0$, then $|\overrightarrow{AB}| = 0$ and A and B coincides.

EQUALITY OF VECTORSS

Two vectors *a* and *b* whose magnitudes are equal i.e. $a = b$ and which are in the same direction regardless of their initial points, are called equal vectors.

It can be seen that for three vectors a, b and c,

$$r = a + b + c = (a + b) + c = a + (b + c) + (a + c) + b$$

This result can be extended to any number of vectors.

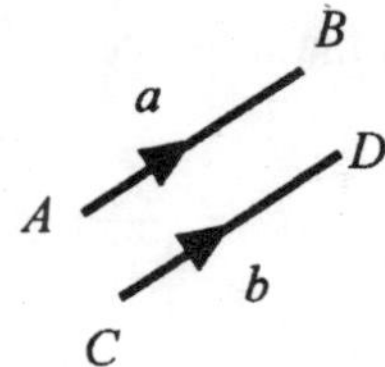

Fig.

Free Vector

A vector which can be shifted along its line of action or parallel to itself and whose initial point can be anywhere, is called a free vector. Free vectors are distinguished only by their length and direction, and not by their position in space.

Unless stated, we always refer a vector as a free vector.

Localized Vector

It is also known as a sliding vector. It can be shifted only along its line of action.

Bounded Vector

It is called as a fixed vector. It must remain at the same point of action or application.

NEGATIVE OF A VECTOR

The negative of a vector u is the vector –u which has the same direction and inclination but is of the opposite sense (of direction).

The negative of the vector from A to B is the vector from B to A. It has the same length as $\overrightarrow{AB}$ but the opposite direction.

Thus the negative of the vector $\overrightarrow{AB}$ is $\overrightarrow{BA}$ and we also write $\overrightarrow{BA} = -\overrightarrow{AB}$.

Fig. The resultant of a system of vectors is the least number of vectors that will replace the given system.

UNIT VECTOR

A vector in a given direction having unit magnitude is called a unit vector in that direction. It is convenient to use vectors of unit length to specify the directions of vector quantities in the various coordinate systems.

It is usually denoted by $\hat{n}, \hat{e}, \hat{a}, \hat{b}$ etc (Read as *n* Cap). Also, note that if $\vec{U}$ is given vector then the unit vector in the direction (or along or parallel) to $\vec{U}$ is given by

$$\hat{U} = \frac{\vec{U}}{|\vec{U}|}$$

in Cartesian coordinates it is typical to use $\hat{i}, \hat{j}$ and $\hat{k}$ to represent unit vectors along $\overline{OX}, \overline{OY}$ and $\overline{OZ}$ respectively.

Position vector $\vec{r} = x\hat{i} + y\hat{j} + z\hat{k}$

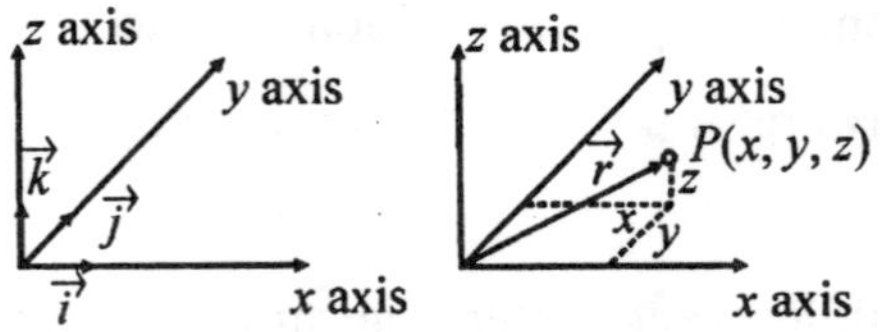

POSITION VECTOR OF A POINT

If *A* and B are any two points then $\overrightarrow{AB}$ is called the position vector of point B with reference to point A.

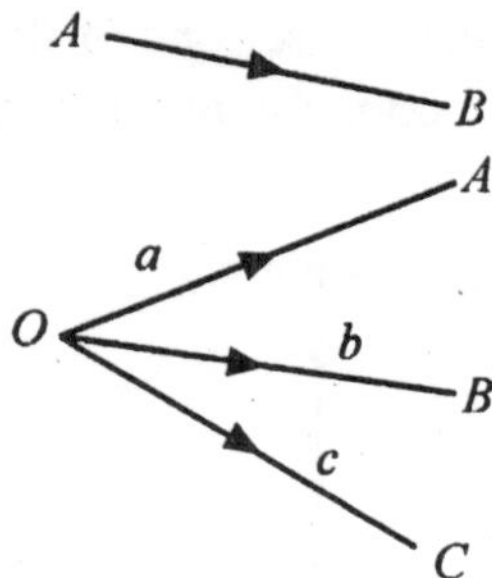

Usually we fix a point *O* and define the position vector of points with reference to *O* which is called the origin of reference.

DEFINITION: POSITION VECTOR

Position vector is a vector that extends from the reference point to the position of the particle.

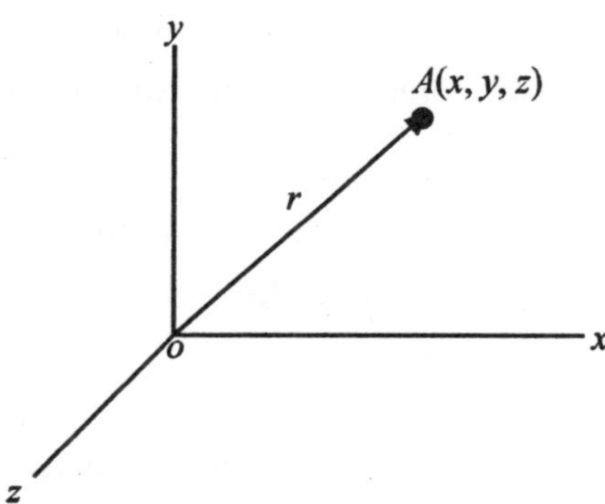

Fig. Position vector is represented by a vector, joining origin to the position of point object.

It is easy to realize that vector representation of position is appropriate, where directional properties of the motion are investigated.

The description of motion involving direction consists of position vector as a function of time. Consider the definitions : the "displacement" is equal to the change in position vector; the "velocity" is equal to the rate of change of position vector with respect to time; and "acceleration" is equal to the rate of change of velocity with respect to time, which, in turn, is the rate of change of position vector. Thus, all directional attributes of motion is based on the processing of position vectors.

VECTOR IN COMPONENT FORM

Recall that scalar components are graphically obtained by dropping two perpendiculars from the ends of the vector to the axes. In the case of position vector, one of the end is the origin itself. As position vector is rooted to the origin, the scalar components of position vectors in three mutually perpendicular directions of the coordinate system are equal to the coordinates themselves. Also, scalar components of position vector, *r*, by definition in the designated directions of the axes are :

$$x = r\cos\alpha,\ y = r\cos\beta,\ z = r\cos\gamma$$

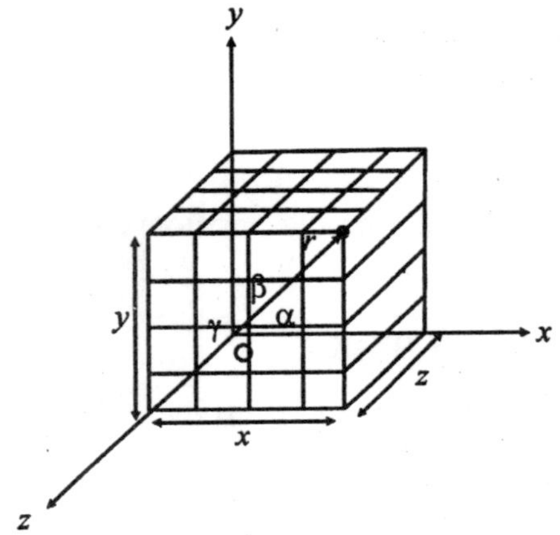

Fig. Scalar components are equal to coordiantes of the position

and position vector in terms of components (coordinates) is :

$$r = xi + yj + zk$$

where *i*, *j* and *k* are unit vectors in *x*, *y* and *z* directions.

The magnitude of position vector is given by :

$$r = |r| = \sqrt{x^2 + y^2 + z^2}$$

Example. Position (in meters) of a moving particle as a function of time (in seconds) is given by :

$$r = (3t^2 - 3)i + (4 - 7t)j + (-t^3)k$$

Find the coordinates of the positions of the particle at the start of the motion and at time $t = 2\ s$. Also, determine the linear distances of the positions of the particle from the origin of the coordinate system at these time instants.

Solution: The coordinates of the position are projection of position vector on three mutually perpendicular axes. Whereas linear distance of the position of the particle from the origin of the coordinate system is equal to the magnitude of the position vector.

Now, When t = 0 (start of the motion)

$$r = (3 \times 0 - 3)\ i + (4 - 7 \times 0)j + (-0)k$$

The coordinates and distance are :

$$x = -1m \text{ and } y = 5m$$

$$r = |r| = \sqrt{(-3)^2 + 4^2} = \sqrt{25} = 5m$$

When t = 2*s*

$$r = (3 \times 2^2 - 3)i + (4 - 7 \times 2)j + (-2^3)k$$

$$= 9i - 10j - 8k$$

The coordinates and distance are :

$$x = 9\text{m}, y = -10\text{m}, \text{ and } z = -8\text{m}.$$

$$r = |r| = \sqrt{9^2 + (-10)^2 + (-8)^2} = \sqrt{245}$$

$$= 15.65m$$

MOTION TYPES

Position is a three dimensional concept, requiring three coordinate values to specify it.

In two dimensional motion, two of the three coordinates change with time. The remaining third coordinate is constant.

In one dimensional motion, only one of the three coordinates is changing with time. Other two coordinates are constant through out the motion. As such, it would be suffice to describe positions of the particle with the values of changing coordinate and neglecting the remaining coordinates.

$$x, y = 2, z = 2$$

A motion along x –axis or parallel to x – axis is, thus, described by x - component of the position vector i.e. x – coordinate of the position as shown in the figure.

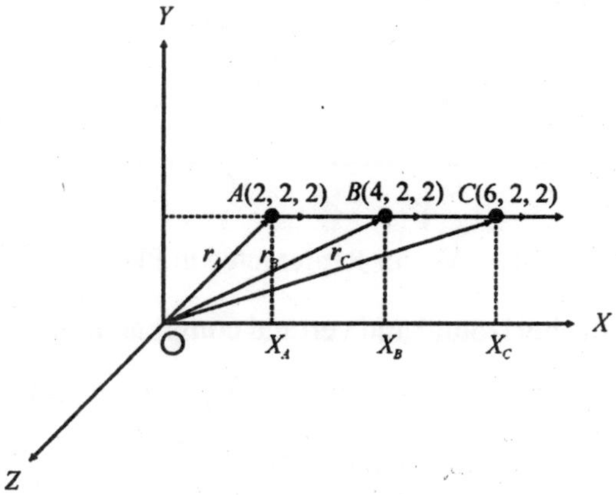

Fig. Motion in one Dimesnion

The description of one dimensional motion is further simplified by shifting axis to the path of motion as shown below.

x, y = 0, z = 0

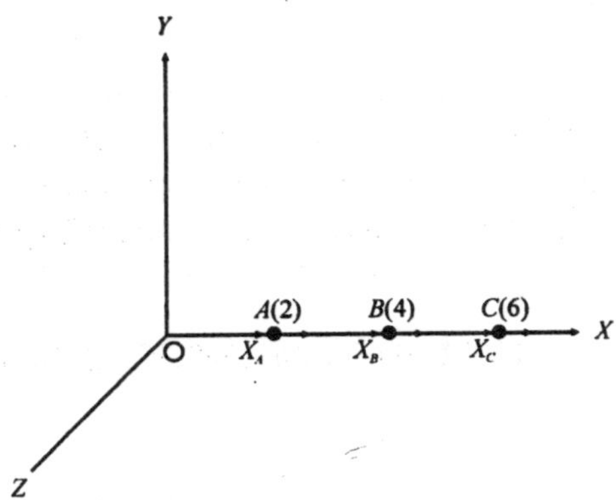

Fig. Motion in One Dimesnion

In this case, position vector itself is along x – axis and, therefore, its magnitude is equal to x – coordinate.

We shall use *a, b, c* etc. to denote position vectors of points *A, B, C* etc. respectively, relative to origin *O*.

REPRESENTATION OF A VECTOR IN A PLANE

Let $P(x, y)$ be any point on the coordinate plane. With $\overrightarrow{OX}$ and $\overrightarrow{OY}$ as coordinate axes and O as the origin. Then $\overrightarrow{OP}$ stands for the position vector of point P relative to origin. Let $P = \overrightarrow{OP}$.

Any such vector can be broken down into two components, a horizontal (along $\overrightarrow{OX}$) and a vertical (along $\overrightarrow{OY}$) component.

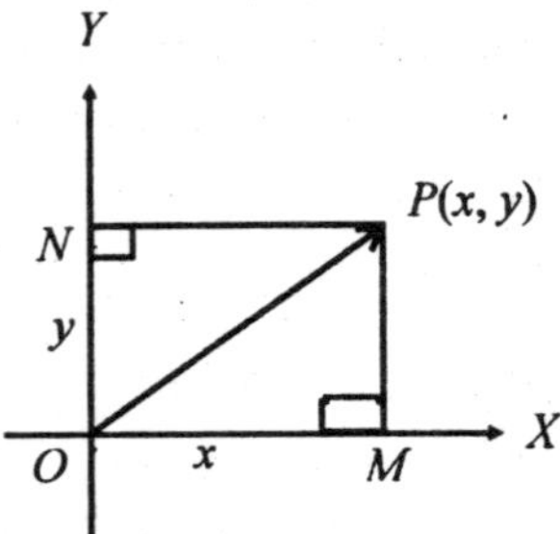

Fig. Vector Represented in Plane.

Let $\overrightarrow{OM}$ and $\overrightarrow{ON}$ be the horizontal and vertical component of $\overrightarrow{OP}$ i.e., P. Now $\overrightarrow{OM}$ and $\overrightarrow{ON}$ are non-collinear vectors but they are collinear with $\hat{i}$ and $\hat{j}$ respectively. So $\overrightarrow{OM} = x\ \hat{i}$ and $\overrightarrow{ON} = y\ \hat{j}$. If we use the tail up rule we have, $\overrightarrow{OP} = \overrightarrow{OM} + \overrightarrow{MP}$. But $\overrightarrow{MP} = \overrightarrow{ON} = y\ \hat{j}$.

$$\therefore \quad \overrightarrow{OP} = x\hat{i} + y\ \hat{j} \text{ i.e., } \mathrm{p} = x\hat{i} + y\ \hat{j}$$

Here x and y are components of p along x-axis and y-axis respectively. This ordered pair (x, y) associated with p is known as the rectangular components of p.

Let $\angle POX = \theta$ then

$$\cos\theta = \frac{|x|}{|p|} \text{ and } \sin\theta = \frac{|y|}{|p|} \text{ so that } p = \sqrt{x^2 + y^2}$$

which is the magnitude of p.

This concept can be extended to higher dimensions such as three dimension.

Also, note that any vector $\overrightarrow{AB}$ can be shifted so that its initial point A is set at 'O', then $\overrightarrow{AB}$ is in 'standard position'. Now $\overrightarrow{AB} = \overrightarrow{OP}$ then $\overrightarrow{OP}$ replaces $\overrightarrow{AB}$.

Thus $\overrightarrow{OP}$ is the standard vector in the plane for all vectors in the plane of the same magnitude and direction as $\overrightarrow{OP}$.

$\therefore \qquad OP = OM + ON$

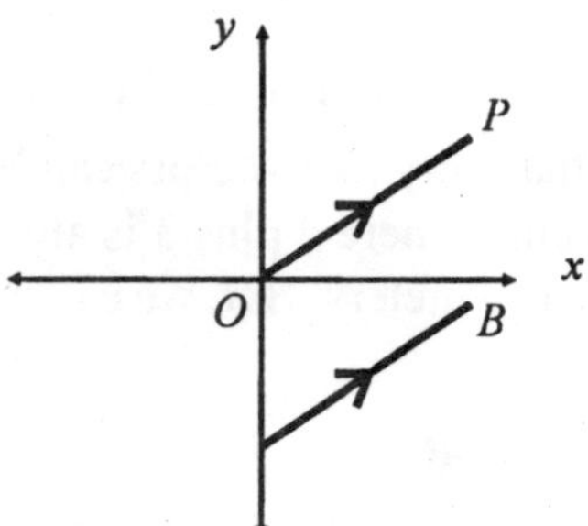

Fig. Vectors of same Magnitude

If coordinates of A are (xa, ya) and that of B are (xb, yb) then shifting A to O and making $B = P$ we have coordinates of P are $(xb - xa, yb - ya)$

$\therefore \overrightarrow{AB}$ or $\overrightarrow{OP} \equiv (xb - xa, yb - ya)$

Thus an algebraic vector is an ordered pair of real numbers that corresponds to a standard geometric vector $\overrightarrow{OP}$.

OPERATIONS ON VECTORS

ADDITION OF VECTORS

Vector addition represents the net effect of the directional quantities

Vectors operate with other scalar or vector quantities in a particular manner. Unlike scalar algebraic operation, vector operation draws on graphical representation to incorporate directional aspect. Basically, vector addition is dealt in two conceptually equivalent ways:

1. Graphical and
2. Algebraic (vector)

The resulting vector after addition is termed as sum or resultant vector. The resultant vector corresponds to the "resultant" or "net" effect of a physical quantity having directional attribute. The effect of a force system on a body, is determined by the resultant force acting on it. The idea of resultant force, in this case, conveys that the resulting force (vector) has the same effect as that of the forces (vectors) being added.

$$R = F_1 + F_2 + F_3 + F_4$$

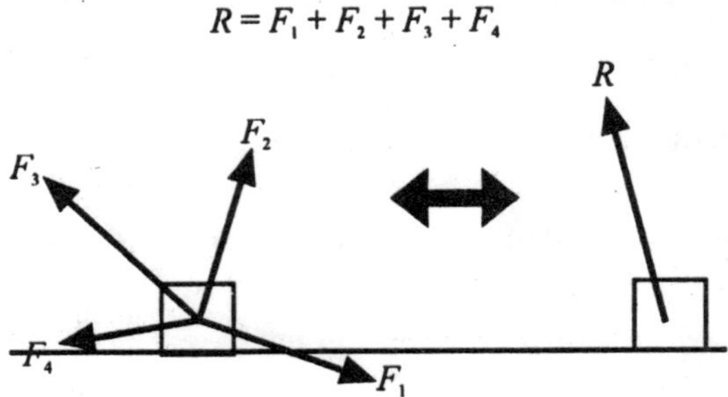

Fig. Resultant Force

Vector addition rule restricts that only like vectors can be added. It is expected also. The requirement is similar to scalar algebra where 2 plus 3 is always 5, but we need to add similar quantity like 2 meters plus 3 meters is 5 meters. But, we can not add, for example, distance and temperature.

Vector Addition: Graphical Method

Let us examine the example of displacement of a person in two different directions. The two displacement vectors, perpendicular to each other, are added to give the "resultant" vector. In this case, the closing side of the right triangle represents the sum (i.e. resultant) of individual displacements *AB* and *BC*.

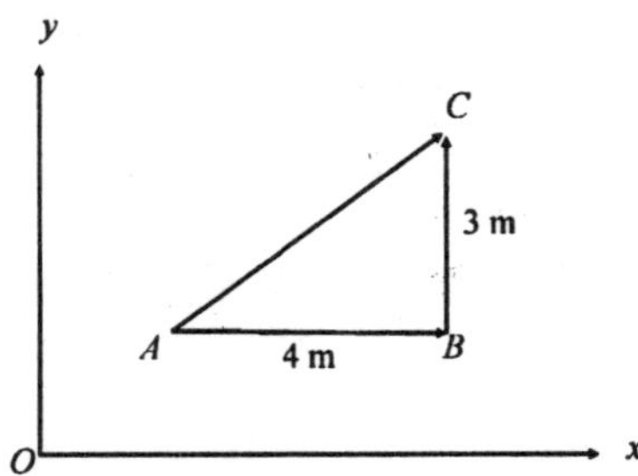

Fig. Displacement $AC = AB + BC$

The method used to determine the sum in this particular case (in which, the closing side of the triangle represents the sum of the vectors in both magnitude and direction) forms the basic consideration for various rules dedicated to implement vector addition.

Triangle Law

In most of the situations, we are involved with the addition of two vector quantities. Triangle law of vector addition is appropriate to deal with such situation.

Definition: *Triangle law of vector addition*

If two vectors are represented by two sides of a triangle in sequence, then third closing side of the triangle, in the opposite direction of the sequence, represents the sum (or resultant) of the two vectors in both magnitude and direction.

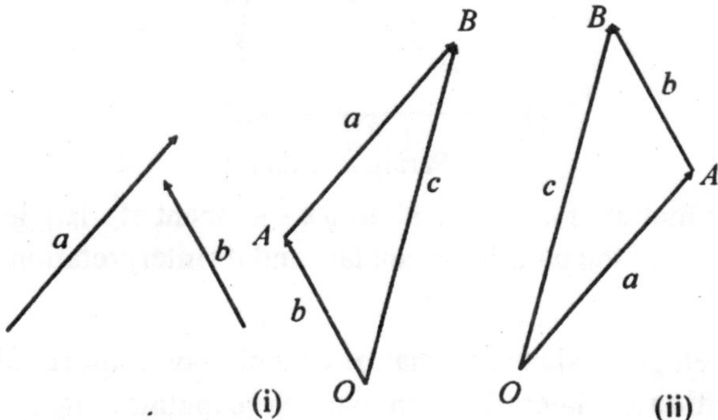

Fig. Triangle Law of Vector Addition

Here, the term "sequence" means that the vectors are placed such that tail of a vector begins at the end of arrow head of the vector placed before it.

The triangle law does not restrict where to start i.e. with which vector to start. Also, it does not put conditions with regard to any specific direction for the sequence of vectors, like clockwise or anti-clockwise, to be maintained. In figure (i), the law is applied starting with vector,b; whereas the law is applied starting with vector, a, in figure (ii). In either case, the resultant vector, c, is same in magnitude and direction.

This is an important result as it conveys that vector addition is commutative in nature i.e. the process of vector addition is independent of the order of addition. Mathematically speaking :

$$a + b = b + a$$

Parallelogram Law

Parallelogram law, like triangle law, is applicable to two vectors.

Definition: *Parallelogram law*

If two vectors are represented by two adjacent sides of a parallelogram, then the diagonal of parallelogram through the common point represents the sum of the two vectors in both magnitude and direction.

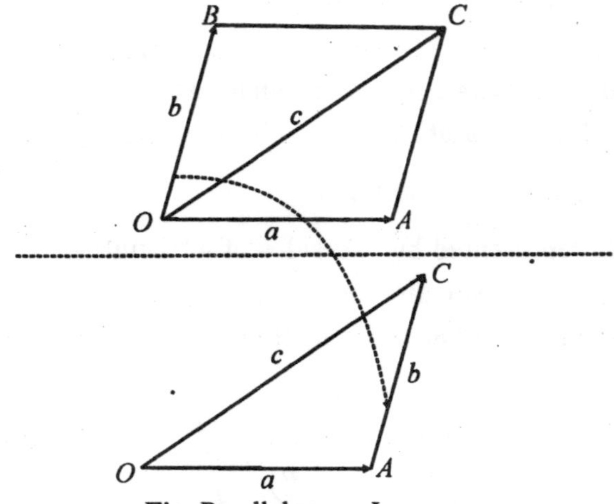

Fig. Parallelogram Law

Parallelogram law, matter of fact, is an alternate statement of triangle law of vector addition. A graphics representation of the parallelogram law and its interpretation in terms of the triangle is shown in the figure :

Converting parallelogram sketch to that of triangle law requires shifting vector, *b*, from the position *OB* to position *AC* laterally as shown, while maintaining magnitude and direction.

Polygon Law

The polygon law is an extension of earlier two laws of vector addition.

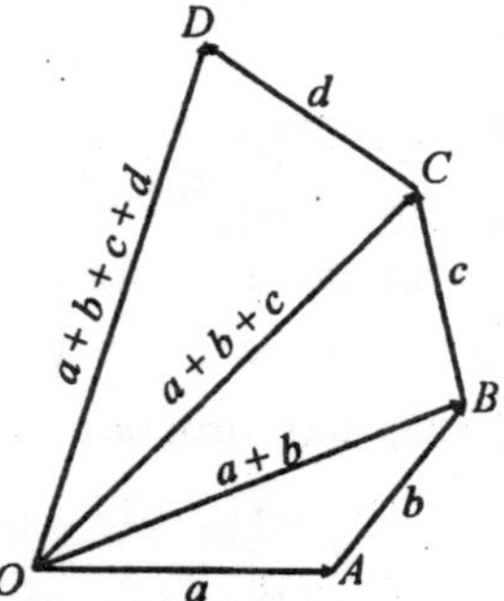

Fig. Successive Application of Triangle Law

Definition: Polygon law

Polygon law of vector addition : If (n –1) numbers of vectors are represented by (n-1) sides of a polygon in sequence, then n^{th} side, closing the polygon in the opposite direction, represents the sum of the vectors in both magnitude and direction.

In the figure shown below, four vectors namely a, b, c and d are combined to give their sum. Starting with any vector, we add vectors in a manner that the subsequent vector begins at the arrow end of the preceding vector. The illustrations in figures 15 i, iii and iv begin with vectors a, d and c respectively.

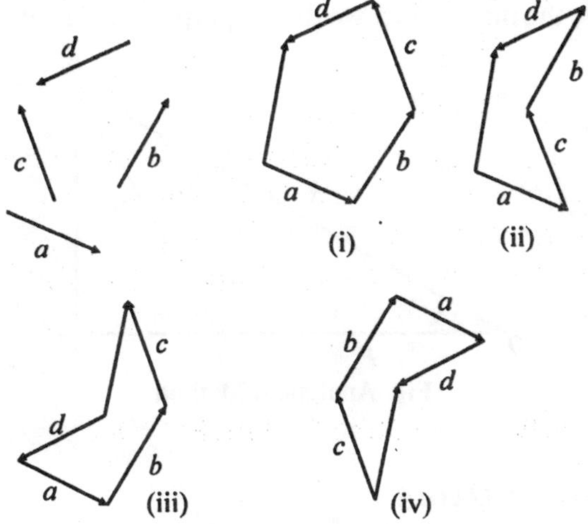

Fig.. Polygon law

Matter of fact, polygon formation has great deal of flexibility. It means that the order of grouping of vectors for addition has no consequence on the result. This characteristic of vector addition is known as "associative" property of vector addition and is expressed mathematically as :

$$(a + b) + c = a + (b + c)$$

ANALYTICAL METHOD

Analytical method, based on geometry, provides a solution in this regard. It allows us to accurately determine the sum or the resultant of the addition, provided velues of magnitudes and angles are supplied.

Here, we shall analyze vector addition in the form of triangle law to obtain the magnitude of the sum of the two vectors. Let *P* and *Q* be the two vectors to be added, which make an angle, with each other. We arrange the vectors in such a manner that two adjacent sides *OA* and *AB* of the triangle *OAB*, represent two vectors *P* and *Q* respectively as shown in the figure.

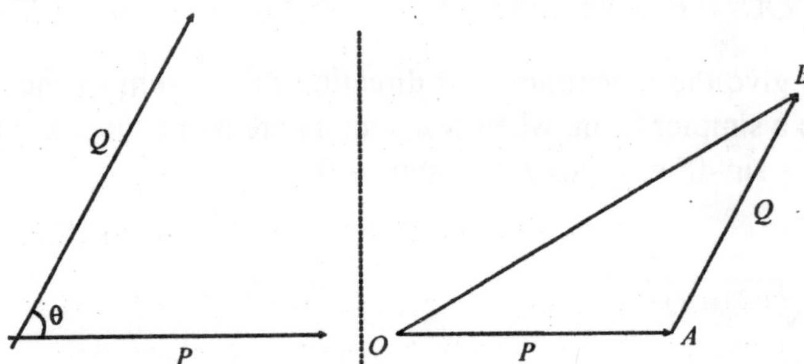

Fig. Analytical method

According to triangle law, the closing side *OB* represent sum of the vectors in both magnitude and direction.

$$OB = OA + AB = P + \text{Q}$$

In order to determine the magnitude, we drop a perpendicular *BC* on the extended line *OC*.

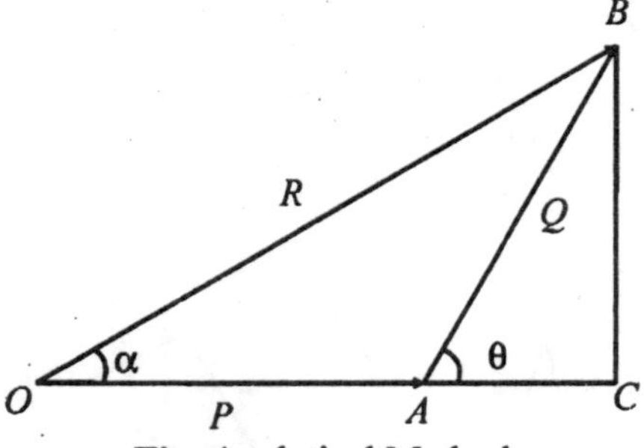

Fig. Analytical Method

In Δ ACB,

$$AC = AB\cos\theta = Q\cos\theta$$

$$BC = AB\sin\theta = Q\sin\theta$$

In right Δ *OCB*, we have:.

$$OB = \sqrt{OC^2 + BC^2} = \sqrt{(OA + AC)_2 + BC^2}$$

Substituting for AC and BC,

$$|OB| = \sqrt{(P + Q\cos\theta)^2 + (Q\sin\theta)^2}$$

$$P = \sqrt{P^2 + Q^2\cos^2\theta + 2PQ\cos\theta + Q^2\sin^2\theta}$$

$$R = OB = \sqrt{P^2 + Q^2 + 2PQ\cos\theta}$$

Let "α" be the angle that line *OA* makes with *OC*, then

$$\tan a = \frac{BC}{OC} = \frac{Q\sin\theta}{P + Q\cos\theta}$$

The equations give the magnitude and direction of the sum of the vectors. The above equation reduces to a simpler form, when two vectors are perpendicular to each other. In that case, $\theta = 90°$; sinq = sin90° = 1; cos $\theta = \cos 90° = 0$

and,

$$OB = \sqrt{P^2 + Q^2}$$

$$\tan\alpha = \frac{Q}{P}$$

These results for vectors at right angle are exactly same as determined, using Pythagoras theorem.

VECTOR ADDITION: ALGEBRAIC METHOD

Graphical method is meticulous and tedious as it involves drawing of vectors on a scale and measurement of angles. We can, however, extend algebraic techniques to vectors, *provided vectors are represented on a rectangular coordinate system.* The representation of a vector on coordinate system uses the concept of unit vector and component.

Components of a Vector

A scalar component, also known as projection, of a vector AB in the positive direction of a straight line $C'C$ is defined as :

$$AC = |AB| \cos\theta = AB \cos\theta$$

where θ, is the angle that vector AB makes with the specified direction C'C as shown in the figure.

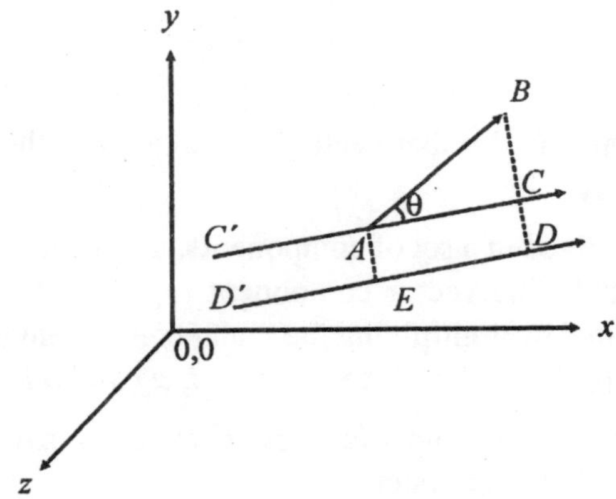

Fig. *Vectors are Represented on a Rectangular Coordinate System.*

It should be understood that the component so determined is not rooted to the common point "*A*" or any specific straight line like *C'C* in a given direction. Matter of fact, component of vector *AB* drawn on any straight lines like *D'D* parallel to *C'C* are same. Note that the component i.e. projection of the vector is graphically obtained by drawing two perpendicular lines from the ends of the vector on the straight line in the specified direction.

$$ED = AC = AB \cos\theta,$$

Following the above definition of component of vector (specifically known as scalar component), we can resolve a given vector in three components in three mutually perpendicular directions of rectangular coordinate system. Note that we measure angle with respect to parallel lines to the axes. By convention, we denote components by using the non-bold type face of the vector symbol with a suffix representing direction (x or y or z).

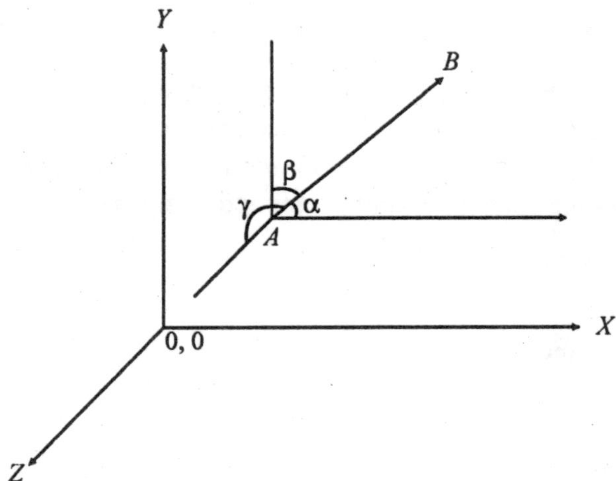

Fig. Components of a Vector

$AB_x = \text{AB} \cos\alpha$

$AB_y = \text{AB} \cos\beta$

$AB_y = \text{AB} \cos\gamma$

where α, β and γ are the angles that vector *AB* makes with the positive directions of *x*, *y* and *z* directions respectively.

As a vector can be resolved in a set of components, the reverse processing of components in a vector is also expected. The vector component is the vector counterpart of the scalar component, which is obtained by multiplying the scalar component with the unit vector in axial direction. The vector components of a vector *a*, are $a_x i$, $a_y j$ and $a_z k$.

A vector *a*, is equal to the vector sum of component vectors in three mutually perpendicular directions of rectangular coordinate system.

$a = normal\ a_x i + normal\ a_y j + normal\ a_z k$

where *i*, *j* and *k* are unit vectors in the respective directions.

Magnitude of the sum of two vectors (a + b) in x and y direction, which are perpendicular to each other, is : $\left|\vec{a}\right| = \sqrt{a_x^{\ 2} + a_y^{\ 2}} = a_x^{\ 2} + a_y^{\ 2}$

and the magnitude of the sum of vectors (a + b) and third vector c, which are perpendicular to each other, is : $\left|\vec{a}\right| = \sqrt{a_x^{\ 2} + a_y^{\ 2}} = \left(a_x^{\ 2} + a_y^{\ 2}\right)^{1/2} \left|\vec{c}\right| = \sqrt{a_x^{\ 2} + a_y^{\ 2} + a_z^{\ 2}}$

VECTOR ADDITION USING COMPONENT OF VECTOR

The frame work for vector addition draws on two important concepts. The first concept is that a vector can be equivalently expressed in terms of three component vectors.

$a = a_x i + a_y j + a_z k$ and $b = b_x i + b_y j + b_z k$

The component vector form has important significance. It ensures that component vectors to be added are restricted to three known directions only. If two vectors are along the same line, then resultant is given as :

When , $\theta = 0°$, $\cos\theta = \cos 0° = 1$ and,

$$\left|\vec{R}\right| = \sqrt{P^2 + 2PQ + Q^2} = \sqrt{(P+Q)^2} = P + Q$$

When θ, $= 180°$, $\cos\theta = \cos 180° = -1$

and, $\left|\vec{R}\right| = \sqrt{P^2 - 2PQ + Q^2} = \sqrt{(P-Q)^2} = P - Q$

Thus, we see that the magnitude of the resultant is equal to algebraic sum of the magnitudes of the two vectors.

Using these two concepts, the addition of vectors is affected as outlined here:

1. Represent the vectors (a and b) to be added in terms of components:
   $a = a_x i + a_y j + a_z k$ and $b = b_x i + b_y j + b_z k$
2. Group components in a given direction:
   $a + b = a_x i + a_y j + a_z k + b_x i + b_y j + b_z k$
   $a + b = (a_x + b_x)\, i + (a_y + b_y)\, j + (a_z + b_z)\, k$
3. Find the magnitude and direction of the sum, using analytical method

Properties of Vector Addition

1. Vector addition is commutative i.e. $a + b = b + a$ Extension to sums of two or more vectors is immediate.
2. Vector addition is associative. Observe the adjoining figure we have $\overrightarrow{OA} = a$, $\overrightarrow{AB} = b$, $\overrightarrow{BC} = c$ and $\overrightarrow{OC} = r$ using tail-tip rule, for triangle OAB, $\overrightarrow{OB} = a + b$, for triangle ABC, $\overrightarrow{AC} = b + c$ for triangle OBC, c i.e. $r = (a + b) + c$. Also, for triangle OAC, $\overrightarrow{OC} = (a + c) + c$ i.e $r = a + (b + c)$.
   Thus, $r = (a + b) + c = a + (b + c)$.

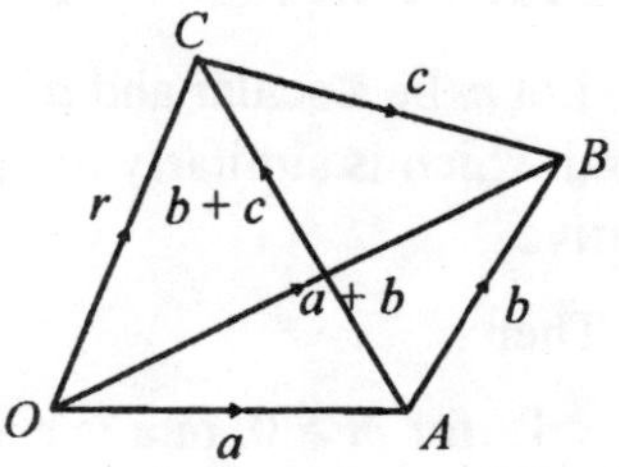

Fig. Vector Addition by Tail-tip Rule

SUBTRACTION OF A VECTOR

It is accomplished by adding the negative of vectors i.e.

$a - b = a + (-b)$. Note also that $-(a + b) = -a - b$.

Subtraction is considered an addition process with one modification that the second vector after reversing its direction is added to the first vector. Let us consider the problem of subtracting vector, b, from, a. Using graphical techniques, we first reverse the direction of vector, b, and obtain the sum applying triangle or parallelogram law.

Symbolically,

$$a - b = a + (-b)$$

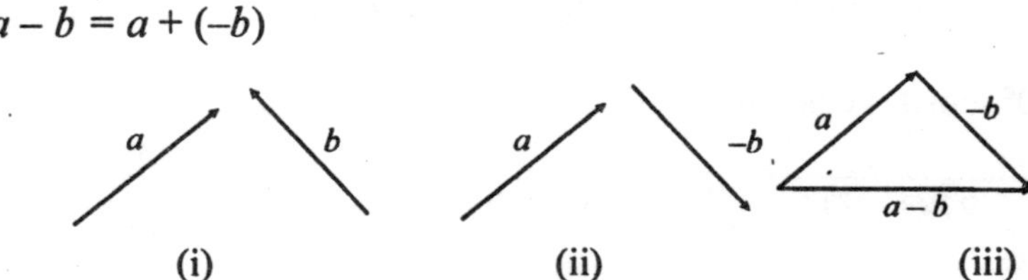

Fig. Subtraction of Vector

COMPOSITION OF VECTORS

It is the process of determining the resultant of a system of vectors. For this we have to draw a vector polygon, placing tail of each vector at the tip of the preceding vector. Then draw a vector from the tail of the first vector to tip of the last vector in the system. Later on we will show that not all vector systems reduce to a single vector. Also note that the order of vectors are drawn is immaterial.

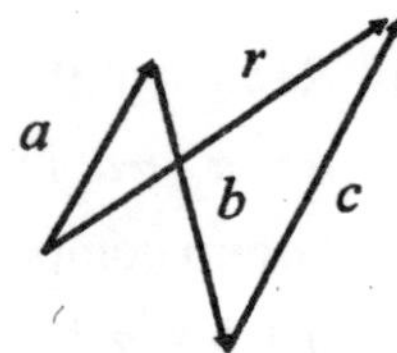

Fig. Vector Composition.

MULTIPLICATION OF VECTORS BY SCALARS

Let *m* be a scalar and *a* be a vector then the product ma whose magnitude is $|m|$ times of $|a|$ and which is similarly or oppositely directed to a depending upon whether m is positive or negative.

Then

1. for $m > 0$, m a is a vector of magnitude m $|a|$ and is in the direction of a.
2. for $m = 0$, ma is a null vector.
3. for $m < 0$, ma is the vector of magnitude $-m\ |a|$ and is exactly opposite to the vector *a*

   From this, it follows that $(-1)\ a = -a$ and $0\ a = 0$

   (B) Other operations with scalars *m* and *n* are

   $(m + n)\ a = m\ \mathrm{a} + n\ \mathrm{a}$

   $m\ (\mathrm{a} + \mathrm{b}) = m\ \mathrm{a} + m\ \mathrm{b}$

   $mn\ (a) = n\ (m\mathrm{a}) = (m\ n\ \mathrm{a})$

LINEAR COMBINATION OF VECTORS

Let a and b be two vectors and x and y be any real numbers. Then the vector $r = x\,a + y\,b$ is called the linear combination of vectors a and b.

COLLINEAR VECTORS

Two vectors a and b are said to be collinear, if they are parallel to one or on the same line. This clearly indicates that 1) they must be coincident or (2) they must be parallel. Also, it must be kept in mind that if two vectors are expressed as the scalar multiple of each other then they must be collinear. i.e. if $a = x\,b$ then a and b are collinear if $x \neq 0$ scalar.

COPLANAR VECTORS

Geometric objects lying in a common plane are said to be coplanar. Three non-collinear points determine a plane and so are trivially coplanar. Four points are coplanar iff the volume of the tetrahedron defined by them is 0,

$$\begin{vmatrix} x_1 & y_1 & z_1 & 1 \\ x_2 & y_2 & z_2 & 1 \\ x_3 & y_3 & z_3 & 1 \\ x_4 & y_4 & z_4 & 1 \end{vmatrix} = 0$$

co-planarity is equivalent to the statement that the pair of lines determined by the four points are not skew, and can be equivalently stated in vector form as

$$(x_3 - x_1) \cdot [(x_2 - x_1) \times (x_4 - x_3)] = 0$$

An arbitrary number of n points $x_1, x_2, \ldots, x_n$ can be tested for co-planarity by finding the point-plane distances of the points $x_1, \ldots, x_n$ from the plane determined by (x_1, x_2, x_3) and checking if they are all zero. If so, the points are all coplanar.

A set of n vectors V is coplanar if the nullity of the linear mapping defined by V has dimension 1, the matrix rank of V (or equivalently, the number of its singular values) is $(n - 1)$.

If vectors are parallel to a given plane, then they are coplanar. Note that if a and b be two non-collinear vectors which can be uniquely expressed as $\mathrm{r} = x\,a + y\,b$, $x, y \neq 0$ scalars, then they must be coplanar.

Parallel lines in three-dimensional space are coplanar, but skew lines are not.

MULTIPLICATION

Vector multiplication provides concise and accurate representation of natural laws, which involve vectors.

The vector multiplication, however, is not an unique mathematical construct like scalar multiplication. The multiplication depends on the nature of quantities (vector or scalar) and on the physical process, necessitating scalar or vector multiplication.

The rules of vector multiplication have been formulated to encapsulate physical processes in their completeness. In order to explore this aspect, let us find out the direction of acceleration in the case of parabolic motion of a particle. There may be two ways to deal with the requirement. We may observe the directions of velocities at two points along the path and find out the direction of the change of velocities. Since we know that the direction of change of velocity is the direction of acceleration, we draw the vector diagram and find out the direction of acceleration. We can see that the direction of acceleration turns out to act in vertically downward direction.

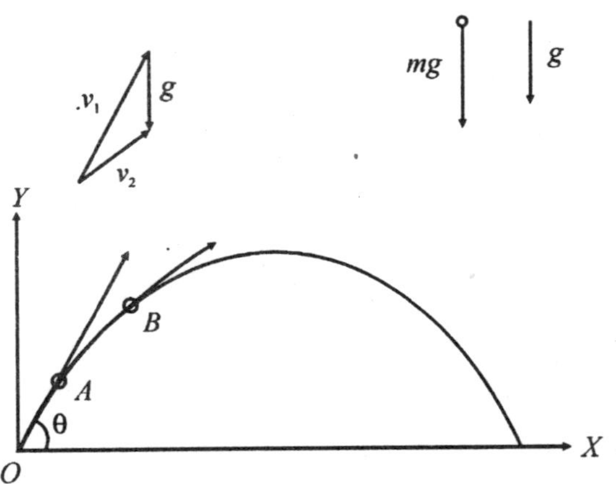

Fig. Direction of Acceleration

The conceptualization of physical laws in vector form, however, provides us with powerful means to arrive at the result in relatively simpler manner. If we look at the flight of particle in parabolic motion, then we observe that the motion of particle is under the force of gravity, which is acting vertically downward. There is no other force (neglecting air resistance). Now, from second law of motion, we know that :

$$F_{\text{Resultant}} = ma$$

This equation reveals that the direction of acceleration is same as that of the resultant force acting on the particle. Thus, acceleration of the particle in parabolic motion is acting vertically downward. We see that this second approach is more elegant of the two methods. Using the fundamental law of nature in vector form involving multiplication of the vector with a scalar in this case, we could arrive at the correct answer in a very concise manner, without getting into the details of the motion.

Multiplication of Vector with Scalar

Multiplication of a vector A, with another scalar quantity, a, results in another vector, B. The magnitude of the resulting vector is equal to the product of the magnitude of vector with the scalar quantity. The direction of the resulting vector, however, is same as that of the original vector.

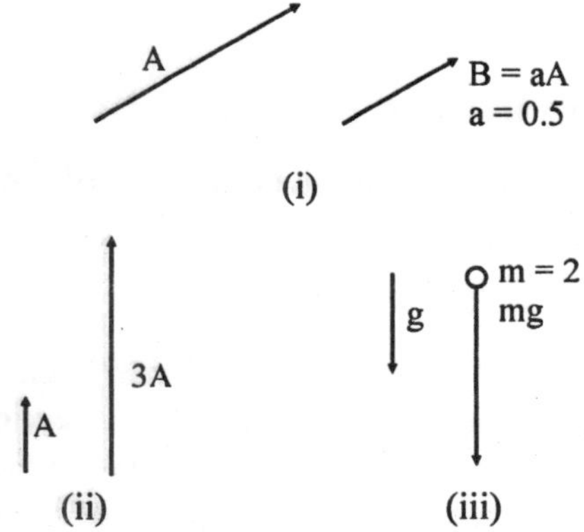

Fig. Multiplication with scalar

We have already made use of this type of multiplication intuitively in expressing a vector in component form.

$$A = A_x i + A_y j + A_z k$$

In this vector representation, each component vector is obtained by multiplying the scalar component with the unit vector. As the unit vector has the magnitude of 1 with a specific direction, the resulting component vector retains the magnitude of the scalar component, but acquires the direction of unit vector.

$$A_x = A_x i$$

PRODUCTS OF VECTORS

Some physical quantities are itself a scalar quantity, but composed from the product of vector quantities. One such example is "work", which is a product of force (a vector) and displacement (a vector), while "work" itself is a scalar quantity. On the other hand, magnetic force is a vector, composed from the product of velocity of a moving charge (a vector) and magnetic field (vector).

Thus, products of vectors are defined in two distinct manner – one resulting in a scalar quantity and the other resulting in a vector quantity.

SCALAR PRODUCT

Scalar product of two vectors a and b is a scalar quantity defined as :

$$a.b = ab\cos\theta,$$

where "a" and "b" are the magnitudes of two vectors and "θ" is the angle between the direction of two vectors. It is important to note that vectors have two angles θ and 2π – θ. We can use either of them as cosine of both "θ" and "2π – θ" are same. However, it is suggested to use the smaller of the enclosed angle as it is required for vector (i.e. cross) product.

The notation "a.b" is important and should be mentally noted to represent a scalar quantity – even though it involves bold faced vectors. It should be noted that all quantities on the right hand side are scalars.

MEANING OF SCALAR PRODUCT

We can read the definition of scalar product in either of the following manners :

$$a.b = a\,(b\cos\theta)$$

$$a.b = b\,(a\cos\theta)$$

Recall that "b cos θ" is the scalar component of vector b along the direction of vector a and "a cos θ" is the scalar component of vector a along the direction of vector b. Thus, we may consider the scalar product of vectors a and b as the product of the magnitude of one vector and the scalar component of other vector along the first vector.

The figure below shows drawing of scalar components. The scalar component of vector b in figure (i) is obtained by drawing perpendicular from the tip of the vector, b, on the direction of vector, a. Similarly, the scalar component of vector a in figure (i) is obtained by drawing perpendicular from the tip of the vector, a, on the direction of vector, b.

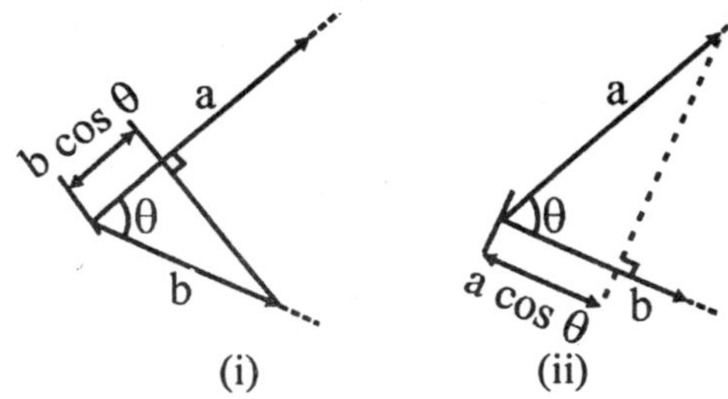

Fig. Scalar Product

The two alternate ways of evaluating dot product of two vectors indicate that the product is commutative i.e. independent of the order of two vectors :

$$a.\ b = b.\ a$$

Values of Scalar Product

The value of dot product is maximum for the maximum value of $\cos\theta$. Now, the maximum value of cosine is cos0°=1. For this value, dot product simply evaluates to the product of the magnitudes of two vectors.

$$(a.b)_{max} = ab$$

For $\theta = 180°, \cos 180° = -1$ and

$$a.b = -ab$$

Thus, we see that dot product can evaluate to negative value as well. This is a significant result as many scalar quantities in physics are given negative value. The work done, for example, can be negative, when the work is done on the system; instead of work being done by the system (positive).

The scalar product evaluates to zero for $\theta = 90^0$ and 270^0 as cosine of these angles are zero. These results have important implication for unit vectors. The dot products of same unit vector evaluates to 1.

$$i.i = j.j = k.k = 1$$

The dot products of combination of different unit vectors evaluate to zero.

$$i.j = j.k = k.i = 0$$

Example. A block of mass "m" moves from point A to B along a smooth plane surface under the action of force as shown in the figure. Find the work done if it is defined as :

$$W = F.\Delta x$$

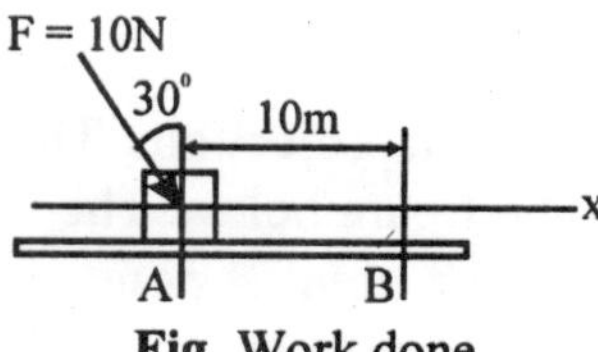

Fig. Work done

where F and Δ x are force and displacement vectors.

Solution. Expanding the expression of work, we have :

Work done

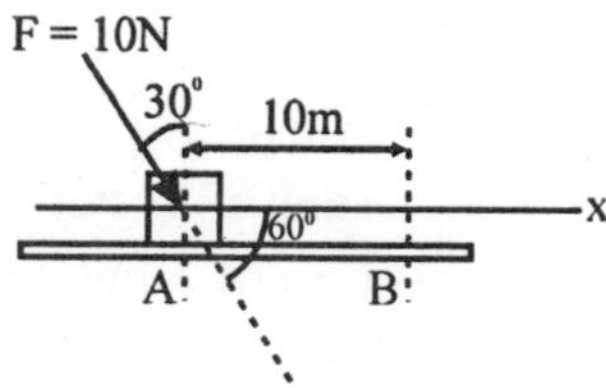

Fig. Work done

$$W = F.\Delta x = F\Delta x\cos,$$

Here, $F = 10\ N$, $\Delta x = 3$ m and $\cos\theta = \cos 60° = 0.5$.

$$W = 10 \times 10 \times 0.5 = 50 \text{ J}$$

SCALAR PRODUCT IN COMPONENT FORM

Two vectors in component forms are written as :

$$a = a_x i + a_y j + a_z k$$

$$b = b_x i + b_y j + b_z k$$

In evaluating the product, we make use of the fact that multiplication of the same unit vectors is 1, while multiplication of different unit vectors is zero. Finally, the dot product evaluates to scalar terms :

$$\vec{a}\cdot\vec{b} = (a_x i + a_y j + a_z k).\ (b_x i + b_y j + b_z k)$$

$$\vec{a}\cdot\vec{b} = a_x i.b_x i + a_y j.b_y j + a_z k.b_z k$$

$$\vec{a}\cdot\vec{b} = a_x b_x + a_y b_y + a_z b_z$$

IMPORTANCE OF DOT PRODUCT

An important use of the dot product is to test whether or not two vectors are orthogonal. Two vectors are orthogonal if the angle between them is 90 degrees. Thus, using we see that the dot product of two orthogonal vectors is zero. Conversely, the only way the dot product can be zero is if the angle between the two vectors is 90 degrees (or trivially if one or both of the vectors is the zero vector). Thus, two non-zero vectors have dot product zero if and only if they are orthogonal.

PROJECTIONS OF A VECTOR

One important use of dot products is in projections. The scalar projection of *b* onto *a* is the *length* of the segment *AB* shown in the figure below. The vector projection of *b* onto *a* is the *vector* with this length that begins at the point *A* points in the same direction (or opposite direction if the scalar projection is negative) as *a*.

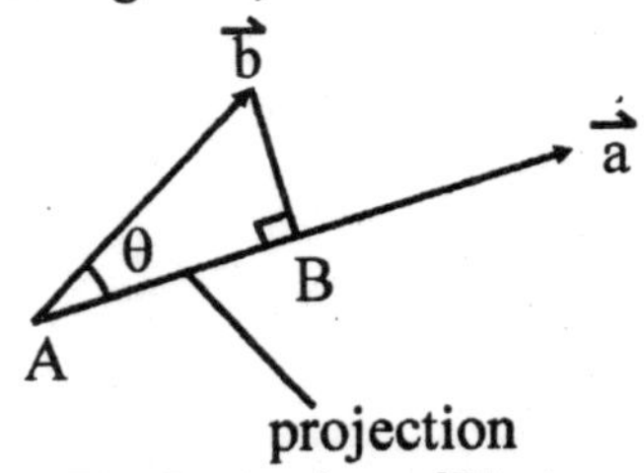

Fig. Projection of Vector

Thus, mathematically, the scalar projection of *b* onto a is $|b| \cos \theta$ (where θ is the angle between *a* and *b*) which from (*) is given by

$$\text{Projection of } b \text{ on } a = \frac{\vec{a}\cdot\vec{b}}{|\vec{a}|}$$

This quantity is also called the component of *b* in the *a* direction (hence the notation comp). And, the vector projection is merely the unit vector $a/|a|$ times the scalar projection of *b* onto *a*:

$$proj_a \vec{b} = \frac{\vec{a}\cdot\vec{b}}{|\vec{a}|}\frac{\vec{a}}{|\vec{a}|} = \frac{\vec{a}\cdot\vec{b}}{|\vec{a}|^2}\vec{a}$$

Thus, the scalar projection of *b* onto *a* is the magnitude of the vector projection of *b* onto *a*.

Example. Suppose you wish to find the work *W* done in moving a particle from one point to another. From physics we know $W = Fd$ where *F* is the magnitude of the force moving the particle and d is the distance between the two points. However, this relation is only valid when the force acts in the direction the particle moves.

Suppose this is not the case. Let the force vector be $F = \langle 2,3,4 \rangle$ and the displacement vector be $d = \langle 1,2,3 \rangle$. In this case, the work is the product of the distance moved (the magnitude of the displacement vector) and the magnitude of the component of the force that acts in the direction of displacement (the scalar projection of *F* onto *d*):

$$W = |\vec{d}||\vec{F}|\cos\theta = \vec{F} \cdot \vec{d}$$

Thus, the work done by the force to displace the particle from say the origin to the point (1,2,3) is

$$W = \vec{F} \cdot \vec{d} = 1(2) + 2(3) + 3(4) = 20 \text{ Joules}$$

Note that this is the easiest way to compute the dot product since the angle between the vectors *F* and *d* is unknown.

VECTOR PRODUCT (CROSS PRODUCT)

The cross product of two vectors a and b is a third vector. The magnitude of the vector product is given by the following expression:

$$|c| = |a \times b| = ab \sin\theta$$

where , is the smaller of the angles between the two vectors. It is important to note that vectors have two angles θ and 2π – θ. We should use the smaller of the angles as sine of θ and 2π – θ are different.

DIRECTION OF VECTOR PRODUCT

The direction of the vector product is perpendicular to the plane containing vectors, *a*, and, *b*. We must understand that two vectors (*a* and *b*) define an unique plane. The vector product is perpendicular to this plane defined by the vectors as shown in the figure. The angle, θ, impacts the magnitude of the cross product and not its direction.

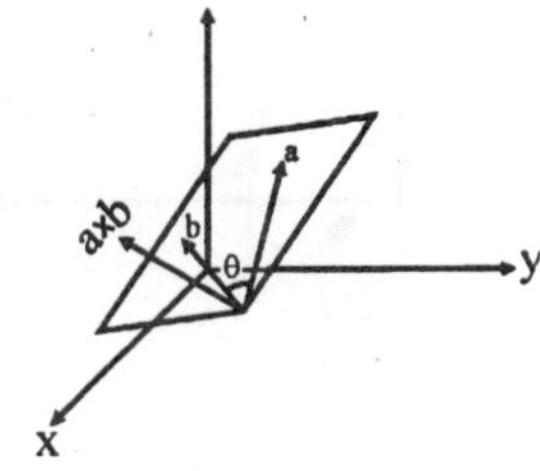

Fig. Direction of vector product

We know that rectangular coordinate system comprises of three planes, which are at right angles to each other. It is, therefore, suggested that we must orient our coordinate system in such a manner that vectors lie in one of the three planes defined by the rectangular coordinate system. This arrangement renders the cross product oriented in the direction of the remaining axis as shown in the figure.

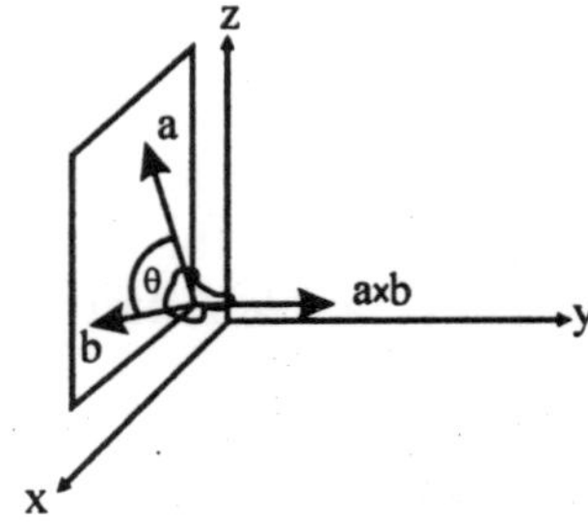

Fig. Direction of Vector Product

We can draw the vector product perpendicular to the plane on either of the two sides. For example, the product can be drawn either along the positive direction of y – axis or along the negative direction of y – axis direction.

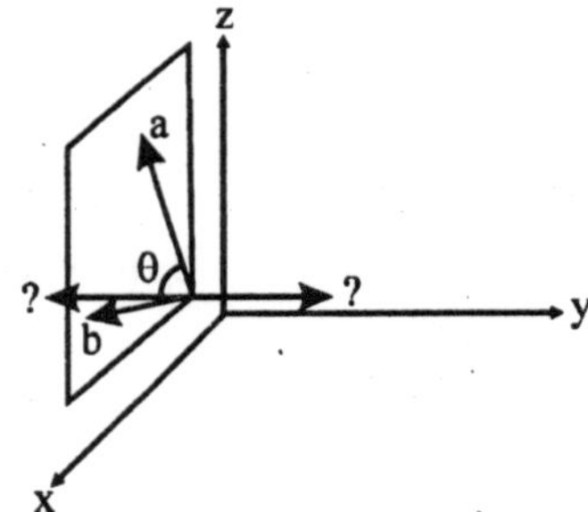

Fig. Direction of Vector Product

The direction of the vector product, including which side of the plane, is determined by right hand rule for vector products. According to this rule, we place right fist such that the curl of the fist follows as we proceed from the first vector, a, to the second vector, b. The stretched thumb, then gives the direction of vector product.

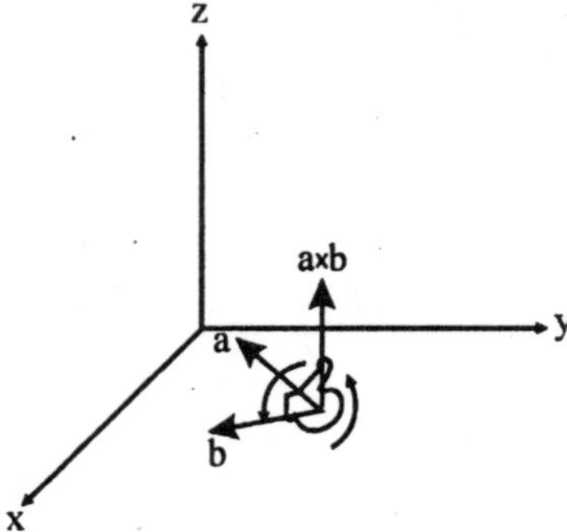

Fig. Direction of Vector Product

When we apply this rule to the case considered before, we find that the vector product is in the positive y – direction as shown below:

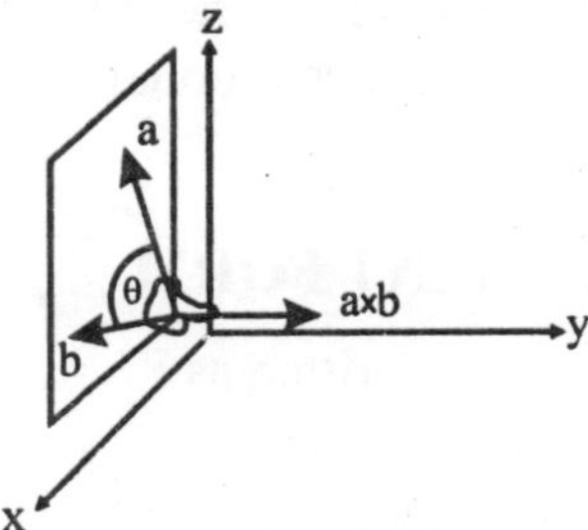

Fig. Direction of Vector Product

Here, we notice that we move in the anti-clockwise direction as we move from vector, a, to vector, b, while looking at the plane formed by the vectors. This fact can also be used to determine the direction of the vector product. If the orientation is anticlockwise, then the vector product is directed towards us; otherwise the vector product is directed away on the other side of the plane.

Further, it is important to note that the direction of cross product can be on a particular side of the plane, depending upon whether we take the product from a to b or from b to a. Implying :

$$a \times b = -b \times a$$

Thus, vector product is not commutative.

VALUES OF CROSS PRODUCT

The value of vector product is maximum for the maximum value of sinθ. Now, the maximum value of $\sin\theta$ is $\sin 90° = 1$. For this value, the vector product evaluates to the product of the magnitude of two vectors multiplied.

Thus, we see that cross product evaluates to negative value, which corresponds to the negative direction of the vector product i.e. other side of the planar formed by the vectors.

$$|a \times b|_{max} = ab$$

For $\theta = 270°$, $\sin 270° = -1$ and

$$|a \times b|_{max} = -ab$$

The vector product evaluates to zero for $\theta = 0°$ and $180°$ as sine of these angles are zero. These results have important implication for unit vectors. The dot products of same unit vector evaluates to 0.

$$i \times i = j \times j = k \times k$$

The cross products of combination of different unit vectors evaluate as :

$$i \times j = k; j \times k = i; k \times i = j$$

$$j \times i = -k; k \times j = -i; i \times k = -j$$

There is a simple rule to determine the sign of the cross product. We write the unit vectors in sequence *i,j,k*. Now, we form pair of vectors as we move from left to right like $i \times j, j \times k$ and right to left at the end like $k \times i$ in cyclic manner. The cross products of these pairs result in the remaining unit vector with positive sign. Cross products of other pairs result in the remaining unit vector with negative sign.

CROSS PRODUCT IN COMPONENT FORM

Two vectors in component forms are written as :

$$a = a_x i + a_y j + a_z k$$
$$b = b_x i + b_y j + b_z k$$

In evaluating the product, we make use of the fact that multiplication of the same unit vectors gives the value of 0, while multiplication of two different unit vectors result in remaining vector with appropriate sign. Finally, the vector product evaluates to vector terms :

$$a \times b = (a_x i + a_y j + a_z k) \times (b_x i + b_y j + b_z k)$$
$$a \times b = a_x i \times b_y j + a_x i \times b_z k + a_y j \times b_x i + a_y j \times b_z k + a_z k \times b_x i + a_z k \times b_y j$$
$$a \times b = a_x b_y k - a_x b_z j \times a_y b_x k + a_y b_z i + a_z b_x j - a_z b_y i$$
$$a \times b = (a_y b_z - a_z b_y)i + (a_z b_x - a_x b_z)j + (a_x b_y - a_y b_x)k$$

Evidently, it is difficult to remember above expression. If we know expanding matrix, then we can write above expression in the matrix form, which is easy to remember.

$$\vec{a} \times \vec{b} = \begin{vmatrix} i & j & k \\ a_x & a_y & c_x \\ b_x & b_y & c_y \end{vmatrix}$$

Example. If $a = 2i + 3j$ and $b = -3i - 2j$, find $a \times b$.

Solution:

$$a \times b = (2i + 3j) \times (-3i - 2j)$$

Neglecting terms involving same unit vectors, we expand the multiplication algebraically as :

$$a \times b = (2i) \times (-2j) + (3j) \times (-3i)$$

$$a \times b = -4k + 9k = 9k$$

Example. Consider the force given as :

$$F = q\,(v \times B)$$

Given $q = 10^{-6}\ C$, $v = (3i + 4j)$ m/s, $B = 1j$ Tesla. Find the force.

Solution:

$$F = q\,(v \times B\,) = 10^{-6}(3i + 4j) \times (-1i)$$

$$F = 4 \times 10^{-6}k$$

PROPERTIES OF THE CROSS PRODUCT

1. The length of the cross product of two vectors is $|\,a \times b\,| = ab \sin\theta$
2. The length of the cross product of two vectors is equal to the area of the parallelogram determined by the two vectors.
3. Anticommutativity: $\vec{a} \times \vec{b} = -\vec{b} \times \vec{a}$
4. Multiplication by scalars: $\left(c\vec{a}\right) \times \vec{b} = c\left(\vec{a} \times \vec{b}\right) = \vec{a} \times \left(c\vec{b}\right)$
5. Distributivity: $\vec{a} \times \left(\vec{b} \times \vec{c}\right) = \vec{a} \times \vec{b} + \vec{a} \times \vec{c}$

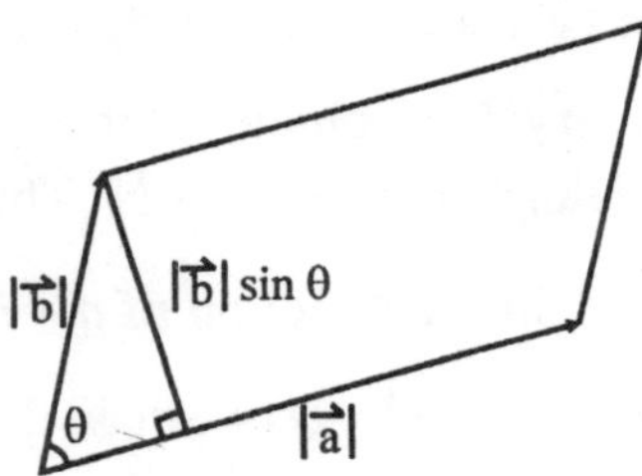

Fig. Cross Product of Vectors.

Note that the result for the length of the cross product leads directly to the fact that two vectors are parallel if and only if their cross product is the zero vector. This is true since two vectors are parallel if and only if the angle between them is 0 degrees (or 180 degrees).

Example. To find the area of the triangle with vertices (1,1,3), (4,-1,1), and (0,1,8), one could find the length of one of the altitudes of the triangle and proceed to find A=1/2(altitude)(base). However, this may not be an easy task in three dimensions. So consider finding the one-half the area of the parallelogram determined by the vector a from (1,1,3) to (4,–1,1) and the vector b from (1,1,3) to (0,1,8). Then a = <2 –1, 2 – (–1), 2 – 3> = <3,–2,–2> and b = <–1,0,5>. Using the result from the previous example and property number two above, we have the length of the cross product (and therefore the parallelogram determined by a and b) as

$$\left|\vec{a} \times \vec{b}\right| = \sqrt{(-10)^2 + (-13)^2 + (-2)^2} = \sqrt{273} = 16.52$$

So, the area of the triangle is one-half this quantity, or 8.26.

The cross product occurs in many formulas in physics, such as curl of a vector field, torque, and many integrals over surfaces.

PRODUCT OF THREE VECTORS

SCALAR TRIPLE PRODUCT

The scalar triple product of three vectors *a*, *b*, and *c* is $(a \times b) \cdot c$. It is a scalar product because it evaluates to a single number (unlike the cross product). The scalar triple product is important because its

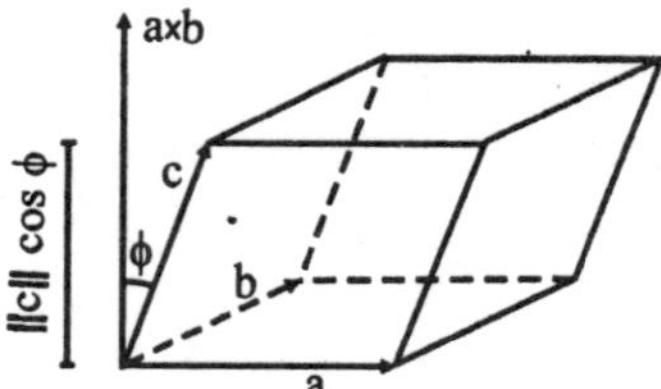

Fig. Triple Product of Vectors.

absolute value $|(a \times b) \cdot c|$ is the volume of the parallelepiped spanned by *a*, b, and *c* (i.e., the parallelepiped whose adjacent sides are the vectors *a*, *b*, and *c*).

This results from simple geometry. The volume is the area of the base times the height. We already know the area of the parallelogram base: $\|a \times b\|$. The height is the component of *c* in the direction normal to the base, i.e., in the direction of $a \times b$. Hence the height is $\left|\|c\| \cos\phi\right|$, where ϕ is the angle between *c* and $a \times b$. (Why do we need the absolute value? $\cos\phi$ would be negative if $\phi > \pi/2$.)

The volume of the parallelepiped is therefore

$$\text{Volume} = \left|\|\vec{a} \times \vec{b}\| \|\vec{c}\| \cos\phi\right| = \left|(\vec{a} \times \vec{b}) \cdot \vec{c}\right|$$

Using the formula for the cross product in components form, we can write the scalar triple product in component form as Write scalar triple product in component form as

$$(\vec{a} \times \vec{b}) \cdot \vec{c} = \begin{vmatrix} a_2 & a_3 \\ b_2 & b_3 \end{vmatrix} c_1 - \begin{vmatrix} a_1 & a_3 \\ b_1 & b_3 \end{vmatrix} c_2 + \begin{vmatrix} a_1 & a_2 \\ b_1 & b_2 \end{vmatrix} c_3$$

$$= \begin{vmatrix} c_1 & c_2 & c_3 \\ a_1 & a_2 & a_3 \\ b_1 & b_2 & b_3 \end{vmatrix}$$

The triple product is also invariant under any cyclic permutation of *a*, *b*, and *c*,

$$a \cdot b \times c = b \cdot c \times a = c \cdot a \times b$$

but any anti-cyclic permutation causes it to change sign,

$$a \cdot b \times c = -b \cdot a \times c$$

The scalar triple product is zero if any two of *a*, *b*, and *c* are parallel, or if *a*, *b*, and *c* are co-planar.

If *a*, *b*, and care non-coplanar, then any vector *r* can be written in terms of them:

$$r = \alpha a + \beta b + \gamma c$$

Forming the dot product of this equation with $b \times c$, we then obtain

$$r \cdot b \times c = \alpha a \cdot b \times c$$

$$\text{so } \alpha = \frac{r \cdot b \times c}{a \cdot b \times c}$$

Analogous expressions can be written for β and γ. The parameters α, and γ are uniquely determined provided $a \cdot b \times c \neq 0$: *i.e.*, provided that the three basis vectors are not co-planar

The scalar triple product has the following properties

$$a \cdot (b \times c) = (b \times c) \cdot a$$
$$a \cdot (b \times c) = c \cdot (a \times b) = (b \times c) \cdot a$$
$$a \cdot (b \times c) = -a \cdot (c \times b)$$

VECTOR TRIPLE PRODUCT

The vector triple product is defined as the cross product of one of the vectors with the cross product of the other two. This is known as the *vector triple product* because it results in a vector.

$$a \times (b \times c) = b(a \cdot c) - c(a \cdot b)$$

For three vectors *a*, *b*, and *c*, the vector triple product is defined $a \times (b \times c)$

1. The brackets are important because $a \times (b \times c) \neq (a \times b) \times c$
2. In fact, it can be demonstrated that

   $$a \times (b \times c) = (a \cdot c)b - (a \cdot b)c$$

   and

   $$(a \times b) \times c = (a \cdot c) - (b \cdot c)a$$

Let us try to prove the first of the above theorems. The left-hand side and the right-hand side are both proper vectors, so if we can prove this result in one particular coordinate system then it must be true in general. Let us take convenient axes such that the x-axis lies along b, and c lies in the x-y plane. It follows that $b = (b_x, 0, 0)$, $c = (c_x, c_y, 0)$ and $a = (a_x, a_y, a_z)$.

The vector $b \times c$ is directed along the z-axis: $b \times c = (0, 0, b_x c_y)$.

It follows that $a\times(b\times c)$ lies in the x-y plane:

$$a\times(b\times c) = (a_y b_x c_y, -a_x b_x c_y, 0).$$

To evaluate the right-hand side, of the above equation we need

$$a \cdot c = a_x c_x + a_y c_y \text{ and } a\cdot b = a_x b_x.$$

It follows that the right-hand side is

$$\text{R.H.S} = ([a_x c_y + a_y c_y]b_x, 0) - (a_x b_x c_y, a_x b_x c_y, 0)$$

$$= (a_x c_y b_x, -a_x b_x c_y, 0) = \text{LH.S}$$

which proves the theorem.

The vector triple product has the properties

$$(a\times b)\times c = -c\times(a\times b) = c\times(b\times a)$$
$$a\times b(b\times c) = -a\cdot(c\times b)$$
$$a\times(b\times c) = (a\cdot c)b - (a\cdot b)c$$

VECTORS AND DIRECTION

A vector quantity is a quantity which is fully described by both magnitude and direction. On the other hand, a scalar quantity is a quantity which is fully described by its magnitude. The emphasis of this unit is to understand some fundamentals about vectors and to apply the fundamentals in order to understand motion and forces which occur in two dimensions.

Examples of vector quantities which have been previously discussed include displacement, velocity, acceleration, and force. Each of these quantities are unique in that a full description of the quantity demands that both a magnitude and a direction are listed. For example, suppose that your teacher tells you that "a bag of gold is located outside the classroom; to find it, displace yourself 20 meters." This statement may provide yourself enough information to pique your interest; yet, not enough information to find the bag of gold.

The displacement required to find the bag of gold has not been fully described. On the other hand, suppose your teacher tells you that "a bag of gold is located outside the classroom; to find it, displace yourself from the center of the classroom door 20 meters in a direction 30 degrees to the west of north." This statement provides a complete description of the displacement vector - it lists both magnitude (20 meters) and direction (30 degrees to the west of north) relative to a reference or starting position (the center of the classroom door). Vector quantities are not fully described unless both magnitude and direction are listed.

Vector quantities are often represented by scaled vector diagrams. Vector diagrams depict a vector by use of an arrow drawn to scale in a specific direction. Vector diagrams were introduced and used in earlier units to depict the forces acting upon an object; such diagrams are known as free-body diagrams. An example of a scaled vector diagram is shown in the diagram at the right to depict a displacement vector. Observe that there are several characteristics of this diagram which make it an appropriately drawn vector diagram.

- A scale is clearly listed
- An arrow (with arrowhead) is drawn in a specified direction; thus, the vector has a head and a tail.
- The magnitude and direction of the vector is clearly labeled; in this case, the diagram shows the magnitude is 20 m and the direction is (30 degrees West of North).

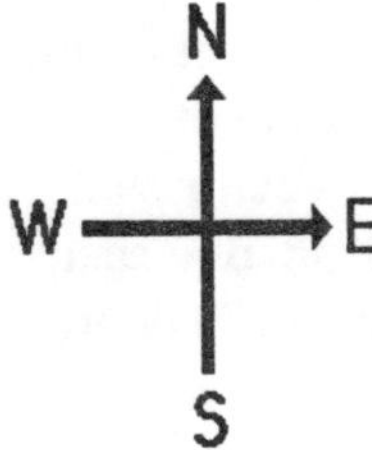

Vectors can be directed due East, due West, due South, and due North. But some vectors are directed northeast (at a 45 degree angle); and some vectors are even directed northeast, yet more north than east. Thus, there is a clear need for some form of a convention for identifying the direction of a vector which is not due East, due West, due South, or due North. There are a variety of conventions for describing the direction of any vector. The two conventions which will be discussed and used in this unit are described below:

1. The direction of a vector is often expressed as an angle of rotation of the vector about its "tail" from either east, west, north, or south. For example, a vector can be said to have a direction of 40 degrees North of West (meaning a vector pointing West has been rotated 40 degrees towards the northerly direction) of 65 degrees East of South (meaning a vector pointing South has been rotated 65 degrees towards the easterly direction).
2. The direction of a vector is often expressed as an counterclockwise angle of rotation of the vector about its "tail" from due East. Using this convention, a vector with a direction of 30 degrees is a vector which has been rotated 30 degrees in a counterclockwise direction relative to due east. A vector with a direction of 160 degrees is a vector which has been rotated 160 degrees in a counterclockwise direction relative to due east. A vector with a direction of 270 degrees is a vector which has been rotated 270 degrees in a counterclockwise direction relative to due east. This is one of the most common conventions for the direction of a vector and will be utilized throughout this unit.

Two illustrations of the second convention (discussed above) for identifying the direction of a vector are shown below.

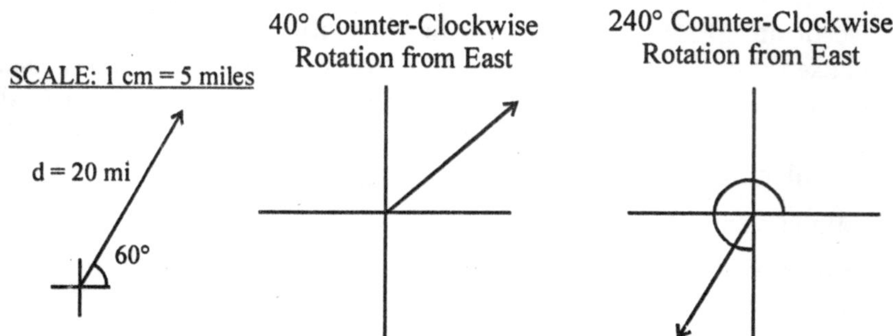

Observe in the first example that the vector is said to have a direction of 40 degrees. This means that the tail of the vector was pinned down, and the vector was rotated an angle of 40 degrees in the counterclockwise direction beginning from due east. Observe in the second example that the vector is said to have a direction of 240 degrees. This means that the tail of the vector was pinned down, and the vector was rotated an angle of 240 degrees in the counterclockwise direction beginning from due east; a rotation of 240 degrees is equivalent to rotating the vector through two quadrants (180 degrees) and then an additional 60 degrees into the third quadrant.

The magnitude of a vector in a scaled vector diagram is depicted by the length of the arrow. The arrow is drawn a precise length in accordance with a chosen scale. For example, the diagram at the right shows a vector with a magnitude of 20 miles. Since the scale used for constructing the diagram is 1 cm = 5 miles, the vector arrow is drawn with a length of 4 cm. That is, 4 cm x (5 miles/1 cm) = 20 miles. Using the same scale (1 cm = 5 miles), a displacement vector which is 15 miles will be represented by a vector arrow which is 3 cm in length. Similarly, a 25 mile displacement vector is represented by a 5 cm long vector arrow. And finally, an 18 mile displacement vector is represented by a 3.6 cm long arrow. See the examples shown below.

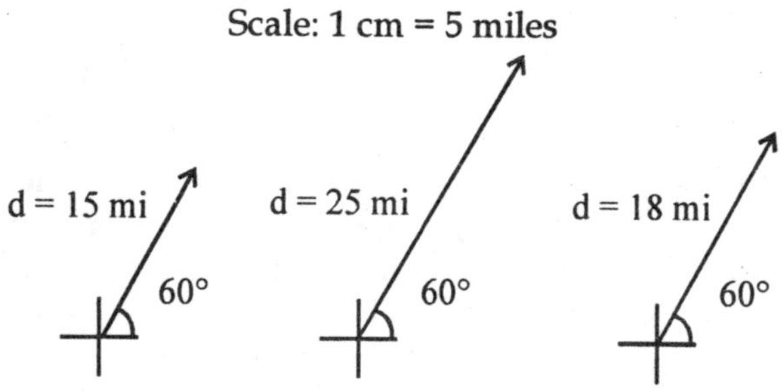

In conclusion, vectors can be represented by use of a scaled vector diagram. On such a diagram, an arrow is drawn to represent the vector; the arrow has an obvious tail and arrowhead. The magnitude of a vector is represented by the length of the arrow; a scale is indicated (such as, 1 cm = 5 miles) and the arrow is drawn the proper length according to the chosen scale.

The arrow points in the precise direction. Directions are described by the use of some convention; the most common convention is that the direction of a vector is the counterclockwise angle of rotation which that vector makes with due East.

In the remainder of this lesson, in the entire unit, and even in future units, scaled vector diagrams and the above convention for the direction of a vector will be commonly used to describe motion and solve problems concerning motion. For this reason, it is critical that you have a comfortable understanding of the means of representing and describing vector quantities.

VECTOR ADDITION

A variety of mathematical operations can be performed with and upon vectors. One such operation is the addition of vectors. Two vectors can be added together to determine the result (or resultant). This process of adding two or more vectors has already been discussed in an earlier unit. Recall in our discussion of Newton's laws of motion, that the net force acting upon an object was determined by computing the vector sum of all the individual forces acting upon that object. That is the net force was the result (or resultant) of adding up all the force vectors. During that unit, the rules for summing vectors (such as force vectors) was kept relatively simple. Observe the following summations of two forces:

$$\xrightarrow{5} + \xrightarrow{5} = \xrightarrow{10}$$

$$\xrightarrow{5} + \xleftarrow{-5} = 0$$

$$\xrightarrow{5} + \xrightarrow{10} = \xrightarrow{15}$$

$$\xrightarrow{5} + \xleftarrow{-10} = \xleftarrow{-5}$$

$$\xrightarrow{5} + \xleftarrow{-15} = \xleftarrow{-10}$$

$$10\uparrow + -5\downarrow = 5\uparrow$$

These rules for summing vectors were applied to free-body diagrams in order to determine the net force (i.e., the vector sum of all the individual forces). Sample applications are shown in the diagram below.

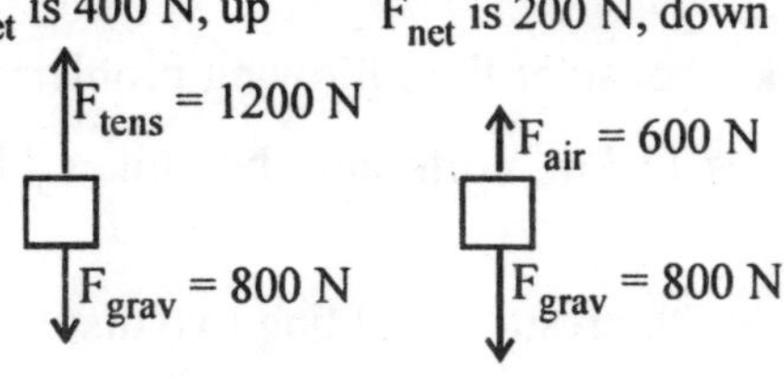

F_{net} is 20 N, left

F_{norm} = 50 N

F_{frict} = 20 N

F_{grav} = 50 N

In this unit, the task of summing vectors will be extended to more complicated cases in which the vectors are directed in directions other vertical and horizontal directions. For example, a vector directed up and to the right will be added to a vector directed up and to the left. For cases such as those shown in the diagrams below, what is the resultant and how can it be determined?

10 N ↗ + ↖ 10 N = ?

10 N ↗ + ← 10 N = ?

10 N ↗ + → 5 N = ?

There are a variety of methods for determining the magnitude and direction of the result of adding two or more vectors. The two methods which will be discussed in this lesson and used throughout the entire unit are:

- The Pythagorean theorem and trigonometric methods
- The head-to-tail method using a scaled vector diagram

The Pythagorean theorem is a useful method for determining the result of adding two (and only two) vectors which make a right angle to each other. The method is not applicable for adding more than two vectors or for adding vectors which are not at 90-degrees to each other. The Pythagorean theorem is a mathematical equation which relates the length of the sides of a right triangle to the length of the hypotenuse of a right triangle.

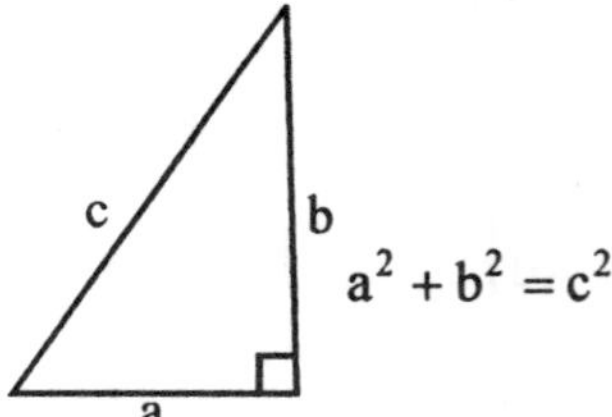

PYTHAGOREAN THEOREM

To see how the method works, consider the following problem:

A hiker leaves camp and hikes 11 km, north and then hikes 11 km east. Determine the resulting displacement of the hiker.

This problem asks to determine the result of adding two displacement vectors which are at right angles to each other. The result (or resultant) of walking 11 km north and 11 km east is a vector directed northeast as shown in the diagram to the right. Since the northward displacement and the eastward displacement are at right angles to each other, the Pythagorean theorem can be used to determine the resultant (i.e., the hypotenuse of the right triangle).

11 km, N + 11 km, E = 11 km, N R

$11^2 + 11^2 = R^2$

$242 = R^2$

$15.6 = R$

The resultant of 11 km, north plus 11 km, east is a vector with a magnitude of 15.6 km. Later, the method of determining the direction of the vector will be discussed. For now, let's test your understanding with the following two practice problems. In each case, use the Pythagorean theorem to determine the magnitude of the resultant. When finished, depress the mouse on the "pop-up menu" to view the answer.

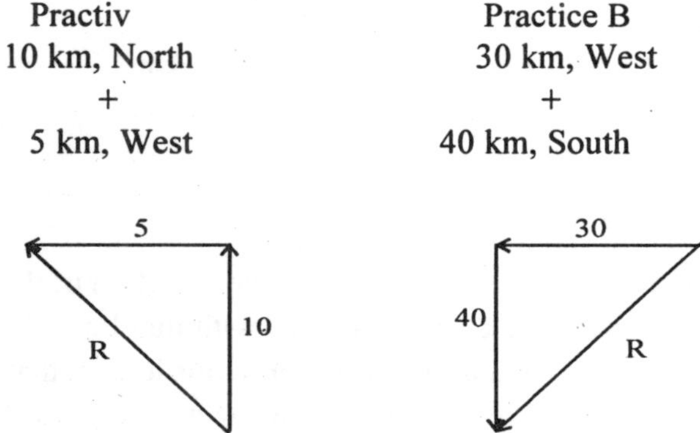

The direction of vector R in the diagram above can be determined by use of three common trigonometric functions - sine, cosine, and tangent functions. These three functions relate the angle of a right triangle to the ratio of the lengths of two of the sides of a right triangle. The sine function relates the sine of an angle to the ratio of the length of the side opposite the angle to the length of the hypotenuse.

The cosine function relates the cosine of an angle to the ratio of the length of the side adjacent the angle to the length of the hypotenuse. The tangent function relates the tangent of an angle to the ratio of the length of the side opposite the angle to the length of the side adjacent to the angle. The three equations below summarize these three functions in equation form.

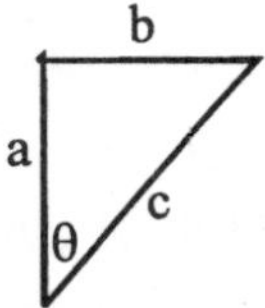

a = side adjacent to angle θ

b = side opposite to angle θ

c = hypotenuse of triangle

$$\sin\theta = \frac{b}{c} \qquad \cos\theta = \frac{a}{c} \qquad \tan\theta = \frac{b}{a}$$

These three trigonometric functions can be applied to the hiker problem in order to determine the direction of the hiker's displacement. The process begins by the selection of one of the two angles (other the right angle) of the triangle. Once the angle is selected, any of the three functions can be used to find the measure of the angle. Write the function and proceed with the proper algebraic steps to solve for the measure of the angle. The work is shown below.

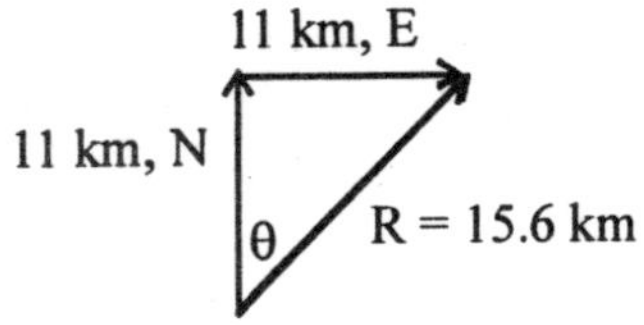

11 km, E

11 km, N

θ R = 15.6 km

$$\sin\theta = \frac{11\text{km}}{15.6\text{km}} = 0.7051 \quad \theta = \sin^{-1}(0.751) = 45°$$

Once the measure of the angle is determined, the direction of the vector can be found. In this case the vector makes an angle of 45 degrees with due East. Thus, the direction of this vector is written as 45 degrees. (Recall from earlier in this lesson, that the direction of a vector is the counterclockwise angle of rotation which the vector makes with due East.)

The measure of an angle as determined is not always the direction of the vector. For example, study the following situation in which the angle of the triangle is determined to be 26.6 degrees, yet the direction of the vector is 206.6 degrees.

10 km, West + 5 km, South

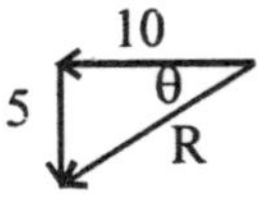

10, 5, θ, R

$$\tan\theta = \frac{5\,\text{km}}{10\,\text{km}} = 0.5$$

$$\theta = \tan^{1}(0.5) = 26.6°$$

Direction of R is 180° + 26.6°

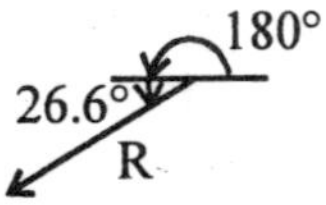

180°, 26.6°, R

Test your understanding of the use of SOH CAH TOA to determine the vector direction by trying the following two practice problems. In each case, use SOH CAH TOA to determine the direction of the resultant. When finished, depress the mouse on the "pop-up menu" to view the answer.

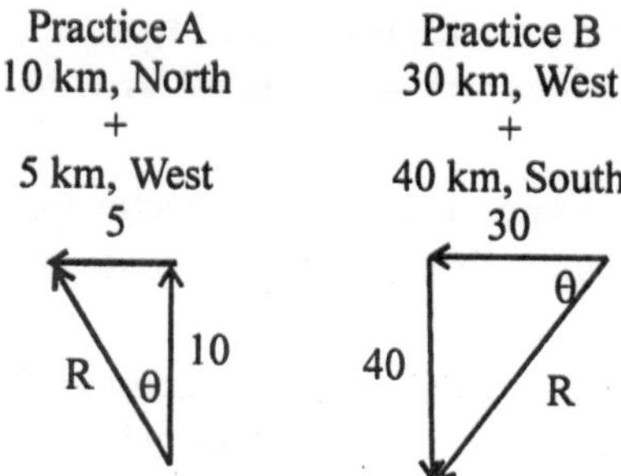

In the above problems, the magnitude and direction of the sum of two vectors is determined using the Pythagorean theorem and trigonometric methods. The procedure is restricted to the addition of two vectors which make right angles to each other. When the two vectors which are to be added do not make right angles to one another, or when there are more than two vectors to add together, we will employ a method known as the head-to-tail vector addition method. This method is described below.

The magnitude and direction of the sum of two or more vectors can also be determined by use of an accurately drawn scaled vector diagram. Using a scaled diagram, the head-to-tail method is employed to determine the resultant. You were given a set of 18 directions to follow, these 18 displacement vectors were added together in consecutive fashion to determine the result of adding the set of 18 directions. Each time, where one measurement ended, the next measurement began. In essence, you were using the head-to-tail method of vector addition.

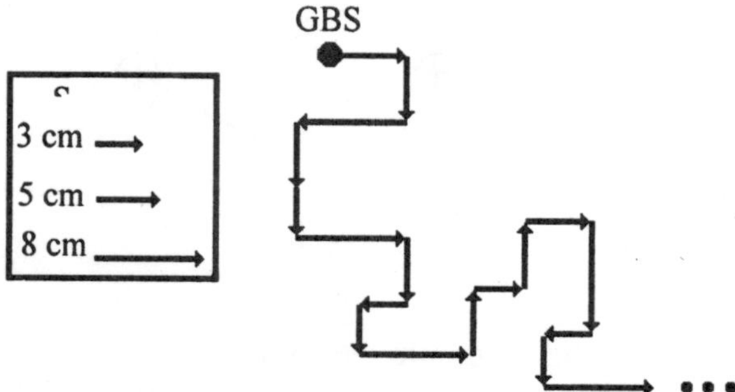

The head-to-tail method involves drawing a vector to scale on a sheet of paper beginning at a designated starting position; where the head of this vector ends the tail of the next vector begins (thus, head-to-tail method). The process is repeated for all vectors which are added. Once all vectors have been added head-to-tail, the resultant is drawn from the tail of the first vector to the head of the last vector; i.e., from start to finish. Once the resultant is drawn, its length can be measured and converted to real units using the given scale. The direction of the resultant can be determined by using a protractor and measuring its counterclockwise angle of rotation from due East.

The following step-by-step method for applying the head-to-tail method to determine the sum of two or more vectors is given below.

1. Choose a scale and indicate it on a sheet of paper (the best choice of scale is one which will result in a diagram which is as large as possible, yet fits on the sheet of paper).
2. Pick a starting location and draw the first vector to scale in the indicated direction. Label the magnitude and direction of the scale on the diagram (e.g., SCALE: 1 cm = 20 m).
3. Starting from where the head of the first vector ends, draw the first vector to scale in the indicated direction. Label the magnitude and direction of the vector on the diagram.
4. Repeat steps 2 and 3 for all vectors which are to be added
5. Draw the resultant from the tail of the first vector to the head of the last vector. Label this vector as "Resultant" or simply "R."
6. Using a ruler, measure the length of the resultant and determine its magnitude by converting to real units using the scale (4.4 cm x 20 m/1 cm = 88 m).
7. Measure the direction of the resultant using the counterclockwise convention discussed earlier in this lesson.

An example of the use of the head-to-tail method is illustrated below. The problem involves the addition of three vectors:

20 m, 45 deg. + 25 m, 300 deg. + 15 m, 210 deg.

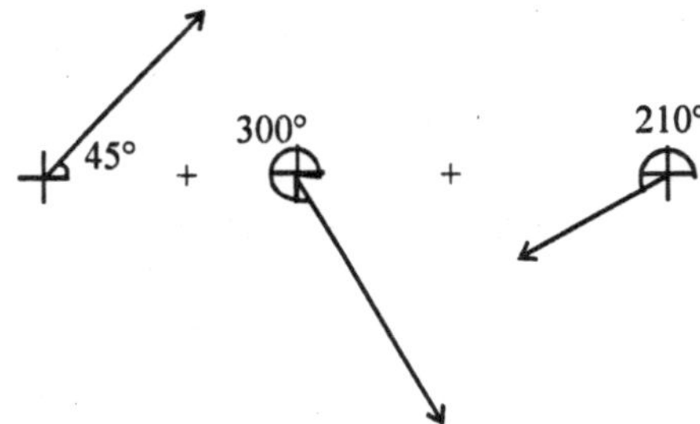

Scale: 1 cm = 5 m

The head-to-tail method is employed as described above and the resultant is determined (drawn in red). Its magnitude and direction is labeled on the diagram.

Scale: 1 cm = 5 m

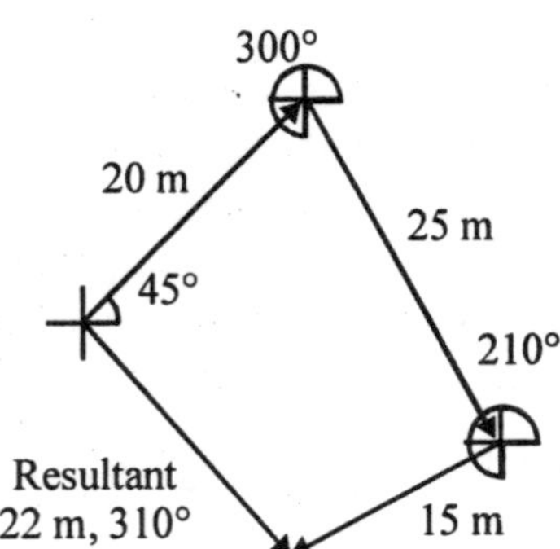

Interestingly enough, the order in which three vectors are added is insignificant; the resultant will still have the same magnitude and direction. For example, consider the addition of the same three vectors in a different order.

15 m, 210 deg. + 25 m, 300 deg. + 20 m, 45 deg.

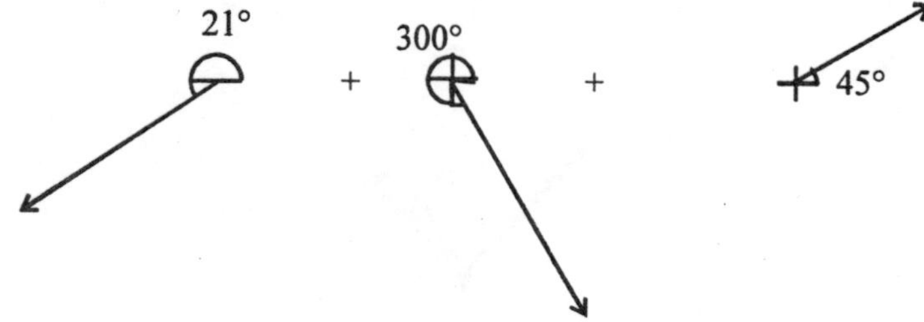

Scale: 1 cm = 5 m

When added together in this different order, these same three vectors still produce a resultant with the same magnitude and direction as before (22 m, 310 deg.). The order in which vectors are added using the head-to-tail method is insignificant.

Scale: 1 cm = 5 m

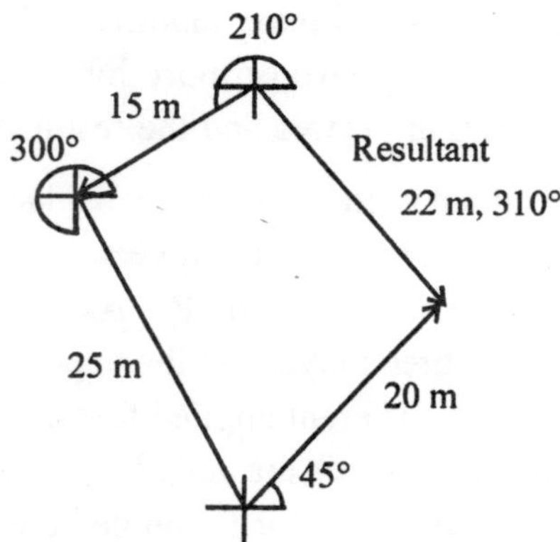

RESULTANTS OF VECTOR ADDITION

The resultant is the vector sum of two or more vectors. It is *the result* of adding vectors together. If displacement vectors A, B, and C are added together, the result will be vector R. As shown in the diagram, vector R can be determined by the use of an accurately drawn, scaled, vector addition diagram.

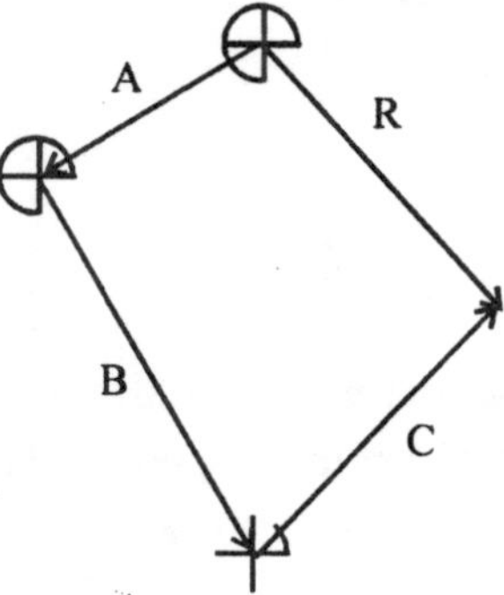

To say that vector R is the resultant displacement of displacement vectors A, B, and C is to say that a person who walked with displacements A, then B, and then C would be displaced by the same amount as a person who walked with displacement R. Displacement vector R gives the same result as displacement vectors A + B + C. That is why it can be said that

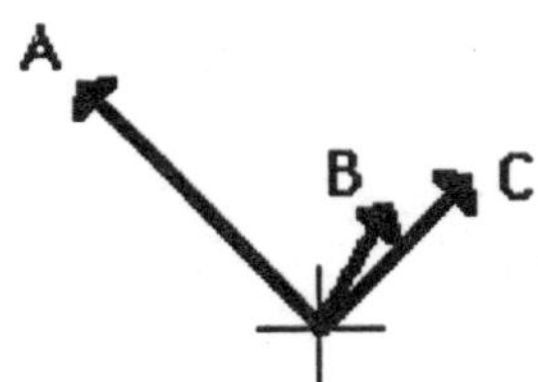

A + B + C = R

The above discussion pertains to the result of adding displacement vectors; when displacement vectors are added, the result is a resultant displacement. But any two vectors can be added as long as they are the same vector quantity. Add two or more velocity vectors and the result is a resultant velocity; add two or more force vectors and the result is a resultant force; add two or more momentum vectors and the result is...

In all such cases, the resultant vector (whether a displacement vector, force vector, velocity vector, etc.) is the result of adding the individual vectors. It is the same thing as adding A + B + C +.... "To do A + B + C is the same as to do R." As an example, consider a football player who gets hit simultaneously by three players on the opposing team (players A. B. and C). The football player experiences three different applied forces. Each applied force contributes to a total or resulting force. If the three forces are added together using methods of vector addition (discussed earlier), then the resultant vector R can be determined. In this case, to experience the three forces A, B and C is the same as experiencing force R. To be hit by players A, B, and C would result in the same force as being hit by one player applying force R. "To do A + B + C is the same as to do R." Vector R is the same result as vector A + B + C!!

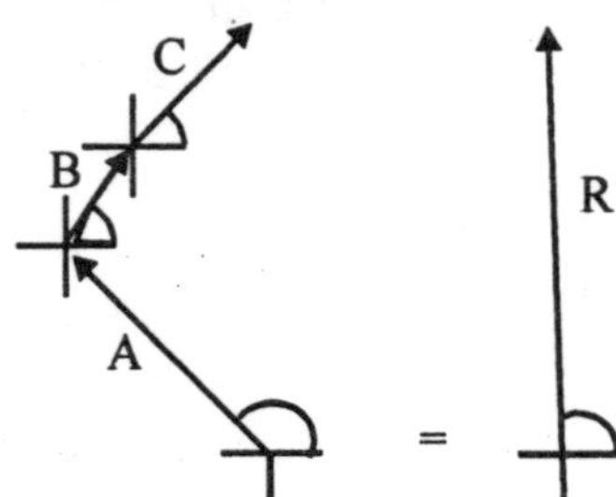

In summary, the resultant is the vector sum of all the individual vectors. The resultant is the result of combining the individual vectors together. The resultant can be determined by adding the individual forces together using vector addition methods.

VECTOR COMPONENTS

A vector is a quantity which has magnitude and direction. Displacement, velocity, acceleration, and force are the vector quantities which we have discussed thus far in our course. In the first couple of units of our course, all vectors which we discussed were simply directed up, down, left or right. When there was a free-body diagram depicting the forces acting upon an object, those forces were directed in one dimension - up, down, left or right. When an object had an acceleration and we described its direction, it was directed in one dimension - up, down, left or right. Now in this unit, we begin to see examples of vectors which are directed in two dimensions - upward and rightward, northward and westward, eastward and southward, etc.

In situations in which vectors are directed at angles to the customary coordinate axes, a useful mathematical trick will be employed to transform the vector into two parts, with each part being directed along the coordinate axes. For example, a vector which is directed Northwest can be thought of as having two parts - a Northward and a Westward part.

A vector which is directed upward and rightward can be thought of as having two parts - an upward and a rightward part.

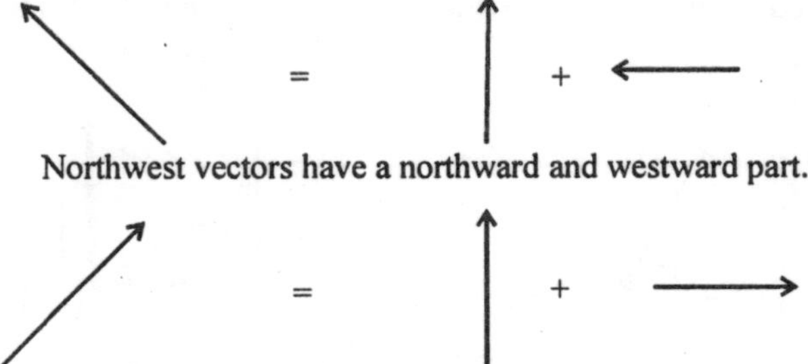

Northwest vectors have a northward and westward part.

An upward and rightward vector has an upward and rightward part.

An upword and rightward vector has an upward and rightward part.

Any vector directed in two dimensions can be thought of as having an influence in two different directions. That is, it can be thought of as having two parts. Each part of a two-dimensional vector is known as a component. The components of a vector depict the influence of that vector in a given direction. The combined influence of the two components is equivalent to the influence of the single two-dimensional vector. The single two-dimensional vector could be replaced by the two components.

If dog chain is stretched upward and rightward and pulled tight by his master, then the tensional force in the chain has two components - an upward and a rightward component. The influence of the chain on dog's body is equivalent to the influence of two chains on his body - one pulling upward and the other pulling rightward. If the single chain were replaced by two chains (each one having the magnitude and direction of the components), the dog would not know the difference. This is not because dog is dumb (a quick glance at his picture reveals that he is certainly not that), but rather because the combined influence of the two components is equivalent to the influence of the single two-dimensional vector.

Consider a picture which is hung to a wall by means of two wires which are stretched vertically and horizontally. Each wire exerts a tensional force upon the picture to support its weight. Since each wire is stretched in two dimensions (both vertically and horizontally), the tensional force of each wire has two components - a vertical and a horizontal component. Focusing on the wire on the left, we could say that the wire has a leftward and an upward component. This is to say that the wire on the left could be replaced by two wires, one pulling leftward and the other pulling upward. If the single wire were replaced by two wires (each one having the magnitude and direction of the components), then there would be no effect upon the stability of the picture. The combined influence of the two components is equivalent to the influence of the single two-dimensional vector.

Consider an airplane which is flying from O'Hare International Airport to a destination in Canada. Suppose that the plane is flying in such a manner that its resulting displacement vector is northwest. If this is the case, then the displacement of the plane has two components - a component in the northward direction and a component in the westward direction. This is to say that the plane would have the same displacement if it were to take the trip into Canada in two segments - one directed due North and the other directed due West. If the single displacement vector were replaced by these two individual displacement vectors, then the passengers in the plane would end up in the same final position. The combined influence of the two components is equivalent to the influence of the single two-dimensional displacement.

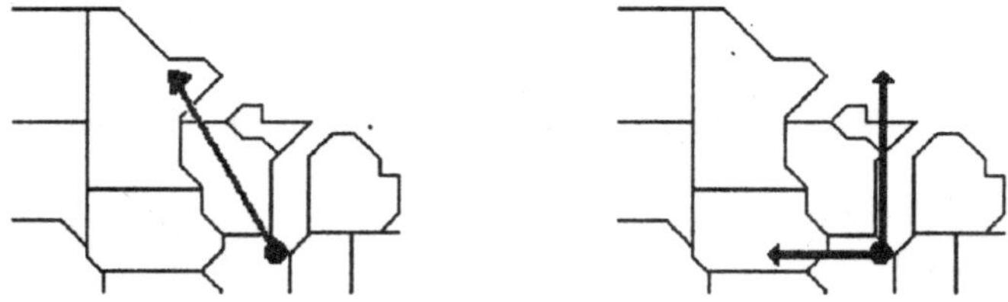

The palne's northwest displacement is equivalent to a northward plus a westward displacement.

Any vector directed in two dimensions can be thought of as having two different components. The component of a single vector describes the influence of that vector in a given direction.

VECTOR RESOLUTION

As mentioned earlier in this lesson, any vector directed at an angle to the horizontal (or the vertical) can be thought of as having two parts (or components). That is, any vector directed in two dimensions can be thought of as having two components. For example, if a chain pulls upward at an angle on the collar of a dog, then there is a tensional force directed in two dimensions. This tensional force has two components: an upward and a rightward component.

As another example, consider an airplane which is displaced northwest from O'Hare International Airport (in Chicago) to a destination in Canada. The displacement vector of the plane is in two dimensions (northwest). Thus, this displacement vector has two components: a northward and a westward component.

In this unit, we learn two basic methods for determining the magnitudes of the components of a vector directed in two dimensions. The process of determining the magnitude of a vector is known as vector resolution. The two methods of vector resolution which we will examine are

1. The parallelogram method
2. The trigonometric method

The parallelogram method of vector resolution involves using an accurately drawn, scaled vector diagram to determine the components of the vector. Briefly put, the method involves drawing the vector to scale in the indicated direction, sketching a parallelogram around the vector such that the vector is the diagonal of the parallelogram, and determining the magnitude of the components (the sides of the parallelogram) using the scale. A step-by-step procedure for using the parallelogram method of vector resolution is:

1. Select a scale and accurately draw the vector to scale in the indicated direction.
2. Sketch a parallelogram around the vector: beginning at the tail of the vector, sketch vertical and horizontal lines; then sketch horizontal and vertical lines at the head of the vector; the sketched lines will meet to form a parallelogram.
3. Draw the components of the vector; the components are the sides of the parallelogram; be sure to place arrowheads on these components to indicate their direction (up, down, left, right).
4. Meaningfully label the components of the vectors with symbols to indicate which component is being represented by which side; a northward force component would be labeled F_{north}; a rightward velocity component might be labeled v_x; etc.
5. Measure the length of the sides of the parallelogram and use the scale to determine the magnitude of the components in real units; label the magnitude on the diagram.

The step-by-step procedure above is illustrated in the diagram below to show how a velocity vector with a magnitude of 50 m/s and a direction of 60 degrees above the horizontal may be resolved into two components. The diagram shows that the vector is first drawn to scale in the indicated direction; a parallelogram is sketched about the vector; the components are labeled on the diagram; and the result of measuring the length of the vector components and converting to m/s using the scale.

Note: Because different computer monitors have different resolutions, the actual length of the vector on your monitor may not be 5 cm.

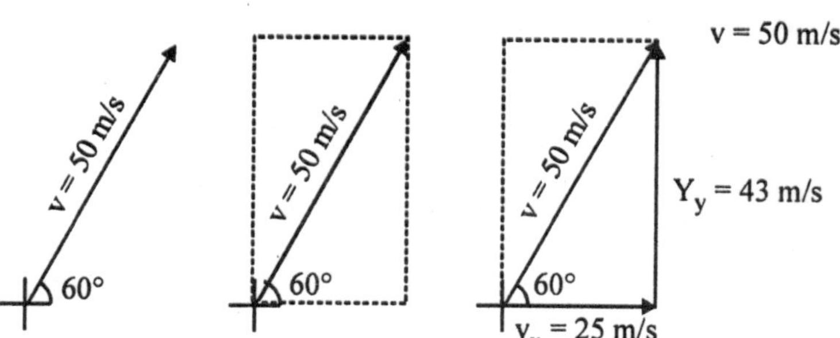

The trigonometric method of vector resolution involves using trigonometric functions to determine the components of the vector. Earlier in lesson 1, the use of trigonometric functions to determine the direction of a vector was described. Now, in this part of lesson 1, trigonometric functions will be used to determine the components of a single vector.

Recall from the earlier discussion that trigonometric functions relate the length of the sides of a right triangle to the angle of a right triangle. As such, trigonometric functions can be used to determine the length of the sides of a right triangle if one angle and the length of one side are known.

The method of employing trigonometric functions to determine the components of a vector are as follows:

1. Construct a sketch (no scale needed) of the vector in the indicated direction; label its magnitude and the angle which it makes with the horizontal.
2. Draw a rectangle about the vector such that the vector is the diagonal of the rectangle; beginning at the tail of the vector, sketch vertical and horizontal lines; then sketch horizontal and vertical lines at the head of the vector; the sketched lines will meet to form a parallelogram.
3. Draw the components of the vector; the components are the sides of the rectangle; be sure to place arrowheads on these components to indicate their direction (up, down, left, right).
4. Meaningfully label the components of the vectors with symbols to indicate which component is being represented by which side; a northward force component would be labeled F-north; a rightward force velocity component might be labeled v-x; etc.
5. To determine the length of the side opposite the indicated angle, use the sine function; substitute the magnitude of the vector for the length of the hypotenuse; use some algebra to solve the equation for the length of the side opposite the indicated angle.
6. Repeat the above step using the cosine function to determine the length of the side adjacent to the indicated angle.

RELATIVE VELOCITY AND RIVERBOAT PROBLEMS

On occasion objects move within a medium which is moving with respect to an observer. For example, an airplane usually encounters a wind - air which is moving with respect to an observer on the ground below. As another example, a motor boat in a river is moving amidst a river current - water which is moving with respect to an observer on dry land. In such instances as this, the magnitude of the velocity of the moving object (whether it be a plane or a motor boat) with respect to the observer on land will not be the same as the speedometer reading of the vehicle.

That is to say, the speedometer on the motor boat might read 20 mi/hr; yet the motor boat might be moving relative to the observer on shore at a speed of 25 mi/hr. Motion is relative to the observer. The observer on land, often named (or misnamed) the "stationary observer" would measure the speed to be different than that of the person on the boat.

To illustrate this principle, consider a plane flying amidst a tailwind. A tailwind is merely a wind which approaches the plane from behind, thus increasing its resulting velocity. If the plane is traveling at a velocity of 100 km/hr with respect to the air, and if the wind velocity is 25 km/hr, then what is the velocity of the plane relative to an observer on the ground below?

The resultant velocity of the plane (that is, the result of the wind velocity contributing to the velocity due to the plane's motor) is the vector sum of the velocity of the plane and the velocity of the wind. This resultant velocity is quite easily determined if the wind approaches the plane directly from behind. As shown in the diagram below, the plane travels with a resulting velocity of 125 km/hr relative to the ground.

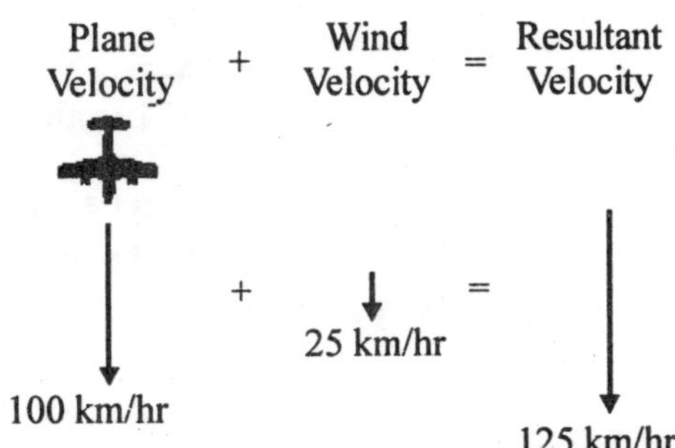

The plane travels with a velocity relative to the ground which is the vector sum of the plane's velocity (relative to the air) plus the wind velocity.

If the plane encounters a headwind, the resulting velocity will be less than 100 km/hr. Since a headwind is a wind which approaches the plane from the front, such a wind would decrease the plane's resulting velocity. Suppose a plane traveling with a velocity of 100 km/hr with respect to the air meets a headwind with a velocity of 25 km/hr. In this case, the resultant velocity would be 75 km/hr; this is the velocity of the plane relative to an observer on the ground. This is depicted in the diagram below.

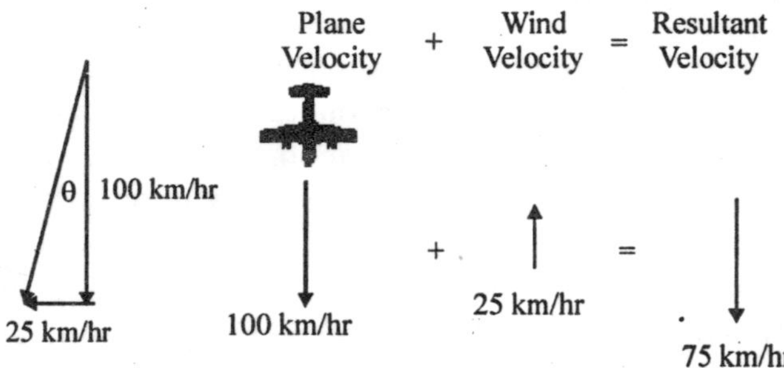

Now consider a plane traveling with a velocity of 100 km/hr, South which encounters a side wind of 25 km/hr, West. Now what would the resulting velocity of the plane be? This question can be answered in the same manner as the previous questions. The resulting velocity of the plane is the vector sum of the two individual velocities. To determine the resultant velocity, the plane velocity (relative to the air) must be added to the wind velocity.

This is the same procedure which was used above for the headwind and the tailwind situations; only now, the resultant is not as easily computed. Since the two vectors to be added - the southward plane velocity and the westward wind velocity - are at right angles to each other, the Pythagorean theorem can be used. This is illustrated in the diagram below.

Plane Velocity + Wind Velocity = Resultant Velocity

100 km/hr + 25 km/hr = R

R = 103.1 km/hr

In this situation of a side wind, the southward vector can be added to the westward vector using the usual methods of vector addition. The magnitude of the resultant velocity is determined using Pythagorean theorem. The algebraic steps are as follows:

$$(100 \text{ km/hr})^2 + (25 \text{ km/hr})^2 = R^2$$

$$10\,000 \text{ km}^2/\text{hr}^2 + 625 \text{ km}^2/\text{hr}^2 = R^2$$

$$10\,625 \text{ km}^2/\text{hr}^2 = R^2$$

$$\Rightarrow \quad R = \sqrt{10625} \text{ km/hr} = 103.1 \text{ km/hr}$$

The direction of the resulting velocity can be determined using a trigonometric function. Since the plane velocity and the wind velocity form a right triangle when added together in head-to-tail fashion, the angle between the resultant and the southward vector can be determined using either the sine, cosine, or tangent functions. The tangent function can be used; this is shown below:

$\tan(\theta) = (25/100)$

$\theta = \tan^{-1}(25/100)$

$\theta = 14.0$ degrees

If the resultant velocity of the plane makes a 14.0 degree angle with the southward direction (theta in the above diagram),then the direction of the resultant is 256 degrees. Like any vector, the resultant's direction is measured as a counterclockwise angle of rotation from due East.

The effect of the wind upon the plane is similar to the effect of the river current upon the motor boat. If a motor boat were to head straight across a river (that is, if the boat were to point its bow straight towards the other side), it would not reach the shore directly across from its starting point. The river current influences the motion of the boat and carries it downstream.

The motor boat may be moving with a velocity of 4 m/s directly across the river, yet the resultant velocity of the boat will be greater than 4 m/s and at an angle in the downstream direction. While the speedometer of the boat may read 4 m/s, its speed with respect to an observer on the shore will be greater than 4 m/s.

Motion of Motor Boat With Current

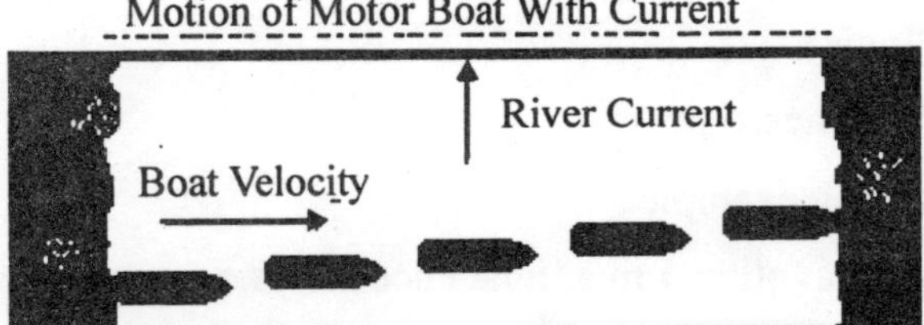

The resultant velocity of the motor boat can be determined in the same manner as was done for the plane. The resultant velocity of the boat is the vector sum of the boat velocity and the river velocity. Since the boat heads straight across the river and since the current is always directed straight downstream, the two vectors are at right angles to each other.

Thus, the Pythagorean theorem can be used to determine the resultant velocity. Suppose that the river was moving with a velocity of 3 m/s, North and the motor boat was moving with a velocity of 4 m/s, East. What would be the resultant velocity of the motor boat (i.e., the velocity relative to an observer on the shore)? The magnitude of the resultant can be found as follows:

$(4.0 \text{ m/s})^2 + (3.0 \text{ m/s})^2 = R^2$

$16 \text{ m}^2/\text{s}^2 + 9 \text{ m}^2/\text{s}^2 = R^2$

R

3 m/s

$\Rightarrow$ R = m/s θ

4 m/s

$5.0 \text{ m/s} = R$

The direction of the resultant is the counterclockwise angle of rotation which the resultant vector makes with due East. This angle can be determined using a trigonometric function as shown below.

$\tan\theta = (3/4)$

$\theta = \tan^{-1}(3/4)$

$\theta = 36.9$ degrees

Given a boat velocity of 4 m/s, East and a river velocity of 3 m/s, North, the resultant velocity of the boat will be 5 m/s at 36.9 degrees.

Motor boat problems such as these are typically accompanied by three separate questions:

1. What is the resultant velocity (both magnitude and direction) of the boat?
2. If the width of the river is X meters wide, then how much time does it take the boat to travel shore to shore?
3. What distance downstream does the boat reach the opposite shore?

The first of these three questions was answered above; the resultant velocity of the boat can be determined using the Pythagorean theorem (magnitude) and a trigonometric function (direction). The second and third of these questions can be answered using the average speed equation (and a lot of logic).

- Average speed = distance/time

Consider the following example.

Example. A motor boat traveling 4 m/s, East encounters a current traveling 3.0 m/s, North.

1. What is the resultant velocity of the motor boat?
2. If the width of the river is 80 meters wide, then how much time does it take the boat to travel shore to shore?
3. What distance downstream does the boat reach the opposite shore?

The solution to the first question has already been shown in the above discussion. The resultant velocity of the boat is 5 m/s at 36.9 degrees. We will start in on the second question.

The river is 80-meters wide; that is, the distance from shore to shore as measured straight across the river is 80 meters. The time to cross this 80-meter wide river can be determined by rearranging and substituting into the average speed equation.

$$\text{time} = \text{distance}/(\text{Average speed})$$

The distance is 80 m can be substituted into the numerator. But what about the denominator? What value should be used for average speed? Should 3 m/s (the current velocity), 4 m/s (the boat velocity), or 5 m/s (the resultant velocity) be used as the average speed value for covering the 80 meters? With what average speed is the boat traversing the 80 meter wide river? Most students want to use the resultant velocity in the equation since that is the actual velocity of the boat with respect to the shore. Yet the value of 5 m/s is the speed at which the boat covers the diagonal dimension of the river. And the diagonal distance across the river is not known in this case.

If one knew the distance C in the diagram below, then the average speed C could be used to calculate the time to reach the opposite shore. Similarly, if one knew the distance B in the diagram below, then the average speed B could be used to calculate the time to reach the opposite shore. And finally, if one knew the distance A in the diagram below, then the average speed A could be used to calculate the time to reach the opposite shore.

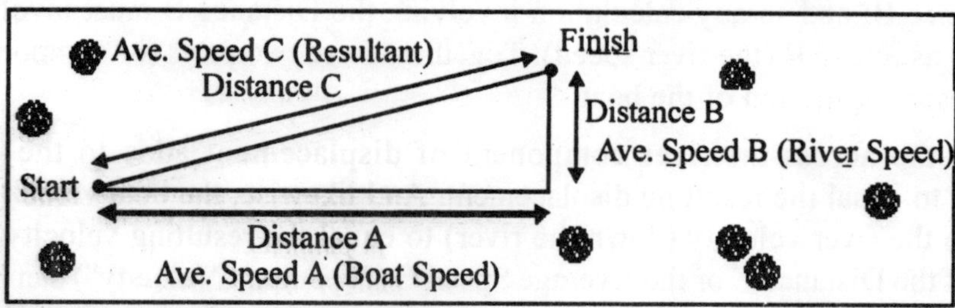

In our problem, the 80 m corresponds to the distance A, and so the average speed of 4 m/s (average speed in the direction straight across the river) should be substituted into the equation to determine the time.

time = (80 m)/(4 m/s) = 20 s

It requires 20 s for the boat to travel across the river. During this 20 s of crossing the river, the boat also drifts downstream. Part c of the problem asks "What distance downstream does the boat reach the opposite shore?" The same equation must be used to calculate this downstream distance. And once more, the question arises, which one of the three average speed values must be used in the equation to calculate the distance downstream? The distance downstream corresponds to Distance B on the above diagram.

The speed at which the boat covers this distance corresponds to Average Speed B on the diagram above (i.e., the speed at which the current moves - 3 m/s). And so the average speed of 3 m/s (average speed in the downstream direction) should be substituted into the equation to determine the time.

distance = ave. speed * time = (3 m/s) * (20 s)

distance = 60 m

The boat is carried 60 meters downstream during the 20 seconds it takes to cross the river.

The mathematics of the above problem is no more difficult than dividing or multiplying two numerical quantities by each other. The mathematics is easy! The difficulty of the problem is conceptual in nature; the difficulty lies in deciding which numbers to use in the equations. That decision emerges from one's conceptual understanding (or unfortunately, one's misunderstanding) of the complex motion which is occurring. The motion of the river boat can be divided into two simultaneous parts - a motion in the direction straight across the river and and a motion in the downstream direction.

These two parts (or components) of the motion occur simultaneously for the same time duration (which was 20 seconds in the above problem). The decision as to which velocity value or distance value to use in the equation must be consistent with the diagram above. The boat's motor is what carries the boat across the river the Distance A; and so any calculation involving the Distance A must involve the speed value labeled as Speed A (the boat speed relative to the water). Similarly, it is the current of the river which carries the boat downstream for the Distance B; and so any calculation involving the Distance B must involve the speed value labeled as Speed B (the river speed). Together, these two parts (or components) add up to give the resulting motion of the boat.

That is, the across-the-river component of displacement adds to the downstream displacement to equal the resulting displacement. And likewise, the boat velocity (across the river) adds to the river velocity (down the river) to equal the resulting velocity. And so any calculation of the Distance C or the Average Speed C ("Resultant Velocity") can be performed using the Pythagorean theorem.

Now to illustrate an important point, let's try a second example problem which is similar to the first example problem. Make an attempt to answer the three questions and then use the pop-up menu to check your answer.

Example. A motor boat traveling 4 m/s, East encounters a current traveling 7.0 m/s, North.

1. What is the resultant velocity of the motor boat?
2. If the width of the river is 80 meters wide, then how much time does it take the boat to travel shore to shore?
3. What distance downstream does the boat reach the opposite shore?

An important concept emerges from the analysis of the two example problems above. In Example 1, the time to cross the 80-meter wide river (when moving 4 m/s) was 20 seconds. This was in the presence of a 3 m/s current velocity. In Example 2, the current velocity was much greater - 7 m/s - yet the time to cross the river remains unchanged. In fact, the current velocity itself has no effect upon the time required for a boat to cross the river. The river moves downstream parallel to the banks of the river; as such, there is no way that the current is capable of assisting a boat in crossing a river.

While the increased current may effect the resultant velocity - making the boat travel with a greater speed with respect to an observer on the ground - it does not increase the speed in the direction across the river. The increased resultant velocity is only a direction pointing down the river. It is often said that "perpendicular components of motion are independent of each other." As applied to river boat problems, this would mean that an across-the-river variable would be independent of (i.e., not be effected by) a downstream variable. The time to cross the river is dependent upon the velocity at which the boat crosses the river.

It is only the component of motion directed across the river (i.e., the boat velocity) which effects the time to travel the distance directly across the river (80 m in this case). The component of motion perpendicular to this direction - the current velocity - only effects the distance which the boat travels downstream. This concept of perpendicular components of motion will be investigated in more detail in the next part of Lesson 1.

EXERCISE

1. Show that the vectors $a = i + 2k$, $b = 2i - j + 3k$ and $c = i - 3j + k$ are position vectors of vertices of a right angled triangle.
2. $a = 3i + 7j + 6k$, $b = i + 2j + 9k$, $c = i + 95j$, find
   (i) 3a –2 b + 7c
   (ii) $|2a + 3b + 4c|$
   (iii) Unit vector petrpendicular to 3a + 2b + 4c.
3. Using dot product, find the angle between the lines AB and AC where A, B, C are the points (6, 4, 1), (1, 1, 1) and (5, 1, 1) respectively.
5. Given two vectors $a = i + j + 5k$, $b = 3i + 5j + 7k$, find $a. b$
6. Find the area of the parallelogram whose two adjacent sides are represented by the vectors
   $a = 4i + 7j$, $b = 2i - 9j - 6k$.
7. If $a = 3i - 4j - 6k,\ b = -5i - 6j + 4k,\ i + j + 6k$, find $(a \times b) \cdot c$
8. Find the volume of parallelopiped whose coterminous edges are represented by the following vectors:
   (i) $a = i - j - k,\ b = -2 - 3j + 6k,\ c = 4i - 6j - 7k$
   (ii) $a = 4i + 5j + 6k,\ b = 12 + 9j + 11k,\ c = i + j - 9k$
9. Find μ so that the vectors
   $a = 7i + \mu j - k,\ b = i + j - k,\ c = \mu i - 8j + 12k$ are coplanar.
10. Show that $[a - b\ b - c\ c - a] = 0$
   (*a*) A plane can travel with a speed of 80 mi/hr with respect to the air. Determine the resultant velocity of the plane (magnitude only) if it encounters a
      (i) 10 mi/hr headwind.
      (ii) 10 mi/hr tailwind.
      (iii) 10 mi/hr crosswind.
      (iv) 60 mi/hr crosswind.
   (*b*) A motor boat traveling 5 m/s, East encounters a current traveling 2.5 m/s, North.
      (*i*) What is the resultant velocity of the motor boat?
      (*ii*) If the width of the river is 80 meters wide, then how much time does it take the boat to travel shore to shore?
      (*iii*) What distance downstream does the boat reach the opposite shore?

(*c*) A motor boat traveling 5 m/s, East encounters a current traveling 2.5 m/s, South.

(*i*) What is the resultant velocity of the motor boat?

(*ii*) What distance downstream does the boat reach the opposite shore?

(*iii*) If the width of the river is 80 meters wide, then how much time does it take the boat to travel shore to shore?

(*d*) A motor boat traveling 6 m/s, East encounters a current traveling 3.8 m/s, South.

(*i*) What is the resultant velocity of the motor boat?

(*ii*) If the width of the river is 120 meters wide, then how much time does it take the boat to travel shore to shore?

(*iii*) What distance downstream does the boat reach the opposite shore?

(*d*) If the current velocity in question #4 were increased to 5 m/s, then

(*i*) how much time would be required to cross the same 120-m wide river?

(*ii*) what distance downstream would the boat travel during this time?

VECTOR SPACES

Definition: An R-module is a set M together with an addition operation on M and a function $\mu : R \times M \rightarrow M$, such that the set M under addition forms an abelian group, and moreover, for all $c, d \in R$ and $\alpha, \beta \in M$, we have:

(i) $\mu(c, \mu(d, \alpha)) = \mu(cd, \alpha)$,

(ii) $\mu(c + d, \alpha) = \mu(c, \alpha) + \mu(d, \mu)$,

(iii) $\mu(c, \alpha + \beta) = \mu(c, \alpha) + \mu(c, \beta)$,

(iv) $\mu(1_R, \alpha) = \alpha$.

One may also call an R-module M a module over R, and elements of R are sometimes called scalars. The function μ in the definition is called a scalar multiplication map, and the value $\mu(c, \alpha)$ is called the scalar product of c and α.

Usually, we shall simply write $c\alpha$ (or $c.\alpha$) instead of $\mu(c, \alpha)$. When we do this, properties (i) – (iv) of the definition may be written as follows:

$$c(d\alpha) = (cd)\alpha,\ (c + d)\alpha = c\alpha + d\alpha,\ c(\alpha + \beta) = c\alpha + c\beta,\ 1_R\alpha = \alpha.$$

Note that there are two addition operations at play here: addition in R (such as $c + d$) and addition in M (such as $\alpha + \beta$). Likewise, there are two multiplication operations at play: multiplication in R (such as cd) and scalar multiplication (such as c). Note that by property (i), we may write cd without any ambiguity, as both possible interpretations, $c(d\alpha)$ and $(cd)\alpha$, yield the same value.

For fixed $c \in R$, the map that sends $\alpha \in M$ to $c\alpha \in M$ is a group homomorphism with respect to the additive group operation of M (by property (iii) of the definition); likewise, for fixed $\alpha \in M$, the map that sends $c \in R$ to $c\alpha \in M$ is a group homomorphism from the additive group of R into the additive group of M (by property (ii)). Combining these observations with basic facts about group homomorphisms, we may easily derive the following basic facts about R-modules:

Theorem. If M is a module over R, then for all $c \in R$, $\alpha \in M$, and $k \in \mathbb{Z}$, we have:

(i) $0_R.\ \alpha = 0_M$,
(ii) $c.\ 0_M = 0_M$,
(iii) $(-c)\alpha = -(c\alpha) = c(-\alpha)$,
(iv) $(kc)\alpha = k(c\alpha) = c(k\alpha)$.

Proof: Exercise.

The definition of a module includes the trivial module, consisting of just the zero element 0_M. If R is the trivial ring, then any R-module is trivial, since for every $\alpha \in M$, we have $\alpha = 1_R\ \alpha = 0_R\ \alpha = 0_M$..

Example. If E is any ring containing R as a subring (i.e., E is an extension ring of R), then E is a module over R, with addition and scalar multiplication defined in terms of the addition and multiplication operations of E.

Example. The ring of polynomials $R[X]$ over R forms an R-module in the natural way, with addition and scalar multiplication defined in terms of the addition and multiplication operations of the polynomial ring.

Example. If $M_1,..., M_k$ are R-modules, then so is their direct product $M_1 \times ... \times M_k$, where addition and scalar product are defined component-wise. If $M = M_1 = ... = M_k$, we write this as $M^{\times k}$.

Example. Any abelian group G, written additively, can be viewed as a $\mathbb{Z}$-module, with scalar multiplication defined in terms of the usual integer multiplication map.

Example. If I is an arbitrary set, and M is an R-module, then Map(I, M), which is the set of all functions $f : I \to M$, may be naturally viewed as an R-module, with point-wise addition and scalar multiplication: for $f, g \in$ Map(I, M) and $c \in R$, we define

$$(f + g)(i) := f(i) + g(i) \text{ and } (cf)(i) := cf(i) \text{ for all } i \in I.$$

Example. Let G be any group, written additively, whose exponent divides n. Then we may define a scalar multiplication that maps $[k]_n \in \mathbb{Z}_n$ and $\alpha \in G$ to k. That this map is unambiguously defined follows from the fact that G has exponent dividing n, so that if $k \equiv k'\alpha$ (mod n), we have $k\alpha - k'\alpha = (k - k')\alpha = 0_G$, since $n \mid (k - k')$. It is easy to check that this scalar multiplication map indeed makes G into a $\mathbb{Z}_n$-module.

Example. The set $R^{\times n}$, which consists of all of n-tuples of elements of R, forms an R-module, with addition and scalar multiplication defined componentwise: for $\alpha = (a_1,..., a_n) \in R^{\times n}$ β, $= (b_1,..., bn) \in R^{\times n}$, and $c \in R$, we define

$$\alpha + \beta := (a_1 + b_1,..., a_n + b_n) \text{ and } c\alpha := (ca_1,..., ca_n).$$

Example. Of course, viewing a group as a module does not depend on whether or not we happen to use additive notation for the group operation. If we specialize the previous example to the group $G = \mathbb{Z}^*_p$, where p is prime, then we may view G as a $\mathbb{Z}_{p-1}$-module. However, since the group operation itself is written multiplicatively, the "scalar product" of $[k]_{p-1} \in \mathbb{Z}_{p-1}$ and $\alpha \in \mathbb{Z}^*_p$ is the power k.

Example. The ring R itself can be viewed as an R-module in the obvious way, with addition and scalar multiplication defined in terms of the addition and multiplication operations of R.

MODULE HOMOMORPHISMS AND ISOMORPHISMS

Again, throughout this section, R is a ring. The notion of a group homomorphism extends in the obvious way to R-modules.

Definition: Let M and M' be modules over R. An R-module homomorphism from M to M' is a function $\rho : M \to M'$, such that

(i) ρ is a group homomorphism from M to M', and

(ii) for all $c \in R$ and $\alpha \in M$, we have $\rho(c\alpha) = c\rho(\alpha)$.

An R-module homomorphism is also called an R-linear map. We shall use this terminology from now on. Expanding the definition, we see that a map $\rho : M \to M'$ is an R-linear map if and only if $\alpha(\alpha + \beta) = \rho(\alpha) + \rho(\alpha)$ and $\rho(c\alpha) = c\rho(\alpha)$ for all $\alpha, \beta \in M$ and all $c \in R$.

Example. Let E be an extension ring of R. As we saw in Example 13.5, E may be viewed as an R-module in a natural way. Let $\alpha \in E$, and consider the -multiplication map on E, which sends $\beta \in E$ to $\alpha\beta \in E$. Then it is easy to see that this is an R-linear map.

Example. Let M be an R-module, and let k be an integer. Then the k-multiplication map on M is not only a group homomorphism, but it is also easily seen to be an R-linear map. Its image is the submodule kM, and its kernel the submodule$M\{k\}$.

Example. Let M be an R-module, and let c be an element of R. The map

$$\rho : M \to M$$

$$\alpha \mapsto c\alpha$$

is called c-multiplication map on M, and is easily seen to be an R-linear map whose image is the submodule cM, and whose kernel is the submodule $M\{c\}$. The set of all $c \in R$ for which $cM = \{0_M\}$ is called the R-exponent of M, and is easily seen to be an ideal of R.

Let M be an R-module, and let be an element of$M$. The map

$$\rho : R \to M$$

$$c \mapsto c\alpha$$

is easily seen to be an R-linear map whose image is the submodule R (i.e., the submodule generated by α). The kernel of this map is called the R-order of α, and is easily seen to be an ideal of R.

Example. If N is a submodule of an R-module M, then the inclusion map $i : N \to M$ is obviously an R-linear map.

Example. Generalizing the previous example, let M be an R-module, and let $\alpha_1,..., \alpha_k$ be elements of M. The map

$$\rho : R^{\times k} \to M$$
$$(c_1,..., c_k) \mapsto c_1\alpha_1 +... + c_k\alpha_k$$

is easily seen to be an R-linear map whose image is the submodule $R\alpha_1 +... + R\alpha_k$ (i.e., the submodule generated by $\alpha_1,..., \alpha_k$).

Example. Suppose N is a submodule of an R-module$M$. Then the natural map

$$\rho : M \to M/N$$
$$\alpha \mapsto [\alpha]_N$$

is not just a group homomorphism, it is also easily seen to be an R-linear map.

Example. Suppose that $M_1,..., M_k$ are submodules of an R-module M.

Then the map

$$\rho : M_1 \times ... \times M_k \to M$$
$$(\alpha_1,..., \alpha_k) \mapsto \alpha_1 +... + \alpha_k$$

is easily seen to be an R-linear map whose image is the submodule $M_1 +... + M_k$.

Theorem. If ρ is an R-module isomorphism of M with M', then the inverse function ρ^{-1} is an R-module isomorphism of M' with M.

Theorem: Let $\rho : M \to M'$ be an R-linear map.

(i) For every submodule N of M, $\rho(N)$ is a submodule of M'; in particular (setting $N := M$), Im ρ is a submodule of M'.

(ii) For every submodule N' of M', $\rho^{-1}(N')$ is a submodule of M'; in particular (setting $N' := \{0_{M'}\}$), Ker ρ is a submodule of M.

Internal Direct Product

Theorem. Let M be an R-module with submodules N_1, N_2, where $N_1 \cap N_2 = \{0_M\}$. Then we have an R-module isomorphism

$$N_1 \times N_2 \cong N_1 + N_2$$

given by the map

$$\rho : N_1 \times N_2 \to N_1 + N_2$$
$$(\alpha_1, \alpha_2) \mapsto \alpha_1 + \alpha_2.$$

Theorem. Let $\rho_i : M \to M_i^{'}$, for $i = 1,..., k$, be R-linear maps. Then the map

$$\rho : M \to M_1^{'} \times ... \times M_k^{'}$$
$$\alpha \mapsto (\rho_1(\alpha),..., \rho_k(\alpha))$$

is an R-linear map.

If an R-linear map $\rho : M \to M'$ is bijective, then it is called an R-module isomorphism of M with M'. If such an R-module isomorphism ρ exists, we say that M is isomorphic to M', and write$M \cong M'$. Moreover, if $M = M'$, then ρ is called an R-module automorphism on M.

Theorem. If $\rho : M \to M'$ and $\rho' : M' \to M''$ are R-linear maps, then so is their composition $\rho' \circ \rho : M \to M''$.

First Isomorphism Theorem

Theorem. Let $\rho : M \to M'$ be an R-linear map with kernel K and image N'. Then we have an R-module isomorphism

$$M/K \cong N'.$$

Specifically, the map

$$\bar{\rho} : M/K \to M'$$

$$[\alpha]_K \mapsto \rho(\alpha)$$

is an injective R-linear map whose image is N'.

Theorem. Let M and M' be R-modules, and consider the R-module of functions Map(M, M'). Then

$$\text{Hom}_R(M, M') := \{\sigma \in \text{Map}(M, M') : \sigma \text{ is an } R\text{-linear mapg}$$

is a submodule of Map(M, M').

Theorem. Let $\rho : M \to M'$ be an R-linear map. Then for every submodule N of M with $N \subseteq \text{Ker}\,\rho$, we may define an R-linear map

$$\bar{\rho} : M/N \to M'$$

$$[\alpha]_N \mapsto \rho(\alpha).$$

Moreover, $\text{Im}\,\bar{\rho} = \text{Im}\,\rho$, and $\bar{\rho}$ is injective if and only if $N = \text{Ker}\,\rho$.

EXERCISE

1. Let $\rho : M \to M'$ be a surjective R-linear map. Let S be the set of all submodules of M that contain Ker ρ, and let S' be the set of all submodules of M'. Show that the sets S and S' are in one-to-one correspondence, via the map that sends $N \in S$ to $\rho(N) \in S'$.
2. Verify that the "is isomorphic to" relation on R-modules is an equivalence relation; that is, for all R-modules M_1, M_2, M_3, we have:
   (a) $M_1 \cong M_1$;
   (b) $M_1 \cong M_2$ implies $M_2 \cong M_1$;
   (c) $M_1 \cong M_2$ and $M_2 \cong M_3$ implies $M_1 \cong M_3$.
3. Let $\rho : M \to M'$ be an R-linear map with kernel K. Let N be a submodule of M. Show that we have an R-module isomorphism $M/(N + K) \cong \rho(M)/\rho(N)$.

4. Let $\rho : M \to M'$ be an R-linear map, and let $c \in R$. Show that $\rho(cM) = c\rho(M)$.
5. Show that if $M = M_1 \times M_2$ for R-modules M_1 and M_2, and N_1 is a subgroup of M_1 and N_2 is a subgroup of M_2, then we have an R-module isomorphism$M/(N_1 \times N_2) \cong M_1/N_1 \times M_2/N_2$.
6. Let M be an R-module with submodules N_1 and N_2. Show that we have an R-module isomorphism $(N_1 + N_2)/N_2 \cong N_1/(N_1 \cap N_2)$.
7. Let $\rho : M \to M'$ be an R-linear map. Let N be a submodule of M, and let $\tau : N \to M'$ be the restriction of ρ to N. Show that τ is an R-linear map and that Ker τ = Ker $\rho \cap N$.
8. Suppose $M_1,..., M_k$ are R-modules. Show that for each $i = 1,..., k$, the projection map $\pi_i : M_1 ... M_k \to M_i$ that sends $(\alpha_1,..., \alpha_k)$ to α_i is a surjective R-linear map.
9. Let $\rho_i : M_i \to M_i'$, for $i = 1,..., k$, be R-linear maps. Show that the map

$$\rho : M_1 \times ... \times M_k \to M_1' \times ... \times M_k'$$

$$(\alpha_1,..., \alpha_k) \mapsto (\rho_1(\alpha_1),..., \rho_k(\rho_k))$$

is an R-linear map.

10. Let M be an R-module with submodules N_1, N_2, and A, where $N_2 \subseteq N_1$. Show that $(N_1 \cap A)/(N_2 \cap A)$ is isomorphic to a submodule of N_1/N_2.

SUBMODULES AND QUOTIENT MODULES [MH]

Again, throughout this section, R denotes a ring. The notions of subgroups and quotient groups extend in the obvious way to R-modules.

Definition: Let M be an R-module. A subset N of M is a submodule (over R) of M if

(*i*) N is closed under scalar multiplication; that is, for all $c \in R$ and $\alpha \in N$, we have $c \in N$.

(*ii*) N is a subgroup of the additive group M.

It is easy to see that a submodule N of an R-module M is also an R-module in its own right, with addition and scalar multiplication operations inherited from M.

Expanding the above definition, we see that a non-empty subset N of M is a submodule if and only if for all $c \in R$ and all α, $\beta \in N$, we have

$$\alpha + \beta \in N, -\alpha \in N, \text{ and } c\alpha \in N.$$

Observe that the condition $-\alpha \in N$ is redundant, as it is implied by the condition $c \in N$ with $c = -1_R$.

Clearly, $\{0_M\}$ and M are submodules of M. For $k \in \mathbb{Z}$, it is easy to see that not only are kM and $M\{k\}$ subgroups of M, they are also submodules of M. Moreover, for $c \in R$,

$$cM := \{c\alpha : \alpha \in M\} \text{ and } M\{c\} := \{\alpha \in M : c\alpha = 0_M\}$$

are also submodules of M. Further, for $\alpha \in M$,

$$R\alpha := \{c\alpha : c \in R\}$$

is a submodule of M. Finally, if N_1 and N_2 are submodules of M, then $N_1 + N_2$ and $N_1 \cap N_2$ are not only subgroups of M, they are also submodules of M. We leave it to the reader to verify all these facts: they are quite straightforward.

Let $\alpha_1, \ldots, \alpha_k \in M$. The submodule

$$R\alpha_1 + \ldots + R\alpha_k$$

is called the submodule (over R) generated by $\alpha_1, \ldots, \alpha_k$. It consists of all R-linear combinations

$$c_1\alpha_1 + \ldots + c_k\alpha_k,$$

where the c_i's are elements of R, and is the smallest submodule of M that contains the elements $\alpha_1, \ldots, \alpha_k$. We shall also write this submodule as $\langle \alpha_1, \ldots, \alpha_k \rangle_R$. As a matter of definition, we allow $k = 0$, in which case this submodule is $\{0_M\}$. We say that M is finitely generated (over R) if $M = \langle \alpha_1, \ldots, \alpha_k \rangle_R$ for some $\alpha_1, \ldots, \alpha_k \in M$.

Example. Let G be an abelian group. We can view G as a $\mathbb{Z}$-module in a natural way. Subgroups of G are just the same thing as submodules of G, and for $a_1, \ldots, a_k \in G$, the subgroup $\langle a_1, \ldots, a_k \rangle$ is the same as the submodule $\langle a_1, \ldots, a_k \rangle \mathbb{Z}$.

Example. Suppose E is an extension ring of R, and I is an ideal of E. Viewing E as an R-module, I is a submodule of E, and hence the quotient ring E/I may naturally be viewed as an R-module, with scalar multiplication defined by $c. [\alpha]_I := [c]_I$ for $c \in R$ and $\alpha \in E$. Example 13.4 is a special case of this, applied to the extension ring $R[X]$ and the ideal (f).

Example. For a given integer $\ell \geq 0$, define $R[X]_{<\ell}$ to be the set of polynomials of degree less than ℓ. The reader may verify that $R[X]_{<\ell}$ is a submodule of the R-module $R[X]$, and indeed, is the submodule generated by $1, X, \ldots, X^{\ell-1}$. If $\ell = 0$, then this submodule is the trivial submodule $\{0_R\}$.

EXERCISE

1. Show that if R is non-trivial, then $R[X]$ is not finitely generated.
2. Let M_1 and M_2 be R-modules, and let N_1 be a submodule of M_1 and N_2 a submodule of M_2. Show that $N_1 \times N_2$ is a submodule of $M_1 \times M_2$.

VECTOR SPACES AND DIMENSION

Throughout this section, F denotes a field.

A module over a field is also called a vector space. In particular, an F-module is called an F-vector space, or a vector space over F.

For vector spaces over F, one typically uses the terms subspace and quotient space, instead of (respectively) submodule and quotient module; likewise, one usually uses the terms F-vector space homomorphism, isomorphism and automorphism, as appropriate.

We now develop the basic theory of dimension for finitely generated vector spaces. Recall that a vector space V over F is finitely generated if

$$V = \langle \alpha_1, ..., \alpha_n \rangle_F$$

for some $\alpha_1,..., \alpha_n$ of V. The main results here are that

- every finitely generated vector space has a basis,
- and all bases have the same number of elements.

Throughout the rest of this section, V denotes a vector space over F. We begin with a technical fact that will be used several times throughout this section:

Theorem. If V is spanned by some family of n elements of V, then every family of $n + 1$ elements of V is linearly dependent.

Proof: We prove this by induction on n. If $n = 0$, the theorem is clear, so assume that $n > 0$. Let $\{\alpha_i\}_{i=1}^{n}$ be a family that spans V, and let $\{\beta_i\}_{i=1}^{n+1}$ be an arbitrary family of elements of V. We wish to show that $\{\beta_i\}_{i=1}^{n+1}$ is linearly dependent.

We know that β_{n+1} is a linear combination of the β_i's, say,

$$\beta_{n+1} = c_1\beta_1 + ... + c_n\beta_n.$$

If all the c_i's were zero, then we would have $\beta_{n+1} = 0_V$, and so trivially, $\{\beta_i\}_{i=1}^{n+1}$ is linearly dependent. So assume that some c_i is non-zero, and for concreteness, say $c_n \neq 0$. Dividing equation through by c_n, it follows that α_n is an F-linear combination of $\alpha_1,..., \alpha_{n-1}, \beta_{n+1}$. Therefore,

$$\langle \alpha_1, ..., \alpha_{n-1}, \beta_{n+1} \rangle_F \supseteq \langle \alpha_1, ..., \alpha_{n-1} \rangle_F + \langle \alpha_n \rangle_F = V.$$

Now consider the subspace $W = \langle \beta_{n+1} \rangle_F$ and the quotient space V/W. Since the family of elements $\alpha_1,..., \alpha_{n-1}, \beta_{n+1}$ spans V, it is easy to see that $\{[\alpha_i]_W\}_{i=1}^{n-1}$ spans V/W ; therefore, by induction, $\{[\beta_i]_W\}_{i=1}^{n}$ is linearly dependent. This means that there exist $d_1,..., dn \in F$, not all zero, such that $d_1\beta_1 + ... + d_n\beta_n \equiv 0 \pmod{W}$, which means that for some $d_{n+1} \in F$, we have $d_1\beta_1 + ... + d_n\beta_n = d_{n+1}\, \beta_{n+1}$. That proves that $\{\beta_i\}_{i=1}^{n+1}$ is linearly dependent.

Theorem. Suppose $\{\alpha_i\}_{i=1}^{n}$ is a family of elements that spans V. Then for some subset $J \subseteq \{1,..., n\}$, the subfamily $\{\alpha_j\}_{j\in J}$ is a basis for V.

Proof: We prove this by induction on n. If $n = 0$, the theorem is clear, so assume $n > 0$. Consider the subspace W of V spanned by $\{\alpha_i\}_{i=1}^{n-1}$. By the induction hypothesis, for some $K \subseteq \{1,..., n - 1\}$, the subfamily $\{\alpha_k\}_{k\in K}$ is a basis for W. There are two cases to consider.

Case 1. $\alpha_n \in W$. In this case, $W = V$, and the theorem clearly holds with $J := K$.

Case 2. $\alpha_n \notin W$. We claim that setting $J := K \cup \{n\}$, the subfamily $\{\alpha_j\}_{j\in J}$ is a basis for V. Indeed, since $\{\alpha_k\}_{k\in K}$ is linearly independent, and $\alpha_n \notin W$, Theorem immediately implies that $\{\alpha_j\}_{j\in J}$ is linearly independent. Also, since $\{\alpha_k\}_{k\in K}$ spans W, it is clear that $\{\alpha_j\}_{j\in J}$ spans $W + \langle \alpha_n \rangle_F = V$.

Theorem. If V is finitely generated, then any two bases have the same size.

Theorem. Suppose that $\{\alpha_i\}_{i=1}^{n}$ is a linearly independent family of elements that spans a subspace $W \subseteq V$, and that $\alpha_{n+1} \in V \setminus W$. Then $\{\alpha_i\}_{i=1}^{n+1}$ is also linearly independent.

Proof: Suppose we have a linear relation

$$0_V = c_1\alpha_1 + ... + c_n\alpha_n + c_{n+1}\,\alpha_{n+1},$$

where the c_i's are in F. We want to show that all the c_i's are zero. If $c_{n+1} \neq 0$, then we have

$$\alpha_{n+1} = -c_{n+1}^{-1}\,(c_1\alpha_1 + ... + c_n\alpha_n) \in W,$$

contradicting the assumption that $\alpha_{n+1} \notin W$. Therefore, we must have $c_{n+1} = 0$, and the linear independence of $\{\alpha_i\}_{i=1}^{n+1}$ implies that $c_1 = ... = c_n = 0$.

The next theorem says that every finitely generated vector space has a basis, and in fact, any family that spans a vector space contains a subfamily that is a basis for the vector space.

Definition: If V is finitely generated, the common size of any basis is called the dimension of V, and is denoted $\dim_F(V)$.

Note that from the definitions, we have $\dim_F(V) = 0$ if and only if V is the trivial vector space (i.e., $V = \{0_V\}$). We also note that one often refers to a finitely generated vector space as a finite dimensional vector space. We shall give preference to this terminology from now on.

To summarize the main results in this section up to this point: if V is finite dimensional, it has a basis, and any two bases have the same size, which is called the dimension of V.

Theorem: Suppose that V is finite dimensional and W is a subspace of V. Then W is also finite dimensional, with $\dim_F(W) \leq \dim_F(V)$. Moreover, $\dim_F(W) = \dim_F(V)$ if and only if $W = V$.

Proof: Suppose $\dim_F(V) = n$. Consider the set S of all linearly independent families of the form $\{\alpha_i\}_{i=1}^{m}$, where $m \geq 0$ and each α_i is in W. The set S is certainly non-empty, as it contains the empty family. Moreover, by Theorem, every member of S must have at most n elements. Therefore, we may choose some particular element $\{\alpha_i\}_{i=1}^{m}$ of S, where m is as large as possible. We claim that this family $\{\alpha_i\}_{i=1}^{m}$ is a basis for W. By definition, $\{\alpha_i\}_{i=1}^{m}$ is linearly independent and spans a subspace of W. If this is a proper subspace $W' \subseteq W$, we can choose an element $\alpha_{m+1} \in W \setminus W'$, and by Theorem, the family $\{\alpha_i\}_{i=1}^{m+1}$ is linearly independent, and therefore, this family also belongs to S, contradicting the assumption that m is as large as possible.

That proves that W is finite dimensional with $\dim_F(W) \leq \dim_F(V)$. It remains to show that these dimensions are equal if and only if $W = V$. Now, if $W = V$, then clearly $\dim_F(W) = \dim_F(V)$. Conversely, if $\dim_F(W) = \dim_F(V)$, then by Theorem, any basis for W must already span V.

Theorem. If V is finite dimensional, then every linearly independent family of elements of V can be extended to form a basis for V.

Proof: One can prove this by generalizing the proof of Theorem 13.18. Alternatively, we can adapt the proof of the previous theorem. Let $\{\beta_i\}_{j=1}^{m}$ be a linearly independent family of elements that spans a subspace W of V. As in the proof of the previous theorem, if $\{[\alpha_i]_W\}_{i=1}^{\ell}$ is a basis for the quotient space V/W, then the elements

$$\alpha_1, \ldots, \alpha_\ell, \beta_1, \ldots, \beta_m$$

form a basis for V.

Example. Suppose that F is finite, say $|F| = q$, and that V is finite dimensional, say $\dim_F(V) = n$. Then clearly $|V| = q^n$. If W is a subspace with $\dim_F(W) = m$, then $|W| = q^m$, and by Theorem 13.24, $\dim_F(V/W) = n - m$, and hence $|V/W| = q^{n-m}$. Just viewing V and W as additive groups, we know that the index of W in V is $[V : W] = |V/W| = |V|/|W| = q^{n-m}$, which agrees with the above calculations.

We next consider the relation between the notion of dimension and linear maps. First, observe that by Theorem, if two finite dimensional vector spaces have the same dimension, then they are isomorphic.

Theorem. If V is finite dimensional, and W is a subspace of V, then the quotient space V/W is also finite dimensional, and

$$\dim_F(V/W) = \dim_F(V) - \dim_F(W).$$

Proof: Suppose that $\{\alpha_i\}_{i=1}^{n}$ spans V. Then it is clear that $\{[\alpha_i]_W\}_{i=1}^{n}$ spans V/W.

By Theorem 13.18, we know that V/W has a basis of the form $\{[\alpha_i]_W\}_{i=1}^{\ell}$, where $\ell \leq n$ (renumbering the α_i's as necessary). By Theorem 13.23, we know that W has a basis, say $\{\beta_i\}_{j=1}^{m}$. The theorem will follow immediately from the following:

Claim. The elements

$$\alpha_1, \ldots, \alpha_\ell, \beta_1, \ldots, \beta_m$$

form a basis for V.

To see that this family spans V, consider any element γ of V. Then since $\{[\alpha_i]_W\}_{i=1}^{\ell}$ spans $V/$

W, we have $\gamma \equiv \sum_i c_i\alpha_i$ (mod W) for some $c_1,\ldots, c_\ell \in F$. If we set $\beta := \gamma - \sum_i c_i\alpha_i \in W$, then since $\{\beta_j\}_{j=1}^m$ spans W, we have $\beta = \sum_j d_j\beta_j$ for some $d_1,\ldots, d_m \in F$, and hence $\gamma = \sum_i c_i\alpha_i + \sum_j d_j\beta_j$. That proves that the family of elements spans V. To prove this family is linearly independent, suppose we have a relation of the form $\sum_i c_i\alpha_i + \sum_j d_j\beta_j = 0_V$, where $c_1,\ldots, c_\ell \in F$ and $d_1,\ldots, d_m \in F$. If any of the c_i's were non-zero, this would contradict the assumption that $\{[\alpha_i]_W\}_{i=1}^{\ell}$ is linearly independent. So assume that all the c_i's are zero. If any of the d_j's were non-zero, this would contradict the assumption that $\{\beta_i\}_{j=1}^m$ is linearly independent. Thus, all the c_i's and d_j's must be zero, which proves that the family of elements is linearly independent. That proves the claim.

Theorem. Suppose that $\dim_F(V) = n$, and that $\{\alpha_i\}_{i=1}^n$ is a family of n elements of V. The following are equivalent:

(i) $\{\alpha_i\}_{i=1}^n$ is linearly independent;

(ii) $\{\alpha_i\}_{i=1}^n$ spans V;

(iii) $\{\alpha_i\}_{i=1}^n$ is a basis for V.

Proof: Let W be the subspace of V spanned by $\{\alpha_i\}_{i=1}^n$.

First, let us show that (i) implies (ii). Suppose $\{\alpha_i\}_{i=1}^n$ is linearly independent. Also, by way of contradiction, suppose that $W \subsetneq V$, and choose $\alpha_{n+1} \in V \setminus W$. Then Theorem implies that $\{\alpha_i\}_{i=1}^{n+1}$ is linearly independent. But then we have a linearly independent family of $n + 1$ elements of V, which is impossible.

Second, let us prove that (ii) implies (i). Let us assume that $\{\alpha_i\}_{i=1}^n$ is linearly dependent, and prove that $W \subsetneq V$. By Theorem we can find a basis for W among the α_i's, and since $\{\alpha_i\}_{i=1}^n$ is linearly dependent, this basis must contain strictly fewer than n elements. Hence, $\dim_F(W) < \dim_F(V)$, and therefore, $W \subsetneq V$.

The theorem now follows from the above arguments, and the fact that, by definition, (iii) holds if and only if both (i) and (ii) hold.

We next examine the dimension of subspaces of finite dimensional vector spaces.

Theorem. If V is of finite dimension n, and V is isomorphic to V', then V' is also of finite dimension n.

Proof: If $\{\alpha_i\}_{i=1}^n$ is a basis for V, then by Theorem, $\{\rho(\alpha_i)\}_{i=1}^n$ is a basis for V'.

Thus, two finite dimensional vector spaces are isomorphic if and only if they have the same dimension.

We next illustrate one way in which the notion of dimension is particularly useful. In general, if we have a function $f: A \to B$, injectivity does not imply surjectivity, nor does surjectivity imply injectivity. If A and B are finite sets of equal size, then these implications do indeed hold. The following theorem gives us another important setting where these implications hold, with finite dimensionality playing the role corresponding to finite cardinality.

Theorem. If V is finite dimensional, and $\rho : V \to V'$ is an F-linear map, then Im ρ is a finite dimensional vector space, and

$$\dim_F(V) = \dim_F(\text{Im } \rho) + \dim_F(\text{Ker } \rho).$$

Intuitively, one way to think of Theorem is as a "law of conservation" for dimension: any "dimensionality" going into ρ that is not "lost" to the kernel of ρ must show up in the image of ρ.

Theorem. If $\rho : V \to V'$ is an F-linear map, and if V and V' are finite dimensional with $\dim_F(V) = \dim_F(V')$, then we have:

ρ is injective if and only if ρ is surjective.

Proof: Let $\{\alpha_i\}_{i=1}^n$ be a basis for V. Then

ρ is injective $\Leftrightarrow \{\rho(\alpha_i)\}_{i=1}^n$ is linearly independent

$\Leftrightarrow \{\rho(\alpha_i)\}_{i=1}^n$ spans V'

$\Leftrightarrow \rho$ is surjective

This theorem may be generalized as follows:

EXERCISE

1. Show that if V is a finite dimensional vector space over F with subspaces W_1 and W_2, then

   $\dim_F(W_1 + W_2) = \dim_F(W_1) + \dim_F(W_2) - \dim_F(W_1 \cap W_2)$.

2. Show that if $V_1, \ldots, V_n$ are finite dimensional vector spaces over F, then $V_1 \times \ldots \times V_n$ has dimension $\sum_{i=1}^n \dim_F(V_i)$.

3. From the previous exercise, one might be tempted to think that a more general "inclusion/exclusion principle" for dimension holds. Determine if the following statement is true or false: if V is a finite dimensional vector space over F with subspaces W_1, W_2, and W_3, then

$$\dim_F (W_1 + W_2 + W_3) = \dim_F (W_1) + \dim_F (W_2) + \dim_F (W_3)$$
$$- \dim_F (W_1 \cap W_2) - \dim_F (W_1 \cap W_3) - \dim_F (W_2 \cap W_3)$$
$$+ \dim_F (W_1 \cap W_2 \cap W_3).$$

4. Let V be a vector space over F with basis $\{\alpha_i\}_{i=1}^n$. Let S be a finite, non-empty subset of F, and define

$$B := \left\{\sum_{i=1}^{n} c_i\alpha_i : c_1,\ldots,c_n \in S\right\}.$$

Show that if W is a subspace of V, with $W \subseteq V$, then $| B \cap W | \leq | S |^{n-1}$.

5. Suppose that V and W are vector spaces over F, V is finite dimensional, and $\{\alpha_i\}_{i=1}^k$ is a linearly independent family of elements of V. In addition, let $\beta_1,\ldots, \beta_k$ be arbitrary elements of W. Show that there exists an F-linear map $\rho : V \to W$ such that $\rho(\alpha_i) = \beta_i$ for $i = 1,\ldots, k$.

6. The theory of dimension for finitely generated vector spaces is quite elegant and powerful. There is a theory of dimension (of sorts) for modules over an arbitrary, non-trivial ring R, but it is much more awkward and limited. This exercise develops a proof of one aspect of this theory: if an R-module M has a basis at all, then any two bases have the same size. To prove this, we need the fact that any non-trivial ring has a maximal ideal. Let n, m be positive integers, let $\alpha_1,\ldots, \alpha_m$ be elements of $R^{\times n}$, and let I be an ideal of R.

   (a) Show that if $\{\alpha_i\}_{i=1}^m$ spans $R^{\times n}$, then every element of $I^{\times n}$ can be expressed as $c_1\alpha_1 +\ldots + c_m\alpha_m$, where $c_1,\ldots, c_m$ belong to I.

   (b) Show that if $m > n$ and I is a maximal ideal, then there exist $c_1,\ldots, c_m \alpha R$, not all in I, such that $c_1\alpha_1 +\ldots + c_m\alpha_m \in I^{\times n}$.

   (c) From (a) and (b), deduce that if $m > n$, then $\{\alpha_i\}_{i=1}^m$ cannot be a basis for $R^{\times n}$.

   (d) From (c), conclude that any two bases for a given R-module M must have the same size.

LINEAR INDEPENDENCE AND BASES

Throughout this section, R denotes a ring.

Definition. Let M be an R-module, and let $\{\alpha_i\}_{i=1}^n$ be a family of elements of M. We say that $\{\alpha_i\}_{i=1}^n$

(i) is linearly dependent (over R) if there exist $c_1,..., c_n \in R$, not all zero, such that $c_1\alpha_1 +... + c_n\alpha_n = 0_M$;

(ii) is linearly independent (over R) if it is not linearly dependent;

(iii) spans M (over R) if for every $\alpha \in M$, there exist $c_1,..., c_n \in R$ such that $c_1\alpha_1 +... + c_n\alpha_n = \alpha$;

(iv) is a basis for M (over R) if it is linearly independent and spans M.

The family $\{\alpha_i\}_{i=1}^n$ always spans some submodule of$M$, namely, the submodule N generated by $\alpha_1,..., \alpha_n$. In this case, we may also call N the submodule (over R) spanned by $\{\alpha_i\}_{i=1}^n$.

The family $\{\alpha_i\}_{i=1}^n$ may contain duplicates, in which case it is linearly dependent (unless R is trivial). Indeed, if, say, $\alpha_1 = \alpha_2$, then setting $c_1 := 1$, $c_2 := -1$, and $c_3 :=... := c_n := 0$, we have the linear relation $\sum_{i=1}^n c_i\alpha_i = 0_M$.

If the family $\{\alpha_i\}_{i=1}^n$ contains 0_M, then it is also linear dependent (unless R is trivial). Indeed, if, say, $\alpha_1 = 0_M$, then setting $c_1 := 1$ and $c_2 :=... := c_n := 0$, we have the linear relation $\sum_{i=1}^n c_i\alpha_i = 0_M$.

The family $\{\alpha_i\}_{i=1}^n$ may also be empty (i.e., $n = 0$), in which case it is linearly independent, and spans the submodule $\{0_M\}$.

In the above definition, the ordering of the elements $\alpha_1,..., \alpha_n$ makes no difference. As such, when convenient, we may apply the terminology in the definition to any family $\{\alpha_i\}_{i\in I}$, where I is an arbitrary, finite index set.

Example. Consider the R-module $R^{\times n}$. Define $\alpha_1,..., \alpha_n \in R^{\times n}$ as follows:

$$\alpha_1 := (1, 0,..., 0),\ \alpha_2 := (0, 1, 0,..., 0),...,\ \alpha_n := (0,..., 0, 1);$$

that is, α_i has a 1 in position i and is zero everywhere else. It is easy to see that $\{\alpha_i\}_{i=1}^n$ is a basis for $R^{\times n}$. Indeed, for all $c_1,..., cn \in R$, we have

$$c_1\alpha_1 +... + c_n\alpha_n = (c_1,..., c_n),$$

from which it is clear that $\{\alpha_i\}_{i=1}^n$ spans $R^{\times n}$ and is linearly independent. The family $\{\alpha_i\}_{i=1}^n$ is called the standard basis for $R^{\times n}$.

Example. Consider the $\mathbb{Z}$-module $\mathbb{Z}^{\times 3}$. In addition to the standard basis, which consists of the tuples

$$(1, 0, 0), (0, 1, 0), (0, 0, 1),$$

the tuples

$$\alpha_1 := (1, 1, 1), \alpha_2 := (0, 1, 0), \alpha_3 := (2, 0, 1)$$

also form a basis. To see this, first observe that for all $c_1, c_2, c_3, d_1, d_2, d_3 \in \mathbb{Z}$, we have

$$(d_1, d_2, d_3) = c_1\alpha_1 + c_2\alpha_2 + c_3\alpha_3$$

if and only if

$$d_1 = c_1 + 2c_3, d_2 = c_1 + c_2, \text{ and } d_3 = c_1 + c_3.$$

If holds with $d_1 = d_2 = d_3 = 0$, then subtracting the equation $c_1 + c_3 = 0$ from $c_1 + 2c_3 = 0$, we see that $c_3 = 0$, from which it easily follows that $c_1 = c_2 = 0$. This shows that the family $\{\alpha_i\}_{i=1}^3$ is linearly independent. To show that it spans $\mathbb{Z}^{\times 3}$, the reader may verify that for any given $d_1, d_2, d_3 \in \mathbb{Z}$, the values

$$c_1 := -d_1 + 2d_3, c_2 := d_1 + d_2 - 2d_3, c_3 := d_1 - d_3$$

satisfy.

The family of tuples

$$(1, 1, 1), (0, 1, 0), (1, 0, 1)$$

is not a basis, as it is linearly dependent: the third tuple is equal to the first minus the second.

The family of tuples (1, 0, 12), (0, 1, 30), (0, 0, 18) is linearly independent, but does not span $\mathbb{Z}^{\times 3}$: the last component of any $\mathbb{Z}$-linear combination of these tuples must be divisible by gcd (12, 30, 18) = 6. However, this family of tuples is a basis for the $\mathbb{Q}$-module $\mathbb{Q}^{\times 3}$.

Example. Consider again the submodule $R[X]_{<\ell}$ of $R[X]$, where $\ell \geq 0$, consisting of all polynomials of degree less than ℓ. Then $\{X^{i-1}\}_{i=1}^{\ell}$ is a basis for $R[X]_{<\ell}$ over R.

Example. Consider again the ring $E = R[X]/(f)$, where $f \in R[X]$ with $\deg(f) = \ell \geq 0$ and $\mathrm{lc}(f) \in R^*$. As in Example 13.4, we may naturally view E as a module over R. From the observations in Example, it is clear that $\{\xi^{i-1}\}_{i=1}^{\ell}$ is a basis for E over R, where $\xi := [X]_f \in E$.

The next theorem highlights a critical property of bases:

Theorem. If $\{\alpha_i\}_{i=1}^n$ is a basis for an R-module M, then the map

$$\varepsilon : R^{\times n} \to M$$

$$(c_1, \ldots, c_n) \mapsto c_1\alpha_1 + \ldots + c_n\alpha_n$$

is an R-module isomorphism. In particular, every element of M can be expressed in a unique way as $c_1\alpha_1 + ... + c_n\alpha_n$, for $c_1, ..., c_n \in R$.

Proof: We already saw that ε is an R-linear map in Example 13.21. Since $\{\alpha_i\}_{i=1}^n$ is linearly independent, it follows that the kernel of ε is trivial, so that ε is injective. That ε is surjective follows immediately from the fact that $\{\alpha_i\}_{i=1}^n$ spans M.

The following is an immediate corollary of this theorem:

Theorem. Any two R-modules that have bases of the same size are isomorphic.

The following theorem develops an important connection between bases and linear maps.

Theorem. Let $\{\alpha_i\}_{i=1}^n$ be a basis for an R-module M, and let $\rho : M \to M'$ be an R-linear map.

(i) ρ is surjective if and only if $\{\rho(\alpha_i)\}_{i=1}^n$ spans M'.

(ii) ρ is injective if and only if $\{\rho(\alpha_i)\}_{i=1}^n$ is linearly independent.

(iii) ρ is an isomorphism if and only if $\{\rho(\alpha_i)\}_{i=1}^n$ is a basis for M'.

Proof:

(*i*) By the previous theorem, we know that every element of M can be written uniquely as $\sum_i c_i\alpha_i$, where the c_i's are in R. Therefore, every element in Im ρ can be expressed as $\rho\left(\sum_i c_i\alpha_i\right) = \sum_i c_i\rho(\alpha_i)$. It follows that Im ρ is equal to the subspace of M' spanned by $\{\rho(\alpha_i)\}_{i=1}^n$. From this, (i) is clear.

(*ii*) consider a non-zero element $\sum_i c_i\alpha_i$ of M, so that not all c_i's are zero. Now, $\sum_i c_i\alpha_i \in$ Ker ρ if and only if $\sum_i c_i\rho(\alpha_i) = 0_{M'}$, and thus, Ker ρ is non-trivial if and only if $\{\rho(\alpha_i)\}_{i=1}^n$ is linearly dependent. That proves (ii).

(*iii*) follows from (i) and (ii). 2

EXERCISE

1. Let M be an R-module. Suppose $\{\alpha_i\}_{i=1}^n$ is a linearly independent family of elements of M. Show that for every $J \subseteq \{1, ..., n\}$, the subfamily $\{\alpha_j\}_{j\in J}$ is also linearly independent.

2. Suppose $\rho : M \to M'$ is an R-linear map. Show that if $\{\alpha_i\}_{i=1}^n$ is a linearly dependent family of elements of M, then $\{\rho(\alpha_i)\}_{i=1}^n$ is also linearly dependent.

3. Suppose $\rho : M \to M'$ is an injective R-linear map and that $\{\alpha_i\}_{i=1}^{n}$ is a linearly independent family of elements of M. Show that $\{\rho(\alpha_i)\}_{i=1}^{n}$ is linearly independent.

4. Suppose that $\{\alpha_i\}_{i=1}^{n}$ spans an R-module M and that $\rho : M \to M'$ is an R-linear map. Show that:

   (a) ρ is surjective if and only if $\{\rho(\alpha_i)\}_{i=1}^{n}$ spans M';

   (b) if $\{\rho(\alpha_i)\}_{i=1}^{n}$ is linearly independent, then ρ is injective.